AF358696

Innovative Research Approaches & Practices towards Sustainable Land Management, Preservation & Restoration

Innovative Research Approaches & Practices towards Sustainable Land Management, Preservation & Restoration

Editors

Kleomenis Kalogeropoulos
Andreas Tsatsaris
Nikolaos Stathopoulos
Demetrios E. Tsesmelis
Nilanchal Patel
Xiao Huang

Basel • Beijing • Wuhan • Barcelona • Belgrade • Novi Sad • Cluj • Manchester

Editors

Kleomenis Kalogeropoulos
University of West Attica
Athens
Greece

Andreas Tsatsaris
University of West Attica
Athens
Greece

Nikolaos Stathopoulos
BEYOND Centre of EO
Research & Satellite Remote
Sensing
Athens
Greece

Demetrios E. Tsesmelis
Hellenic Open University
Patras
Greece

Nilanchal Patel
Birla Institute of Technology
Mesra
Ranchi
India

Xiao Huang
University of Arkansas
Fayetteville, AR
USA

Editorial Office
MDPI
St. Alban-Anlage 66
4052 Basel, Switzerland

This is a reprint of articles from the Special Issue published online in the open access journal *Land* (ISSN 2073-445X) (available at: https://www.mdpi.com/journal/land/special_issues/sustainable_land).

For citation purposes, cite each article independently as indicated on the article page online and as indicated below:

Lastname, A.A.; Lastname, B.B. Article Title. *Journal Name* **Year**, *Volume Number*, Page Range.

ISBN 978-3-7258-0339-2 (Hbk)
ISBN 978-3-7258-0340-8 (PDF)
doi.org/10.3390/books978-3-7258-0340-8

Cover image courtesy of Demetrios E. Tsesmelis

Contents

About the Editors

Kleomenis Kalogeropoulos

Dr. Kleomenis Kalogeropoulos is a surveying engineer (PhD, MSc, and BSc) and an adjunct lecturer teaching GIS and Thematic Cartography at the Department of Surveying and Geoinformatics Engineering at the University of West Attica, Greece. Dr. Kalogeropoulos also teaches about Geographical Information Systems at the Greek National Centre for Public Administration and Local Government. His research interests include GIScience, spatial data infrastructures, spatial analysis, natural disaster modeling, geodemography, geoarchaeology, digital cartography, and geography. Dr. Kalogeropoulos has made substantial contributions to his field, having authored over one hundred articles in peer-reviewed journals and international conferences and eighteen chapters in respective collective volumes. He is a co-Editor of a book on Geoinformatics and a co-Editor of six Special Issues in scientific journals. He has also participated in numerous national and international research programs. He is a certified evaluator of the General Secretariat in Research and Innovation and an expert in Earth sciences, GIS, and Remote Sensing applications for the Institute of Educational Policy of the Greek Ministry of Education.

Andreas Tsatsaris

Professor Dr. Andreas Tsatsaris is a rural and surveying engineer (PhD, MSc, and BSc) and has been the Head of the Surveying and Geoinformatics Engineering Department at the University of West Attica (UniWA), Greece, since 2019. He has also been, on a number of occasions since 2013, elected as the Director of the Research Laboratory "GAEA". He teaches GIS, Thematic Cartography, Spatial Epidemiology, and Medical Geography, in both the undergraduate and postgraduate study programs of the Faculty of Engineering at UniWA, as well as in those of the Faculties of Medicine at the University of Crete, the Aristotle University of Thessaloniki, and the National and Kapodistrian University of Athens. He has published more than 110 articles in prestigious scientific journals and international conferences, with several references being made to his work. He has participated in national and international research programs (fewer than 30) and currently supervises seven doctoral theses.

Nikolaos Stathopoulos

Dr. Nikolaos Stathopoulos is a Geotechnical–Environmental Engineer who graduated from the School of Mining and Metallurgical Engineering (MME) at the National Technical University of Athens, Greece (NTUA), majoring in Geotechnology (2006). He holds two Master's degrees, one in the "Science and Technology of Water Resources" from the School of Civil Engineering at NTUA (2008), and one in "Geoinformatics" from Harokopion University, Greece (2017). In 2019, he completed his Doctoral research at the MME School on the topic "Research Methods for Geoenvironmental Hazards & Water Resources, via Geographic Information Systems & Remote Sensing", and was awarded a PhD in Engineering from NTUA. Today, he works as a Researcher and Scientific Technical Project Manager in the Operational Unit "BEYOND Centre of Earth Observation Research and Satellite Remote Sensing" at the Institute for Astronomy, Astrophysics, Space Applications and Remote Sensing at the National Observatory of Athens (IAASARS/NOA), focusing on natural hazards and disasters, soil erosion and land degradation, and water resources. He has authored more than fifty peer-reviewed articles in scientific journals and conferences and has participated in more than fifteen national and international projects. His scientific interests are in the fields of Earth Observation (mainly Satellite Remote Sensing), Applied Spatial Analysis, Natural

Hazards/Disasters (modeling, vulnerability, and risk assessment), Land Degradation (soil erosion, desertification, etc.), Natural Resources Engineering and Management (mainly water and soil), and Climate Change impacts on the Human–Nature Ecosystem.

Demetrios E. Tsesmelis

Demetrios E. Tsesmelis is an agronomist (BSc, MSc, and PhD) and an Associate Adjunct Professor of Environmental Management at the School of Applied Arts and Sustainable Design at the Hellenic Open University, Greece. He has more than 15 years of teaching experience in the fields of natural resources, sustainability, and environmental management. He has also made significant contributions to his research field, having authored about forty-five articles in peer-reviewed journals, international conferences, and books. He has been a Guest Editor of four Special Issues in scientific journals and is the referee for eighty papers published in twenty-two international journals. He specializes in drought management, and has conducted extensive research in the fields of water resources, natural hazards, land degradation, drought indices, forecasting, and vulnerability.

Nilanchal Patel

Dr. Nilanchal Patel has been a Professor in the Department of Remote Sensing at the Birla Institute of Technology, Mesra, Ranchi, India, for 23 years. He received his Ph.D. from the Indian Institute of Technology in Kanpur. His research interests include various applications of remote sensing, GIS, and digital image processing. He has served as a member of multiple editorial boards, such as those of the *Arabian Journal of Geosciences*, the *Journal of Forestry Research*, the *Journal of Mountain Science*, the *International Journal of Sustainable Development and World Ecology*, and of many MDPI Special Issues. He has also published over seventy papers in international and national journals.

Xiao Huang

Dr. Xiao Huang is an Assistant Professor in the Department of Environmental Sciences at Emory University. He holds a Ph.D. in Geography from the University of South Carolina and a Master's degree in Geographic Information Science and Technology from the Georgia Institute of Technology. Dr. Huang's research expertise include human–environment interactions, computational social sciences, urban informatics, disaster mapping and mitigation, GeoAI, and disaster remote sensing. Dr. Huang has contributed extensively to his field, authoring over 150 peer-reviewed journal articles and seven book chapters, and has played a pivotal role in the editing of five books. In his professional capacity, he serves as an Associate Editor for *Computational Urban Science* and is a member of the Editorial Board for several prestigious journals, including *Big Earth Data*, the *International Journal of Digital Earth*, *Frontiers Remote Sensing*, *Nature Scientific Reports*, the *Journal of Remote Sensing*, *Current Social Sciences*, and *PLoS ONE*. Dr. Huang's research has garnered significant attention, receiving coverage in renowned media outlets such as Nature News, NASA, NBC, and Fox. His work has attracted substantial funding from NSF, NASA, the National Academies, and the Centers for Medicare and Medicaid Services. Furthermore, Dr. Huang is a respected reviewer for NASA and NSF grants and has reviewed manuscripts for over fifty-two international and national journals.

Preface

The intricate interplay between human activity and the environment has underscored the urgency for innovative research approaches and practices to address the challenges of sustainable land management, preservation, and restoration. The profound impact of these issues on ecosystems, biodiversity, and human well-being demands a concerted effort in exploring novel solutions and in contributing to the global discourse on environmental sustainability. This Special Issue, titled "Innovative Research Approaches & Practices towards Sustainable Land Management, Preservation & Restoration," presents an array of cutting-edge research articles that investigate various aspects of land management and conservation. The papers featured offer insights into a number of topics, ranging from geospatial approaches to assessing ecological restoration sites in post-fire landscapes to the effects of agricultural policies on sustainable rural development in European agriculture. These studies exemplify the multidisciplinary nature of contemporary research by addressing issues such as soil erosion, water loss, carbon sequestration, and the restoration of degraded landscapes.

The contributors to this Special Issue employ advanced geospatial technologies, sustainable agricultural practices, and policy analyses to develop comprehensive solutions that aim to strike a balance between the needs of human societies and those of the natural environment. By leveraging these diverse approaches, these authors pave the way for sustainable land management practices that are environmentally conscious and beneficial to future generations. This preface serves as an introduction to a collection of works that not only highlights the challenges we face, but also underscores the innovative solutions researchers are devising to navigate the complexities of sustainable land management. As we navigate these uncharted territories, we also extend our gratitude to the authors for their invaluable contributions and hope that this Special Issue will inspire further collaborative efforts to safeguard our planet for future generations.

Kleomenis Kalogeropoulos, Andreas Tsatsaris, Nikolaos Stathopoulos, Demetrios E. Tsesmelis, Nilanchal Patel, and Xiao Huang
Editors

 *land*

Article

A Geospatial Approach to Identify and Evaluate Ecological Restoration Sites in Post-Fire Landscapes

Stefanos Dosis [1], George P. Petropoulos [1,*] and Kleomenis Kalogeropoulos [2]

[1] Department of Geography, Harokopio University of Athens, El. Venizelou 70, 17676 Athens, Greece; stefdosis@gmail.com

[2] Department of Surveying and Geoinformatics Engineering, University of West Attica, Ag. Spyridonos Str., 12243 Athens, Greece; kkalogeropoulos@uniwa.gr

* Correspondence: gpetropoulos@hua.gr

Abstract: Wildfires are a pervasive natural phenomenon in Mediterranean forest ecosystems, causing significant ecological imbalances that demand immediate restoration efforts. The intricacy of reinstating the ecological balance necessitates a proactive approach to identifying and assessing suitable restoration sites. The assessment and investigation of the most suitable restoration sites is of particular importance both for the relevant authorities and for planning and decision making by the state. This study proposes the development of a user-friendly model for evaluating and identifying the most suitable restoration sites immediately after a fire, using geoinformation technologies. For the purposes of demonstrating the method's applicability, the 2016 fire of "Prinos", Thasos, Greece, an area that has been repeatedly affected by forest fires, was chosen as a case study. The methodology evaluation was carried out by applying the weighted multicriteria decision analysis method (MC-DAM) and was based on a number of variables. The analysis, processing and extraction of the results were performed using primarily remote sensing datasets in a geographical information system (GIS) environment. The methodology proposed herein includes the classification of the individual criteria and their synthesis based on different weighting factors. In the final results, the restoration suitability maps are presented in five suitability zones based on two different scenarios. Based on this study, the integration of geospatial and remote sensing data offers a valuable and cost-effective means for promptly assessing post-fire landscapes, with the aim of identifying suitable restoration sites.

Keywords: wildfires; forest restoration; geographical information systems; multicriteria analysis; normalized burn ratio index; remote sensing; earth observation

Citation: Dosis, S.; Petropoulos, G.P.; Kalogeropoulos, K. A Geospatial Approach to Identify and Evaluate Ecological Restoration Sites in Post-Fire Landscapes. *Land* **2023**, *12*, 2183. https://doi.org/10.3390/land12122183

Academic Editor: Dianfeng Liu

Received: 14 October 2023
Revised: 1 December 2023
Accepted: 3 December 2023
Published: 18 December 2023

1. Introduction

Forest ecosystems, as natural formations, are vulnerable to a range of natural and anthropogenic threats, which have significant impacts on ecosystem health and are often directly related to their degradation [1–4]. Wildfires are one of the threats that bring devastating results at environmental, economic and social levels [5,6]. Although it is a natural phenomenon that occurs extensively in all Mediterranean forest ecosystems, the frequency and magnitude of its occurrence in recent decades highlight the magnitude of the problem. According to the European Commission's 2020 technical report on wildfires [7], climate change is directly related to the increase in incidents, both at the level of fire initiation and in the indirect change in vegetation and fuel characteristics. The EU Strategy for Adaptation to Climate Change recognized the severity of the problem and promotes a series of measures in different sectors and activities, including reforestation initiatives, vegetation management, community engagement and awareness, early warning systems [7]. Thus, climate change with prolonged periods of drought, combined with an increase in average temperatures, especially during the summer months, has resulted in more frequent, but also more severe, fires. If all of this is combined with the abandonment of the Greek countryside, the increase in urbanization in recent decades and the lack of management of forest ecosystems, along with a consequent increase in fuel use, it is concluded that the impacts of wildfires are even more devastating [8–10].

The Mediterranean landscape, with its special characteristics, the varied topography, the mountain ranges and the abundance of flora and habitats, has been significantly affected by fires [11]. Fire, as an integral part of the Mediterranean landscape, is therefore a factor that has shaped and influenced the terrestrial ecosystems of the Mediterranean, especially the Mediterranean vegetation zone and, to a lesser extent, the Mediterranean vegetation zone (vegetation zones as defined by the Braun–Blanquet classification system) [12]. These areas are dominated by species such as evergreen shrubs and trees, low shrubs and toadstools, herbaceous plants and geophytes that mainly flower in spring and autumn. In Greece, for example, there are mainly pure forests of *coniferous conifers, pine halepensis* (*Pinus halepensis*) and *Pinus brutia*, while, to a lesser extent, there are stands *of Pinus pinea, Cupressus sempervirens*, etc. Similarly, evergreen broad-leaved shrubs are spreading in the southern and island regions, but also widely throughout the country, creating forests and woodlands with species such as holly (*Quercus coccifera*), *schisin* (*Pistacia lentiscus*), *heather* (*Erica arborea*), etc. These areas and forest species are primarily subject to the largest number of wildfires occurring annually in the country during the fire season, from May to October. These ecosystems and their species, in order to cope, have developed mechanisms of adaptation to fires (such as resprouting or epicormic regeneration, thick bark, serotiny, fire-adapted seeds, etc.) and are therefore considered to be fire-loving or fire-resistant species [13]. However, even these fire-adapted ecosystems, when affected by repeated fires or subjected to particularly severe destructive effects, are often unable to cope. As a result, they are at immediate risk from soil erosion and landslides, but also in the long term from loss of forest vegetation and the possibility of desertification [14].

The change in climatic conditions and the occurrence of increasingly frequent recurrent fires in the same areas of thermophilic Mediterranean species, as well as in areas of higher altitudes with cold-tolerant conifers, have resulted in these ecosystems being irreversibly affected. In these cases, restoring ecological balance is particularly difficult and requires immediate restoration measures, including protection of soil capital, erosion control, flood control and artificial regeneration (reforestation) [15–18].

A government authority is not in a position to carry out reforestation in the entire affected area mainly because of the high economic costs of reforestation works. However, it must know which areas will be selected for intervention and which are most at risk after the fire has passed. For this reason, it is particularly important to assess and investigate the most suitable locations for implementing restoration measures. Moreover, according to a Greek ministerial decision, it is foreseen that, within 15 days from the outbreak of a fire, the competent services should be able to propose the areas and their parts that need immediate restoration/reforestation. The ability to select suitable restoration sites in a timely manner is important both for the competent authorities and for planning and decision making by the Greek Government.

The advances in modern technologies and, in particular, of Geoinformation have played a key role in decision-making systems and on the solution of complex problems. The development of specialized and easy-to-use software, the increase in the power and capabilities of computer systems have brought significant changes in spatial analysis and in the field of wildfires compared to earlier analog methods of analysis and processing. The advantages of these new Geoinformation technologies include the possibility of easy recording and observation; the use of open data; the possibility of monitoring ecological and socioeconomic parameters; and the evaluation, design and management of a decision-making system [19–23].

The use of geographic information systems (GISs) as part of Geoinformation technologies is of great significance in the planning and analysis of complex spatial problems [24]. These are used to transform geographic data, process them, analyze and visualize results [25–27]. In particular, through GIS it is possible to achieve the identification of areas which carry a combination of features, the application of multicriteria parameter analysis to solve complex phenomena, statistical data processing and the creation of decision making models [28].

At the same time, remote sensing (RS) provides a wealth of high-resolution satellite data available in recent years, and technological advances in science offer the possibility of timely, remote and highly accurate spatial information, as well as a wide range of possibilities [29–31]. They can contribute to parameter mapping and change detection, as well as assessment and indicator generation [32–35].

The assessment of areas affected by catastrophic wildfires and the selection of areas in need of immediate restoration is a complex and multifactorial process [17]. Through the science of Geoinformation and its advantages, the analysis of the factors and criteria related to the phenomenon becomes more accessible. While Geoinformation provides valuable insights, it is crucial to recognize the inherent limitations, especially in cases where ecosystems are exposed to novel climatic conditions. This approach enhances our understanding and helps us draw informed conclusions, acknowledging the potential impact of imperfect ecological knowledge [36]. The development of an easy-to-use tool for the direct, targeted and low-cost assessment of areas requiring further post-fire rehabilitation is possible through modern technologies. The combination of geospatial and RS data to investigate potential reforestation sites can provide useful and meaningful results [37,38]. These results, with the help of modern internet technologies and web-based GIS (WebGIS), can be easily available and accessible to the general public. In recent years, there has been a rapid advancement of the internet which has followed the general development of the IT sector. This development has not been lacking in the field of GIS, where WebGIS services have been developed for the processing, analysis and dissemination of geographic information. These capabilities were based on a series of services and standards such as WMSs (Web Map Services) and WFSs (Web Feature Services) based on the specifications of the OGC (Open Geospatial Consortium). These services allow for easy and fast sharing of geographic data with the public at large, as well as the possibility of editing by remote users [39,40].

GIS and RS together have been used extensively in recent years to exploit and analyze spatial information, to disseminate results in map form in an understandable way and to facilitate decision making (e.g., [17,29]). They have been exploited in a multitude of studies and are widely applied in the environmental field, particularly in wildfires and post-fire succession and restoration [17,41–43].

Already in the previous decades, GIS had started to be used in relevant studies and played an important role in data analysis and processing. In San Bernandino National Forest, California, a fire prediction system developed, taking into account available geographic data and making use of GIS to calculate relevant indicators and the probability of occurrence [44]. Epp and Lanonville [45] developed an intelligent fire management system for Northwestern Canada, using GIS and remote sensing data. They included procedures for fire risk prediction, fire progression and response planning. Later, in another study, Ruiz-Gallardo et al. [46] generated a priority map of intervention after a forest fire in the Southeastern Iberian Peninsula, taking into account fire severity, slope and exposure. They propose a semi-automatic method to identify the areas that are most at risk of erosion, using GIS technology and related remote sensing indicators.

In the contemporary literature, for example, Schulz and Schroder [47] used GIS to evaluate areas with Mediterranean climate characteristics in Central Chile which are facing deforestation and need immediate implementation of restoration programs. Their study applied the spatial multicriteria analysis to identify sites for reforestation that meet multipurpose forestry, using land cover data. The use of such new tools of geoinformatics, combined with data that can be quickly and easily assessed from satellite data, has been applied in several works aimed at evaluating areas for restoration; for example, [35,48,49] studied the rates of new vegetation emergence after a fire in a maritime pine forest in Northwestern Spain, using high-resolution satellite data (WorldView-2). Fernández-Manso et al. [25] used radar and LIDAR datasets, combined with optical and thermal data, to map the destruction and regrowth of Mediterranean pine forests after wildfires. The synergy between the two sciences thus offers a wealth of advantages and opportunities, particularly in the field of wildfires and restoration. In these fields, the study areas cover large areas, the natural environment is dynamic and constantly changing, and, as a result, data collection and field measurements are time-consuming and costly [18]. Therefore, the combination of modern information technologies can provide easy, economical and accurate results to identify and evaluate ecological restoration sites in post-fire landscapes [50].

The aim of this paper is to use geoinformatics technologies to create an easy-to-use model which provides the possibility to evaluate and identify the most suitable sites for restoration after a wildfire. For the purposes of this work, the boundary of the catastrophic fire of 2016 in the area of Prinos, Thasos Island, was chosen as the study area. The selection was made by taking into account all available data and the physical and geographic characteristics of the area. It was also taken into account that this forest ecosystem has been repeatedly affected by wildfire events and is a representative Mediterranean thermophilous coniferous forest.

2. Materials and Methods

2.1. Study Area in General

The island of Thasos (Figure 1) is located in the North Aegean Sea and is considered the northernmost island of Greece (40°41′34″ latitude and 24°39′1.95″ longitude). The coastline of Thasos is about 115 km, and the island's area is 378.8 km². Administratively, it belongs to the Region of Eastern Macedonia and Thrace and the Regional Unit of Thasos, according to the Kallikratis program (the Kallikratis program is a reform of the administrative divisions of Greece). It is about 24 km away from Kavala, but it is also connected by sea transport to the town of Keramoti. The capital of the island is Limenas, and the other main settlements are Limenaria, Prinos, Theologos and Kallirachi. Thasos is a mountainous island with an intense relief, presenting extensive ridges and several trenches. The highest peak is the Ipsario Peak at 1203 m in altitude.

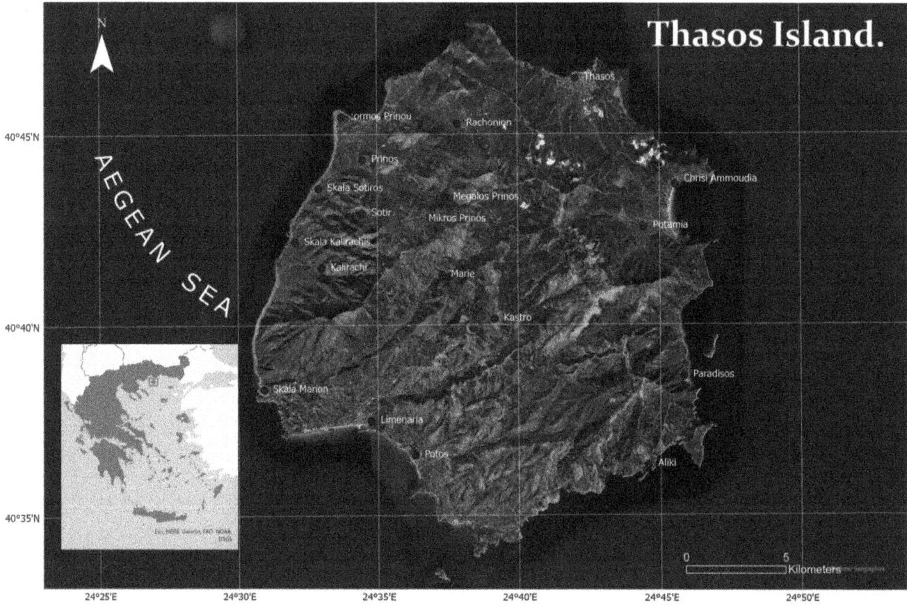

Figure 1. The study area.

The island of Thasos is chronically affected by forest fires. In the Region of Eastern Macedonia and Thrace, the three worst fires for the period 1983–2006 were recorded in the prefecture of Kavala. A large percentage of the areas of Thasos has been affected at least twice by catastrophic fires. Also, about half of the island's surface area was burnt between 1984 and 2000. Specifically, in 1984, 1985, 1989 and 2000, major catastrophic fires occurred in the western and southern part of the island which incinerated almost the half of the total area of the island. Forest fires broke out again in the western part of the island in 2004, 2008 and 2013. One of the worst years for the island of Thasos was 2016, as 68,870 ha was burnt again in three areas of the southern and western parts of the island. In particular, on 10 September 2016, four fire fronts occurred in different parts of the island which developed into the major fires of 2016. The outbreaks were caused by an abundance of lightning recorded in the area, without the presence of rain. This rare phenomenon of "dry thunderstorms" affected Eastern Macedonia and Thrace but mainly occurred on the island of Thasos. The increased dryness of the fuel, due to the absence of rainfall in the previous period, allowed the fire fronts to burn for three consecutive days and cause great damage to the natural environment and infrastructure.

2.1.1. Vegetation

Thasos Island features Pinus brutia forests, alongside stands of *black pine* (*Pinus nigra*) at higher altitudes and some hybrid spruce (*Abies borissi-regis*) in cooler spots. The island also hosts evergreen broadleaves like holly (*Quercus coccifera*), *arbutus unedo*, *Quercus ilex* and *Pistacia lentiscus*. The understory mainly consists of heather (*Erica arborea*), vital for beekeeping, and there are occasional appearances of broad-leaved species like the plane tree (*Platanus orientalis*), *chestnut* (*Castanea sativa*), *poplar* (*Populus tremula*) and willow (*Salix* sp.) Thasos is rich in flora, including endemic and rare species. Thasia is found only in this region. The Mediterranean vegetation zone (*Quercetalia pubescentis*) is found at altitudes between 300 and 800 m and includes the predominant species of the island, rough pine [13]).

2.1.2. Climate–Meteorology

This study used the data from a meteorological station in Thasos, the National Observatory of Athens in Prinos, which was recently put into operation and has meteorological data only after 2020. This station is located next to the study area. The yearly average temperature is 15.3 °C. The highest average monthly temperature is 26.1 °C and was observed in July, while the minimum is 5.6 °C and was observed in January. Similarly, the average minimum monthly minimum temperature of 1.8 °C is observed in January, and the average maximum of 30.2 °C is observed in August. The annual rainfall is 466.2 mm. A higher amount of rainfall occurs in the winter months, and it decreases significantly in the summer months. The highest average rainfall is observed in December, i.e., 78.4 mm.

2.1.3. Geology

Thasos belong to the Rhodope crystal-rich complex of metamorphic rocks. It is dominated by crystalline schistose and exfoliated rocks of the Pre-Paleozoic period and formations of the Mesozoic century, as well as metamorphic limestones. The younger layers of Neogene, Tertiary and Quaternary deposits are found in places. According to the geological map, the center of the island, especially in the more mountainous parts, is dominated by compact semi-metamorphic limestones of amphibolite and gneiss. At lower altitudes around the island, crystalline limestone and marble are found at lower altitudes. To a lesser extent, mainly in the coastal zone, newer layers of sediment, sandy loams and sandstones are restricted to the coastal zone. The dominance of crystalline schistose and metamorphic rocks, especially in mountainous areas, may impact vegetation types and fuel availability, potentially influencing fire behavior. Additionally, the presence of sedimentary formations, such as sandy loams and sandstones in the coastal zone, might contribute to different fire dynamics compared to the compact semi-metamorphic limestones found in the mountainous center of the island.

2.2. Study Datasets

The evaluation of the burnt area of Prinos, as to the optimal restoration sites, will be based on a series of characteristics and elements of the area that need to be analyzed and combined with each other. The data, in this particular case, arise after the review and thorough research of the international literature and are divided into two distinct categories: ecological and economic criteria. The ecological criteria refer to the parameters that favor or hinder the growth of vegetation, based on the climatic and physical–geographical characteristics of the area. The economic criteria refer to the factors that increase or reduce the cost of afforestation, depending on the location where it is carried out. The next table (Table 1) presents the data used for the proposed methodology.

Table 1. The data used.

Data	Type	Source
Vegetation	Vector	Vegetation map (Ministry of Environment)
Soil	Vector	Soil map (Ministry of Agriculture)
Burnt area	Raster	Sentinel 2A (Copernicus Program)
Digital Elevation Model (DEM)	Raster	Greek Cadaster
Settlements	Vector	Google Earth
Road network	Vector	Google Earth

2.3. The Proposed Methodology

The aim of the study is to develop a model for the evaluation of optimal rehabilitation sites based on a set of criteria that will be used in combination. The solution to this which is proposed herein is based on variables such as vegetation, soil, burnt area, digital elevation model (DEM), settlements and road network. The existence of a multitude of criteria that must be combined with each other to evaluate a situation and draw the final conclusions necessitates the use of the multicriteria decision analysis (MCDA). This method incorporates quantitative indicators to determine the importance of each criterion and proceed to the synthesis of the individual criteria [28,51–53]. Since the phenomenon under consideration and the variables of the model have spatial characteristics, the model of the weighted multicriteria analysis with cartographic superposition was adopted [54].

The analysis that follows in the next section, in the first phase, presents the actions required to develop the 7 variables from all available data. The ArcGIS Pro 3.0 and SNAP 8.0 software packages were used for the tasks and data processing needs. The following figure (Figure 2) shows, in diagram form, the workflow and procedures followed to create the individual criteria from the primary data.

The repetition of criteria and variables in this study arises from a deliberate alignment between the chosen spatial parameters and the fundamental components of our evaluation model. These factors were meticulously selected due to their intrinsic importance in post-fire landscape assessment. The term "scenarios" encapsulates the various conditions or situations resulting from the interplay of these variables, forming the core of our work. Each scenario represents a unique combination of spatial parameters, facilitating a comprehensive evaluation of optimal rehabilitation sites. In response to the reviewer's valuable feedback, we provide an enhanced explanation in the manuscript, elucidating the critical role of these variables in shaping different scenarios and their significance in achieving the study's objectives.

In the next phase, after the configuration of the individual variables, all the necessary actions are applied to allow for the synthesis of each individual criterion. In the following (Figure 3), the workflow diagram of the weighted multicriteria analysis for the development of the sites' rehabilitation evaluation model is presented.

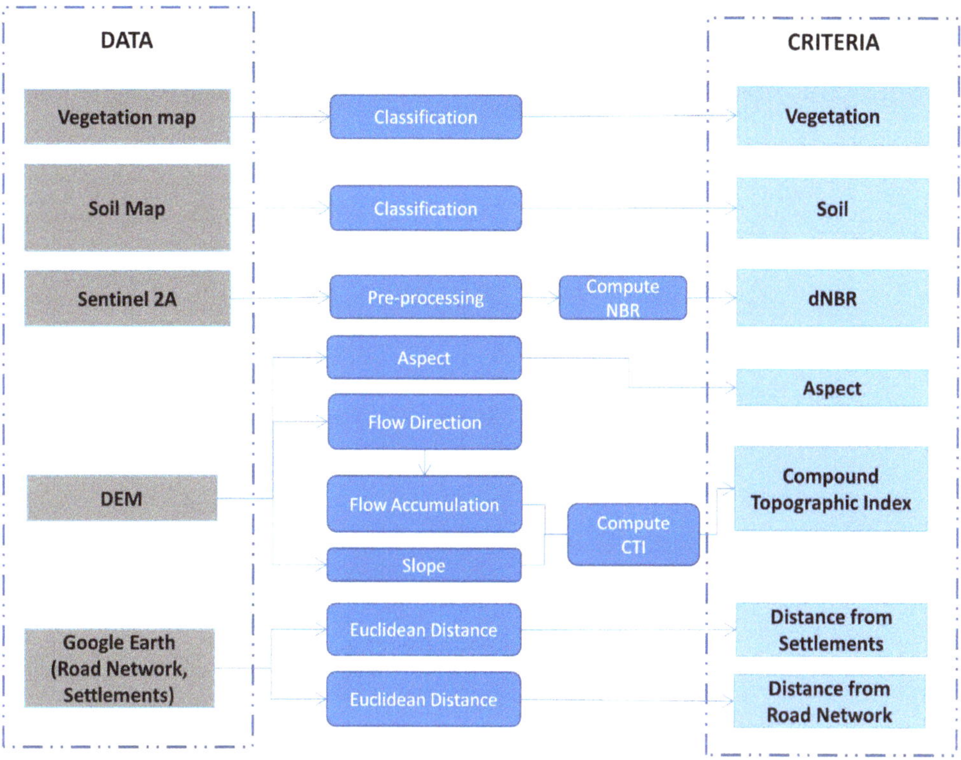

Figure 2. Flowchart summarizing the proposed methodology.

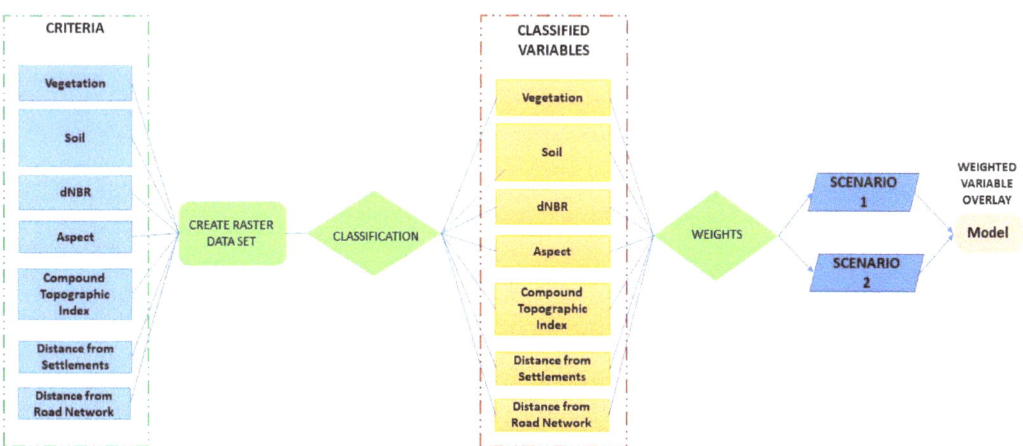

Figure 3. Flowchart of the weighted multicriteria analysis.

The final variables used are as follows:

i. Vegetation

The state and composition of vegetation both play a pivotal role in the recovery process of ecosystems following a wildfire event. GIS technology allows for the detailed mapping and analysis of vegetation types, their density and their health. By considering these factors, restoration planners can make informed decisions about the suitability of a site for restoration efforts. Assessing the pre-fire and post-fire vegetation can help identify areas where native species have been severely impacted or invasive species have taken hold, guiding the prioritization of restoration efforts. Additionally, vegetation data can help determine the ecological significance of a site, its potential for natural regeneration or the need for active intervention like reseeding or planting.

ii. Soil

Wildfires can significantly impact soil properties, including the nutrient levels, composition and stability. GIS technology facilitates the analysis of soil characteristics, helping to assess the extent of damage caused by the fire and identify areas that may require restoration interventions. Understanding soil conditions is essential for determining the suitability of a site for ecological restoration, as it influences the success of plant establishment and overall ecosystem recovery. Soil data can inform decisions about the choice of plant species, erosion control measures and the need for soil amendments. Table 2 presents the different rock types and soil depths identified at the study area's boundary. They are classified into four classes of suitability, based on their ability to accommodate new seedlings and provide the appropriate supplies for their growth. Gneisses are clearly superior soils to limestone for vegetation establishment [55].

Table 2. Soil classification.

Bedrock	Soil Depth	Class
Gneisses	2	I
Gneisses	4	II
Hard limestone	6	III
Gneisses	8	III
Hard limestone	8	IV

iii.   Differenced Normalized Burn Ratio (dNBR)

The calculation of the NBR requires the use of satellite data. It is based on the wavelengths of the electromagnetic spectrum of the near infrared (NIR) and shortwave infrared ($SWIR$), according to the following equation [56,57].

$$BR = \frac{(NIR - SWIR)}{(NIR + SWIR)} \tag{1}$$

The analysis and calculation of the severity index was based on images from the Sentinel 2 satellite. Specifically, images of the Sentinel 2A and Level 1C logger were selected as recorded in the table below. The images of this level were atmospherically corrected, and the values of reflected electromagnetic radiation recorded are affected by the atmosphere. These images (1st image ID: S2A_MSIL1C_20160819T091032_N0204_R050_T34TGL_20160819T091 026119-8-2016, 2nd image ID: S2A_MSIL1C_20160918T090622_N0204_R050_T35TKF_20160 918T090923/18-9-2016) are close to the 10 September event and also show little to no cloud cover in the region of interest, while their spatial resolution is 10–60 m.

The figure (Figure 4) below shows the situation in the northwestern part of Thasos after the devastating fire (Figure 4a), and the situation before the fire on 19 August (Figure 4b). The determination of the difference in the severity index between the two pre-fire and post-fire images had to be developed as a different channel so that it would include the information for the whole area.

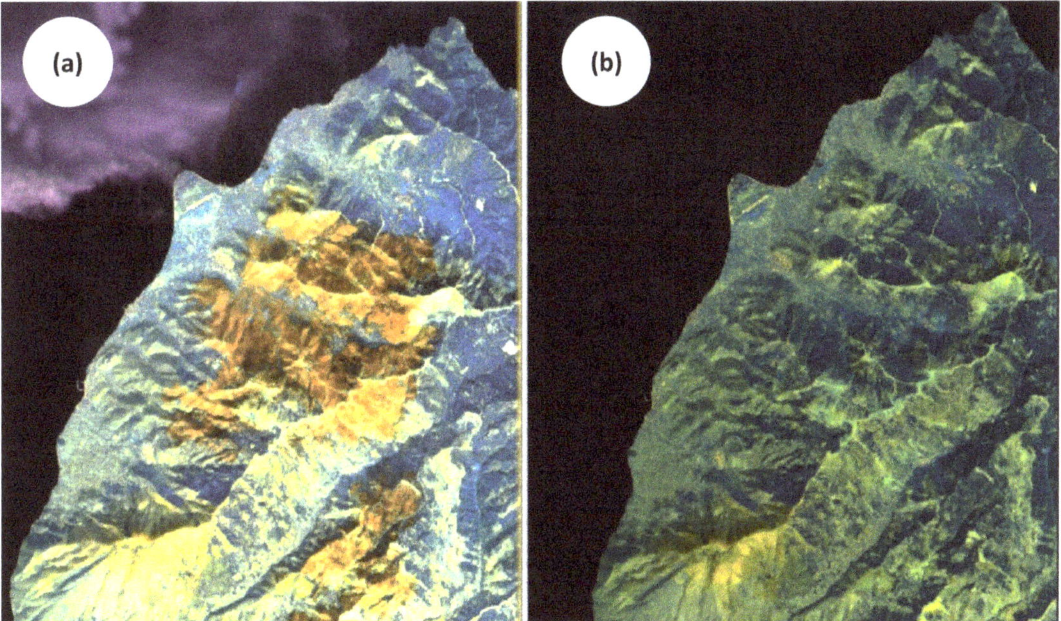

Figure 4. (**a**) Post-fire (19 August 2016) and (**b**) pre-fire condition of the study area (18 August 2016) (Sentinel 2A).

This geometric process was performed by combining the two images into one, taking into account and combining the values of each image cell. The image taken on 19 August was designated as the main image, and the one taken on 18 September as the subordinate. In the new product created, the function is as follows:

$$dNBR = PrefireNBR - PostfireNBR \tag{2}$$

iv.   Aspect

Aspect is a fundamental geospatial parameter that plays a crucial role in assessing and identifying ecological restoration sites in post-fire landscapes within the context GIS. Aspect refers to the compass direction that a slope or landform faces, influencing the amount of sunlight and moisture an area receives and ultimately impacting the distribution of vegetation and ecological conditions. In a geospatial approach to identifying and evaluating restoration sites, aspect data are instrumental for assessing the suitability of potential sites for restoration efforts. It helps in understanding how solar exposure and moisture patterns vary across the landscape, which, in turn, aids in selecting sites that can support the regeneration of vegetation and the restoration of ecological balance in areas affected by wildfires. In this study, aspect was derived from the DEM in a mosaic form, with a resolution of 10 m.

v.   Compound Topographic Index (*CTI*)

The Compound Topographic Index (*CTI*) is an indicator which is related to the accumulation of surface water runoff and the slope of the soil. It is calculated according to the following equation proposed by Beven and Kirkby [58].

$$CTI = Ln\left(\frac{As}{tan\beta}\right) \tag{3}$$

The index is defined as the natural logarithm of A_s times the tangent of β, where As is the runoff accumulation based on the scale of analysis, and β is the slope of the terrain in radians. The calculation of the CTI index was based on the digital elevation model (DEM) of the study area. Runoff accumulation, a variable required to determine the CTI index, is derived from the level of runoff direction. Any point on the ground that has an elevation less than its neighbors is given the value of the sum of the cells contributing to runoff up to that point. The sum of the overlying cells depends on the direction of runoff movement, with the result that the highest values are observed in the stream beds and estuaries.

The denominator of the CTI index equation includes the soil slope criterion, which was calculated based on the DEM. The slope plane generated was converted from degrees to radians, where each cell was assigned a slope value based on the following equation:

$$\beta = slope * \frac{\left(\frac{\pi}{2}\right)}{90} \tag{4}$$

vi.   Distance from Settlements

The distance from settlements is a critical parameter in a geospatial approach to identifying and evaluating ecological restoration sites in post-fire landscapes. As wildfires can pose significant threats to human settlements, it is essential to consider the proximity of restoration sites to inhabited areas. GIS technology enables the analysis of this distance, helping to ensure the safety of both local communities and restoration crews. It also plays a pivotal role in the strategic planning of restoration efforts, as selecting sites far removed from settlements reduces the risk of potential conflicts, resource competition and interference with ongoing recovery efforts.

vii.   Distance from Road Network.

The road network is a crucial parameter in a geospatial approach to identifying and evaluating ecological restoration sites in post-fire landscapes. Roads can have a significant impact on the environment, affecting wildlife habitat, soil erosion and water quality. In this context, GIS technology is invaluable for assessing the accessibility of potential restoration sites in relation to the existing road network. Analyzing the road infrastructure allows restoration planners to make informed decisions regarding site selection. The presence of nearby roads can facilitate logistical aspects of restoration efforts, such as the transportation of equipment and personnel. However, it is essential to strike a balance, as excessive road proximity may result in increased human disturbances and associated negative ecological impacts.

Each variable will be introduced into the model as a different thematic level after the necessary processing. The analysis is based on converting the criteria into a mosaic format so that each point in the space is given a separate value based on its characteristics. This is followed by classifying the individual criteria and their synthesis based on different weighting factors obtained through the Analytical Hierarchy Process [59]. Two different scenarios are applied, assigning each one more weight to the ecological or economic criteria. The spatial resolution to be used is 10 m so that there is extensive detail for the study area to allow for detailed planning by managers during the decision-making process.

After reclassifying the criteria, the subsequent crucial step in the analytical process involved the homogenization of variables to establish a consistent scale spanning from 1 to 5. This harmonization was undertaken with the primary objective of creating a standardized platform for evaluating and comparing all the variables within the study. The process served to eliminate any disparities stemming from varying measurement units or scales employed for different criteria. This standardization not only facilitated direct and equitable comparisons between the variables but also simplified the overall analysis, rendering it more comprehensible and user-friendly for interpretation and decision making. The uniform scale ensured that each variable held an equal footing in the evaluation process, contributing to a more balanced and objective assessment. Furthermore, this standardized 1–5 scale offered an efficient means of conveying the results and recommendations to stakeholders, project team members or decision makers, enhancing the clarity and effectiveness of communication within the context of the project.

The final stage of the methodological approach was the synthesis of the homogenized variables from the previous procedure based on specific weighting factors. For this reason, two scenarios of weighted variable composition were chosen to be applied.

- Scenario 1 prioritizes ecological constraints and favors the importance of optimal vegetation growth conditions.
- Scenario 2 gives greater weight to the economic and social benefits of restoration, favoring the criteria of distance from settlements and roads.

The determination of the weighting coefficients was based on the Analytical Hierarchy Process, as described by Saaty [59]. According to the method, the variables are compared pairwise in an aggregate comparison matrix. The comparison was made on a scale of 1 to 9. The higher the preference for one criterion over another, the higher the score tends toward 9, while the lowest criterion in the comparison pair automatically receives a score below one (e.g., 1/9 = 0.11).

3. Results and Discussion

3.1. The Parameters Used

3.1.1. Vegetation

The vegetation variable was converted to a raster layer, as described by the workflow (Figure 3). The vegetation in the wider area before the catastrophic fires includes herbaceous, bushy and wooded areas. Most of the affected area was covered by black pine. The remaining areas were occupied by bushy species, such as juniper, oak and holly. Barren land and grassland belong to the same herbaceous vegetation category. The areas that are suitable for afforestation include mainly open areas of low vegetation and burnt areas of coniferous species where there is a high probability that they will not regenerate naturally. Shrub and rocky areas should be avoided for afforestation. Moreover, black pine, as a cold-hardy conifer, should be given priority in restoration work over rough pine [1,2]. The next figure presents the initial vegetation map (Figure 5a) and the reclassified one (Figure 5b), which is one of the parameters of the methodology used.

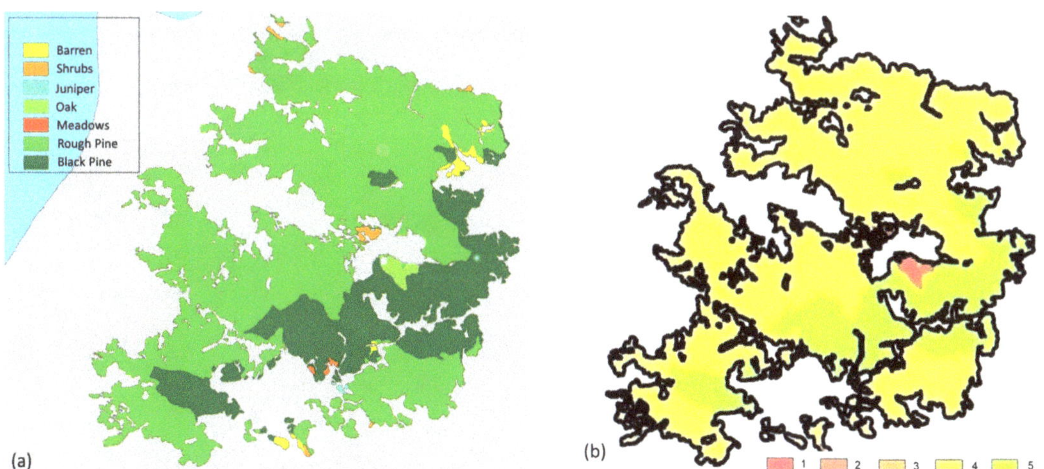

Figure 5. Vegetation parameter (**a**) before the classification and (**b**) after the classification (in terms of afforestation suitability, 1 is the lowest and 5 is the highest suitability).

3.1.2. Soil

The soil characteristics map records the bedrock and soil depth, characteristics that relate to the potential for vegetation growth and the quality of a site. The codes used to describe the type of parent material and soil depth have values from 1 to 9, where 1–3 is deep, 4–6 is deep to shallow, and 7–9 is shallow to rocky soil [60]. The soil classification is presented within the figure that follows (Figure 6).

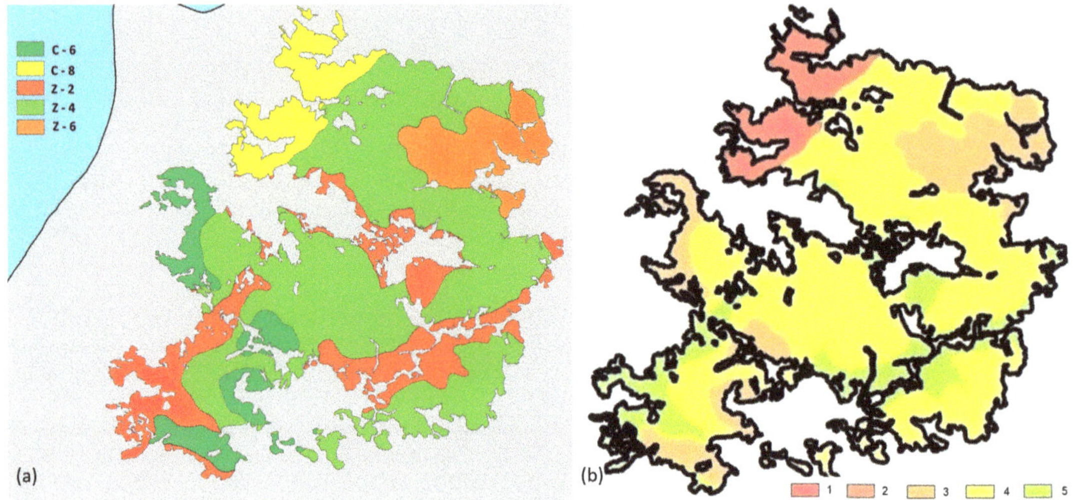

Figure 6. Soil (**a**) before the classification and (**b**) after the classification.

3.1.3. dNBR

The applications of dNBR formula (2) produce, as a result, a new image with a spatial resolution of 10 m that includes the NBRprefire, NBRpostfire and dNBR channels. The final dNBR map and its reclassification are presented in the following figure (Figure 7).

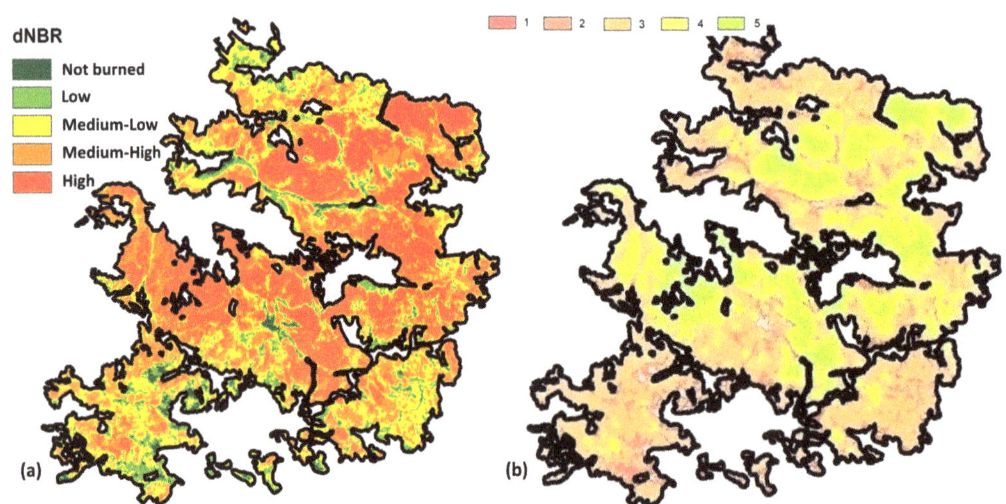

Figure 7. (**a**) dNBR and (**b**) reclassification of dNBR.

3.1.4. Aspect

The creation of the aspect layer was based on the DEM of the study area. The next figure presents the calculation of the aspect (Figure 8a) for the study area and its reclassification (Figure 8b).

Figure 8. (**a**) Aspect and (**b**) reclassification of aspect.

Aspect data are valuable for selecting appropriate planting sites during ecological restoration. These data allow practitioners to match plant species with the prevailing microclimate conditions, increasing the chances of successful revegetation.

3.1.5. Compound Topographic Index (CTI)

The following figure presents the CTI calculation (Figure 9a) for the study area and its reclassification (Figure 9b). The parameter of runoff accumulation, As, which is used in the numerator of the CTI index, was related to the size of the analysis applied. The index values ranged from 1.81 to 18.87, with a mean of 5.09 and a standard deviation of 1.585. Higher values identify, to a greater extent, the drainage network, while lower values are associated with steep slopes, peaks and low water concentration.

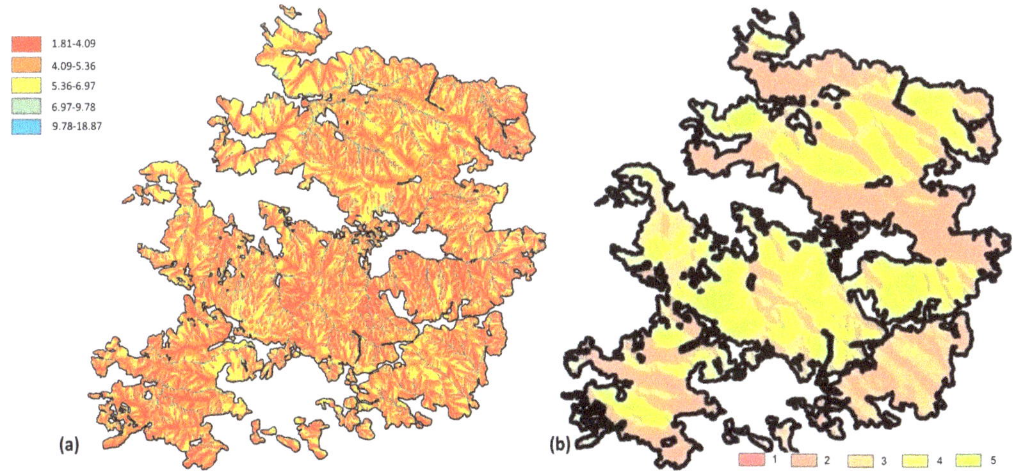

Figure 9. (**a**) CTI and (**b**) reclassification of CTI.

3.1.6. Distance from Settlements

This criterion is related to both economic and aesthetic and protective purposes. Both the establishment and restoration of vegetation around settlement boundaries enhance the landscape and provide relief to the local community. However, they also performs other roles, such as protecting soil from erosion and potential landslides and improving local climatic conditions [61,62]. For the calculation of the distance from the nearest settlement, a maximum distance of 4 km was set as a parameter so that there is overlap between the points and to cover the entire study area. The figure that follows shows the distance from the settlement map (Figure 10a) and its reclassification (Figure 10b).

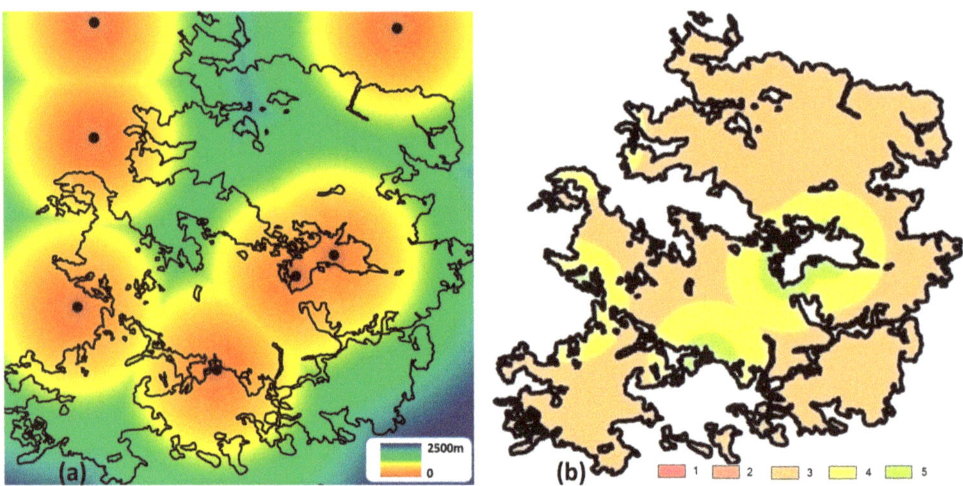

Figure 10. (**a**) Distance from settlements (**b**) reclassification of distance from settlements.

3.1.7. Distance from the Road Network

The areas that are close to the road network, at a short distance from the deck, are more favorable locations for reforestation activities, as they significantly reduce the cost of the individual works. The transport of materials, staff labor and the necessary equipment increases as we move further away from the existing road network. An example is the watering of young seedlings, where the necessary hoses and infrastructure increase dramatically as we move away from a road [1]. For the variable of distance from the road network, three suitability zones of 50, 200 and more than 200 m applied. This classification of the distance from the Road Network (Figure 11a), together with its reclassification (Figure 11b), is shown in the next figure. The red lines in Figure 11a represent the initial Road Network.

3.2. Criteria Synthesis Results

After reclassifying criteria, the analytical process included homogenizing variables to a 1–5 scale. This standardized approach eliminated disparities, enabling equitable comparisons and facilitating clear communication of results and recommendations to stakeholders. The next table (Table 3) presents this homogenization of the variables.

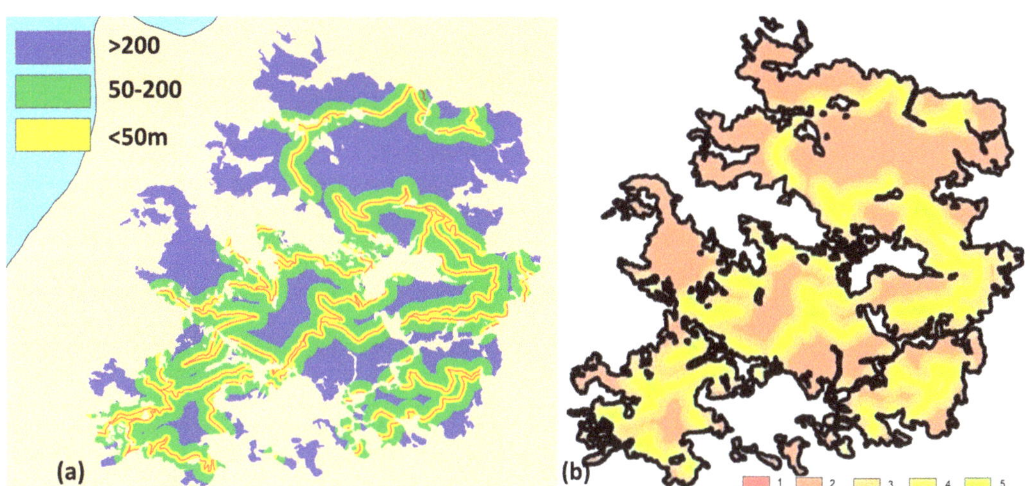

Figure 11. (**a**) Classification of the distance from the Road Network and (**b**) reclassification of distance from the Road Network.

Table 3. Parameters classification.

Variable	Value	Suitability
Vegetation	Cool-weather conifers	5
	Thermophilous conifers	4
	Herbaceous vegetation	4
	Shrubby vegetation	1
Soil	I	4
	II	3
	III	2
	IV	1
dNBR	High	5
	Medium–high	3
	Medium–low	3
	Low	1
	Not burned	0
	0–45°	4
	45–135°	3
	135–225°	2
	225–315°	2
	315–360	4
CTI	High	5
	Medium	4
	Low	2
Distance from settlements	0–500 m	5
	500–1000 m	4
	>1000 m	3
Distance from Road Network	0–50 m	5
	50–200 m	3
	>200 m	2

The variables under consideration encompass a diverse range of both quantitative and qualitative attributes, each possessing distinct characteristics that cannot be readily aggregated due to their inherent disparities (such as the juxtaposition of quantitative measures like distance in meters and qualitative assessments related to vegetation by species). To address this challenge and enable a more cohesive and meaningful analysis, a crucial methodology was employed: the reclassification of each variable onto an ordinal scale spanning from 1 to 5. Within this scale, a value of 1 signifies a condition of very low suitability for reforestation, whereas a value of 5 denotes a state of very high suitability for reforestation. This systematic reclassification process not only harmonized the diverse variables into a common framework but also allowed for a nuanced and standardized assessment of their individual contributions to the overall analysis. By ascribing these values, we not only achieved a clear and consistent means of evaluating each variable's suitability for reforestation independently but also paved the way for more robust and integrated decision-making processes in the context of environmental restoration and land management initiatives.

This ordinal reclassification approach has proven instrumental in distilling the complex array of attributes associated with each variable into a simplified yet informative format. It acknowledges that variables encompass multifaceted dimensions, and by assigning them values on a 1 to 5 scale, we can capture their inherent nuances and relative importance. This structured methodology not only streamlines the assessment process but also lends itself to more effective comparisons and prioritization of variables within the broader context of reforestation and ecological restoration efforts. Furthermore, it fosters a deeper understanding of the intricacies of each variable's role in influencing suitability for reforestation, facilitating more informed and data-driven decision making. In essence, the ordinal reclassification of variables into this 1-to-5 scale serves as a pivotal step toward achieving a holistic and well-informed approach to land management and environmental restoration, ensuring that the unique characteristics of each variable are appropriately considered and integrated into the overarching analysis.

3.3. Weighted Overlay of the Classified Variables

The final step in this study involved combining homogenized variables, using specific weighting factors. Two scenarios were implemented: Scenario 1 prioritizes ecological constraints, focusing on optimal vegetation growth conditions, while Scenario 2 emphasizes economic and social benefits, favoring criteria like the distance from settlements and roads. Weighting coefficients were determined through the Analytical Hierarchy Process [59], where pairwise comparisons on a scale of 1 to 9 informed the significance of each criterion. This approach ensures a balanced assessment, considering both ecological and socioeconomic factors for identifying optimal ecological restoration sites.

The scoring of the criteria was repeated three times for each scenario, each time applying small differences between the variables. The weighting coefficients were then calculated, and the final model results were obtained for each scenario. The test showed that the small changes applied to the values of the variables did not bring about significant changes in the results of the scenarios. The two tables that follow (Tables 4 and 5) are presenting the comparison matrices for the above two scenarios.

Table 4. Pairwise comparison matrix for Scenario 1.

	Vegetation	Soil	dNBR	CTI	Aspect	Roads	Settlements
Vegetation	1.00	2.00	3.00	3.00	5.00	6.00	7.00
Soil	0.50	1.00	3.00	3.00	7.00	6.00	8.00
dNBR	0.33	0.33	1.00	3.00	4.00	6.00	6.00
CTI	0.33	0.33	0.33	1.00	4.00	2.00	2.00
Aspect	0.20	0.14	0.25	0.25	1.00	2.00	2.00
Roads	0.17	0.17	0.17	0.50	0.50	1.00	3.00
Settlements	0.14	0.13	0.17	0.50	0.50	0.33	1.00

Table 5. Pairwise comparison matrix for Scenario 2.

	Roads	Settlements	Vegetation	Soil	dNBR	CTI	Aspect
Roads	1.00	2.00	1.00	2.00	3.00	3.00	6.00
Settlements	0.50	1.00	1.00	2.00	3.00	3.00	6.00
Vegetation	1.00	1.00	1.00	2.00	3.00	4.00	6.00
Soil	0.50	0.50	0.50	1.00	3.00	5.00	5.00
dNBR	0.33	0.33	0.33	0.33	1.00	2.00	3.00
CTI	0.33	0.33	0.25	0.20	0.50	1.00	2.00
Aspect	0.17	0.17	0.17	0.20	0.33	0.50	1.00

The final scoring values and the hierarchy between the variables in the two tables above were considered reasonable, since the CR consistency ratio was less than 0.10 for both scenarios [63]. Specifically, in Scenario 1, the consistency ratio was CR = 0.06, and in Scenario 2, it was CR = 0.03. Based on the averages of the comparison tables, the weighting coefficients of each variable were obtained, as presented in the following table (Table 6). The sum of the weighting coefficients for each scenario is summed to unity, so that the results are on the same classification scale and allow for a comparison between them.

Table 6. The weights for the variables for the two scenarios.

Variable	Weights	
	Scenario 1	Scenario 2
Vegetation	0.319	0.226
Soil	0.275	0.162
Aspect	0.052	0.032
dNBR	0.178	0.079
CTI	0.096	0.056
Distance from the Road Network	0.048	0.245
Distance from settlements	0.032	0.200

The weights derived from the Analytic Hierarchy Process (AHP), pioneered by Saaty, constitute a pivotal component of informed decision-making processes across diverse domains. AHP offers a systematic and structured methodology for evaluating and prioritizing alternatives or criteria within a hierarchical framework. These weights are of paramount importance as they provide a quantitative representation of the relative significance or influence of different elements within the decision-making hierarchy. The AHP process commences with the construction of a hierarchical model, comprising a goal, criteria, sub-criteria and potential alternatives. Stakeholders or experts are then tasked with the crucial step of making pairwise comparisons between elements at each level of the hierarchy. These comparisons serve as the foundation for computing priority vectors, which assign numerical values to the relative importance of each element. The significance of these weights is multifaceted. The final weights are presented in the next table (Table 6).

In Scenario 2, the variable weights reflect a deliberate emphasis on economic and social factors, particularly favoring criteria related to distance from the road network. The seemingly less smooth distribution of values in this scenario is a result of the intentional higher influence assigned to the distance from the Road Network criterion. This emphasis is in line with the scenario's prioritization of economic and social benefits, where road accessibility is a key consideration. While this may result in variations in the weights compared to the smoother distribution in Scenario 1, it aligns with the model's intended emphasis on specific criteria in different scenarios to capture diverse restoration considerations.

3.4. The Results for the Two Scenarios

As mentioned, Scenario 1 favors the ecological characteristics of the reforestation point, while Scenario 2 gives more weight to economic criteria. The final suitability zones of the results were classified and presented on the same regular classification scale used to categorize each of the model variables. As shown in the two maps (Figures 12 and 13), areas with very high suitability for reforestation are depicted in dark green, and the sum of the weighted variables tends toward a value of 5. In these locations, according to the previous analysis of the criteria, the conditions for the establishment of new plants are considered to be particularly favorable. Furthermore, the severity of the fire and the type of vegetation at these sites indicate that these are areas where natural regeneration is unlikely to recover. In contrast, areas in red (Figures 12 and 13) record areas where there is very little suitability for reforestation, where the corresponding sum of the individual weighted variables is close to unity. In these areas, the success of new plantings is considered ineffective, and it is considered that such sites should be avoided for reforestation activities.

Figure 12. Map of suitability for ecological restoration of Scenario 1.

A first approximation of the results shows a concentration of areas with very high suitability in the center of the burnt area near the settlement of Mikro Prinos for both scenarios. These areas range from approximately 400 to 800 m in elevation and have moderate slopes. Most of them are located in northern exposures and near a dense road network. It can be concluded that the proposed areas cover a large part of the requirements for successful reforestation interventions.

However, to make further comparisons between the cartographic results of the two scenarios and to draw conclusions, the area of each suitability zone was calculated. Table 7 compares the values of the two scenarios.

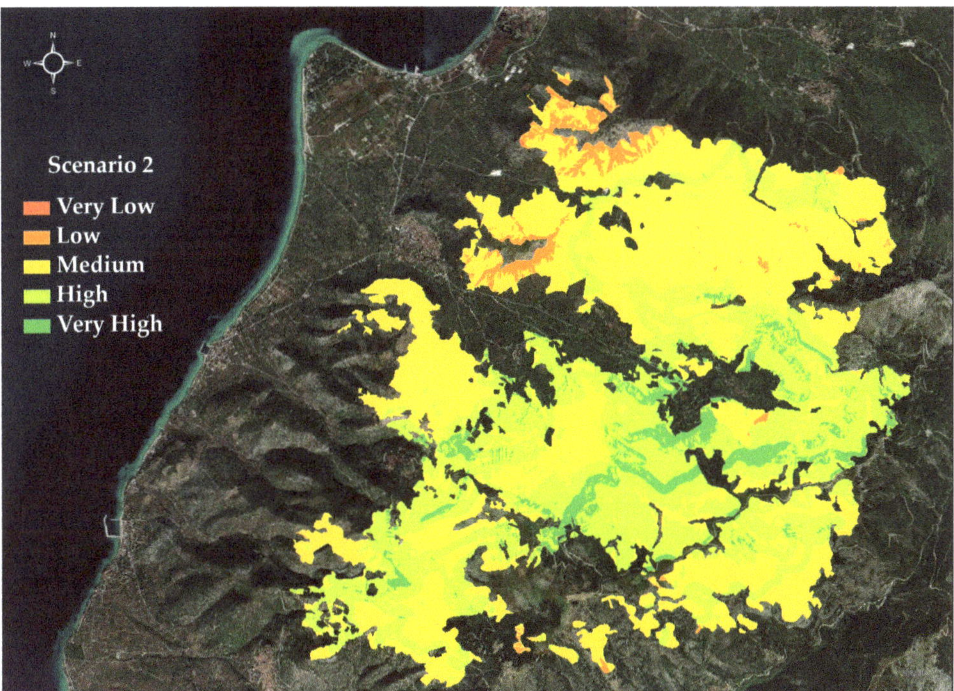

Figure 13. Map of suitability for ecological restoration of Scenario 2.

Table 7. Area and afforestation suitability rates for each scenario.

Suitability	Scenario 1		Scenario 2	
	Area (km^2)	Percentage	Area (km^2)	Percentage
Very low	439	0.17	59	0.02
Low	11,437	4.32	8263	3.12
Medium	85,473	32.27	125,388	47.34
High	151,245	57.10	115,398	43.57
Very high	16,286	6.15	15,772	5.95
Total	264,880	100.00	264,880	100.00

The ability to assess and identify the most suitable sites for restoration immediately after a forest fire is very important for both the affected natural ecosystem and the local community. In this study, a model was developed to evaluate affected areas for suitability for reforestation based on a number of ecological and economic parameters. The final results obtained for the study area and the generated suitability maps show the usefulness of the model and its application in the decision-making process.

It should be emphasized that the evaluation model was based on a set of variables that were evaluated and selected according to a review of the international literature. The assessment of phenomena based on ecological parameters is a complex and sophisticated process. The selection of the variables used is related to the importance of the specific criteria in the restoration processes, and this was based on the hierarchy between them. According to the results of the two scenarios and the small deviations between them, it can be concluded that the hierarchy and classification of the variables were correct. Moreover, rerunning the model with slight variations in the scoring values of the criteria gave similar results.

It should be noted that the contribution of geographical information systems and remote sensing played a key role in the analysis and implementation of the evaluation model, providing a wealth of tools for data processing and preparation. The application of the methodology, the processing of the variables and the extraction of the final results in cartographic format relied on both the capabilities of GIS and remote sensing. The synergy of these two disciplines in this work allowed for the evaluation of the study area and the extraction of detailed and usable results. Otherwise, the absence of these tools would have required time-consuming and costly field work, as well as analogous processing methods that would have complicated and significantly delayed its completion.

It is worth noting that the methodology applied in this work was mainly based on the conversion and processing of mosaic files. The analysis applied to the set of variables was based on a 10×10 m tessellation size. The detail discernible in the cartographic results of the model proves that the choice of this scale was successful. The area manager can identify individual surfaces with great accuracy and plan his operations in detail. After all, reforestation operations are, in practice, carried out on individual areas that are much larger than 10% of the hectare. A resolution of 100 square meters per pixel is more than sufficient to identify suitable areas. One problem identified during processing with mosaic files has to do with the selection of tiles at the edges of the study area. The complex boundary of the fire resulted in several mosaics intersecting with the study area boundary line. These tesserae were taken into account in the subsequent analysis, even though a portion of them were outside the boundary line. The change in the total area of the study area was very small, as the 10×10 m scale applied was very detailed.

According to the results of the table, the largest changes between the two scenarios are seen in the medium and large scales, where the differences are 15.07 and 13.53 (%), respectively. On the other hand, in the very small and very large suitability categories, there are no major changes between the two scenarios. In other words, it is detected that, both at the level of percentages and the level of delimitation, the two categories are significantly similar. The detailed spatial analysis reveals nuanced patterns in the very small and very large suitability categories between Scenario 1 and Scenario 2. In the very small suitability category, where economic and social factors are more heavily weighted in Scenario 2, the spatial distribution appears to reflect an increased emphasis on areas with enhanced accessibility, potentially influenced by proximity to the road network. This shift aligns with the socioeconomic focus of Scenario 2. Conversely, in the very large suitability category, dominated by the ecological considerations of Scenario 1, the spatial distribution tends to exhibit a more extensive pattern, prioritizing areas with optimal ecological conditions for restoration. The observed variations in these categories emphasize the model's sensitivity to the distinct emphases of each scenario and underscore the need for a comprehensive understanding of the trade-offs between ecological and socioeconomic priorities in the decision-making process for identifying optimal ecological restoration sites.

It can therefore be concluded that—at least on the basis of this study—both scenarios are equally effective for identifying areas that are not suitable for afforestation and those that are highly suitable for afforestation. However, more research is required before generalizing this finding, such as conducting the same experiment in other settings. More thorough testing of the results could be achieved by validating the model from actual restoration data. After checking all available data and contacting the competent authorities, it was found that no extensive reforestation program has been implemented for this area or the wider area of Thasos. Verification of the model with field data would therefore require its application to a burnt area for which a restoration program has been carried out, followed by longitudinal monitoring of the evolution and progress of reforestation. Such a task is very time-consuming and difficult to implement in the context of a thesis but could be the subject of a future research project.

4. Conclusions

Identifying and evaluating ecological restoration sites through geospatial technologies involves the systematic analysis of spatial data, such as GIS mapping and remote sensing imagery, to pinpoint areas in need of ecological rehabilitation. By leveraging these advanced technologies, environmental scientists can assess the extent of damage, identify key ecological indicators and prioritize strategic restoration efforts for landscapes affected by various disturbances, including wildfires or habitat degradation.

Regarding the results of the two different scenarios, it is concluded that both delineate the "very large" and "very small" suitability sites in an almost similar way. Despite the assignment of different weighting factors between the two scenarios, both identify and qualify successful sites suitable for reforestation. In this paper, the proposed model for the evaluation of a burnt area for restoration was developed based on a set of variables and applied to the fire in Prinos, Thasos. It would be of interest to further study and investigate additional criteria that may influence the assessment and progression of the restoration process, including research on possible parameters, e.g., soil water potential or climatic conditions, that could be included as variables in the assessment of the area. At the same time, it is proposed to extend the model developed to other areas that may not have the same physiographic characteristics as those of Prinos, Thasos.

Author Contributions: Conceptualization, S.D.; and G.P.P.; methodology, S.D. and G.P.P.; formal analysis, S.D.; investigation, S.D.; writing—original draft preparation, S.D., K.K. and G.P; writing—review and editing, S.D., K.K. and G.P.P.; visualization, S.D., K.K. and G.P.P.; supervision, G.P.P.; project administration, G.P.P. All authors have read and agreed to the published version of the manuscript.

Funding: This research received no external funding.

Conflicts of Interest: The authors declare no conflict of interest.

References

1. Kakouros, P.; Dafis, S. *Guidelines for Restoration of Pinus Nigra Forests Affected by Fires through a Structured Approach*; Greek Biotope Wetland Centre: Thermi, Greece, 2010; p. 27.
2. Martín-Alcón, S.; Coll, L. Unraveling the Relative Importance of Factors Driving Post-Fire Regeneration Trajectories in Non-Serotinous Pinus Nigra Forests. *For. Ecol. Manag.* **2016**, *361*, 13–22. [CrossRef]
3. Khan, S.M.; Shafi, I.; Butt, W.H.; Diez, I.D.L.T.; Flores, M.A.L.; Galán, J.C.; Ashraf, I. A Systematic Review of Disaster Management Systems: Approaches, Challenges, and Future Directions. *Land* **2023**, *12*, 1514. [CrossRef]
4. Karamesouti, M.; Petropoulos, G.P.; Papanikolaou, I.D.; Kairis, O.; Kosmas, K. Erosion Rate Predictions from PESERA and RUSLE at a Mediterranean Site before apnd after a Wildfire: Comparison & Implications. *Geoderma* **2016**, *261*, 44–58. [CrossRef]
5. Birot, Y. *Living with Wildfires: What Science Can Tell Us*; European Forest Institute: Joensuu, Finland, 2009; Volume 15.
6. Vukomanovic, J.; Steelman, T. A Systematic Review of Relationships Between Mountain Wildfire and Ecosystem Services. *Landsc. Ecol.* **2019**, *34*, 1179–1194. [CrossRef]
7. San-Miguel-Ayanz, J.; Durrant, T.; Boca, R.; Maianti, P.; Liberta, G.; ArtesVivancos, T.; Jacome Felix Oom, D.; Branco, A.; De Rigo, D.; Ferrari, D.; et al. *Forest Fires in Europe, Middle East and North Africa 2020*; Publications Office of the European Union: Luxembourg, 2021.
8. Wu, M.; Knorr, W.; Thonicke, K.; Schurgers, G.; Camia, A.; Arneth, A. Sensitivity of Burned Area in Europe to Climate Change, Atmospheric CO_2 Levels, and Demography: A Comparison of Two Fire-vegetation Models. *JGR Biogeosci.* **2015**, *120*, 2256–2272. [CrossRef]
9. Pausas, J.G.; Keeley, J.E. Wildfires as an Ecosystem Service. *Front. Ecol. Environ.* **2019**, *17*, 289–295. [CrossRef]
10. Parissi, Z.M.; Kyriazopoulos, A.P.; Drakopoulou, T.A.; Korakis, G.; Abraham, E.M. Wildfire Effects on Rangeland Health in Three Thermo-Mediterranean Vegetation Types in a Small Islet of Eastern Aegean Sea. *Land* **2023**, *12*, 1413. [CrossRef]
11. Di Castri, F.; Mooney, H.A. (Eds.) *Mediterranean Type Ecosystems: Origin and Structure*; Ecological Studies; Springer: Berlin/Heidelberg, Germany, 1973; Volume 7, ISBN 978-3-642-65522-7.
12. Davis, K.T.; Dobrowski, S.Z.; Higuera, P.E.; Holden, Z.A.; Veblen, T.T.; Rother, M.T.; Parks, S.A.; Sala, A.; Maneta, M.P. Wildfires and Climate Change Push Low-Elevation Forests across a Critical Climate Threshold for Tree Regeneration. *Proc. Natl. Acad. Sci. USA* **2019**, *116*, 6193–6198. [CrossRef]
13. Dafis, S. *Forest Ecology*; Yahoudis-Giapouli Publications: Thessaloniki, Greece, 1986; ISBN 978-618-5092-47-4.
14. Xanthopoulos, G.; Calfapietra, C.; Fernandes, P. Fire Hazard and Flammability of European Forest Types. In *Post-Fire Management and Restoration of Southern European Forests*; Moreira, F., Arianoutsou, M., Corona, P., De Las Heras, J., Eds.; Managing Forest Ecosystems; Springer: Dordrecht, The Netherlands, 2012; Volume 24, pp. 79–92, ISBN 978-94-007-2207-1.

15. Castro, J. Post-Fire Restoration of Mediterranean Pine Forests. In *Pines and Their Mixed Forest Ecosystems in the Mediterranean Basin*; Ne'eman, G., Osem, Y., Eds.; Managing Forest Ecosystems; Springer International Publishing: Cham, Switzerland, 2021; Volume 38, pp. 537–565, ISBN 978-3-030-63624-1.

16. Eleftheriadis, N.; Vergos, S.; Tzortzi, T. *Planning after Forest Fire Disaster*; General Secretariat of Civil Protection: Athens, Greece, 2001; pp. 211–214.

17. Ireland, G.; Petropoulos, G.P. Exploring the Relationships between Post-Fire Vegetation Regeneration Dynamics, Topography and Burn Severity: A Case Study from the Montane Cordillera Ecozones of Western Canada. *Appl. Geogr.* **2015**, *56*, 232–248. [CrossRef]

18. Petropoulos, G.P.; Griffiths, H.M.; Kalivas, D.P. Quantifying Spatial and Temporal Vegetation Recovery Dynamics Following a Wildfire Event in a Mediterranean Landscape Using EO Data and GIS. *Appl. Geogr.* **2014**, *50*, 120–131. [CrossRef]

19. Bonazountas, M.; Kallidromitou, D.; Kassomenos, P.; Passas, N. A Decision Support System for Managing Forest Fire Casualties. *J. Environ. Manag.* **2007**, *84*, 412–418. [CrossRef] [PubMed]

20. Kalabokidis, K.; Xanthopoulos, G.; Moore, P.; Caballero, D.; Kallos, G.; Llorens, J.; Roussou, O.; Vasilakos, C. Decision Support System for Forest Fire Protection in the Euro-Mediterranean Region. *Eur. J. For. Res.* **2012**, *131*, 597–608. [CrossRef]

21. Sakellariou, S.; Sfougaris, A.; Christopoulou, O. Review of Geoinformatics-Based Forest FireManagement Tools for IntegratedFire Analysis. *Pol. J. Environ. Stud.* **2021**, *30*, 5423–5434. [CrossRef]

22. Milczarek, M.; Aleksandrowicz, S.; Kita, A.; Chadoulis, R.-T.; Manakos, I.; Woźniak, E. Object- Versus Pixel-Based Unsupervised Fire Burn Scar Mapping under Different Biogeographical Conditions in Europe. *Land* **2023**, *12*, 1087. [CrossRef]

23. Tselka, I.; Detsikas, S.E.; Petropoulos, G.P.; Demertzi, I.I. Google Earth Engine and Machine Learning Classifiers for Obtaining Burnt Area Cartography: A Case Study from a Mediterranean Setting. In *Geoinformatics for Geosciences*; Elsevier: Amsterdam, The Netherlands, 2023; pp. 131–148, ISBN 978-0-323-98983-1.

24. Tsatsaris, A.; Kalogeropoulos, K.; Stathopoulos, N.; Louka, P.; Tsanakas, K.; Tsesmelis, D.E.; Krassanakis, V.; Petropoulos, G.P.; Pappas, V.; Chalkias, C. Geoinformation Technologies in Support of Environmental Hazards Monitoring under Climate Change: An Extensive Review. *ISPRS Int. J. Geo-Inf.* **2021**, *10*, 94. [CrossRef]

25. Abdullah, A.Y.M.; Masrur, A.; Adnan, M.S.G.; Baky, M.A.A.; Hassan, Q.K.; Dewan, A. Spatio-Temporal Patterns of Land Use/Land Cover Change in the Heterogeneous Coastal Region of Bangladesh between 1990 and 2017. *Remote Sens.* **2019**, *11*, 790. [CrossRef]

26. Eskandari, S.; Pourghasemi, H.R.; Tiefenbacher, J.P. Relations of land cover, topography, and climate to fire occurrence in natural regions of Iran: Applying new data mining techniques for modeling and mapping fire danger. *For. Ecol. Manag.* **2020**, *473*, 118338. [CrossRef]

27. Fragou, S.; Kalogeropoulos, K.; Stathopoulos, N.; Louka, P.; Srivastava, P.K.; Karpouzas, S.; Kalivas, D.P.; Petropoulos, G.P. Quantifying Land Cover Changes in a Mediterranean Environment Using Lands at TM and Support Vector Machines. *Forests* **2020**, *11*, 750. [CrossRef]

28. Malczewski, J. Multiple Criteria Decision Analysis and Geographic Information Systems. In *Trends in Multiple Criteria Decision Analysis*; Ehrgott, M., Figueira, J.R., Greco, S., Eds.; International Series in Operations Research & Management Science; Springer: Boston, MA, USA, 2010; Volume 142, pp. 369–395, ISBN 978-1-4419-5903-4.

29. Petropoulos, G.P.; Kontoes, C.C.; Keramitsoglou, I. Land Cover Mapping with Emphasis to Burnt Area Delineation Using Co-Orbital ALI and Landsat TM Imagery. *Int. J. Appl. Earth Obs. Geoinf.* **2012**, *18*, 344–355. [CrossRef]

30. Knopp, L.; Wieland, M.; Rättich, M.; Martinis, S. A Deep Learning Approach for Burned Area Segmentation with Sentinel-2 Data. *Remote Sens.* **2020**, *12*, 2422. [CrossRef]

31. Bachofer, F.; Quénéhervé, G.; Märker, M.; Hochschild, V. Comparison of SVM and boosted regression trees for the delineation of lacustrine sediments using multispectral ASTER data and topographic indices in the Lake Manyara Basin. *Photogramm. Fernerkund. Geoinf.* **2015**, *2015*, 81–94. [CrossRef]

32. Veraverbeke, S.; Gitas, I.; Katagis, T.; Polychronaki, A.; Somers, B.; Goossens, R. Assessing Post-Fire Vegetation Recovery Using Red–near Infrared Vegetation Indices: Accounting for Background and Vegetation Variability. *ISPRS J. Photogramm. Remote Sens.* **2012**, *68*, 28–39. [CrossRef]

33. Yadav, C.S.; Pradhan, M.K.; Gangadharan, S.M.P.; Chaudhary, J.K.; Singh, J.; Khan, A.A.; Haq, M.A.; Alhussen, A.; Wechtaisong, C.; Imran, H.; et al. Multi-Class Pixel Certainty Active Learning Model for Classification of Land Cover Classes Using Hyperspectral Imagery. *Electronics* **2022**, *11*, 2799. [CrossRef]

34. Matin, M.A.; Chitale, V.S.; Murthy, M.S.R.; Uddin, K.; Bajracharya, B.; Pradhan, S. Understanding Forest Fire Patterns and Risk in Nepal Using Remote Sensing, Geographic Information System and Historical Fire Data. *Int. J. Wildland Fire* **2017**, *26*, 276. [CrossRef]

35. Atun, R.; Kalkan, K.; Gürsoy, Ö. Determining The Forest Fire Risk with Sentinel 2 Images. *Turk. J. Geosci.* **2020**, *1*, 22–26.

36. Fornacca, D.; Ren, G.; Xiao, W. Evaluating the Best Spectral Indices for the Detection of Burn Scars at Several Post-Fire Dates in a Mountainous Region of Northwest Yunnan, China. *Remote Sens.* **2018**, *10*, 1196. [CrossRef]

37. Park, J.; Woodam, S.; Lee, J. Evaluation of Suitable REDD+ Sites Based on Multiple-Criteria Decision Analysis (MCDA): A Case Study of Myanmar. *J. For. Environ. Sci.* **2018**, *34*, 461–471.

38. Christakopoulos, P.; Hatzopoulos, I.; Kalabokidis, K.; Paronis, D.; Filintas, A. Assessment of the response of a Mediterranean-type forest ecosystem to recurrent wildfires and to different restoration practices using Remote Sensing and GIS techniques. In *Advances in Remote Sensing and GIS Applications in Forest Fire Management, Proceedings of the 6th International Workshop of the EARSeL, Thessaloniki, Greece, 27–29 September 2007*; Gitas, I., Carmona-Moreno, C., Eds.; European Commission: Thessaloniki, Greece, 2007; pp. 213–216.
39. Stefanov, S. Model of an Open-Source Web GIS Application for Observation in Case of Wildland Fires. In *New Frontiers in Communication and Intelligent Systems*; Soft Computing Research Society: New Delhi, India, 2021; pp. 547–553, ISBN 978-81-955020-0-4.
40. Adaktylou, N.; Stratoulias, D.; Landenberger, R. Wildfire Risk Assessment Based on Geospatial Open Data: Application on Chios, Greece. *ISPRS Int. J. Geo-Inf.* **2020**, *9*, 516. [CrossRef]
41. Bridges, J.M.; Petropoulos, G.P.; Clerici, N. Immediate Changes in Organic Matter and Plant Available Nutrients of Haplic Luvisol Soils Following Different Experimental Burning Intensities in Damak Forest, Hungary. *Forests* **2019**, *10*, 453. [CrossRef]
42. Arciniegas, G.; Janssen, R. Spatial Decision Support for Collaborative Land Use Planning Workshops. *Landsc. Urban Plan.* **2012**, *107*, 332–342. [CrossRef]
43. Amos, C.; Petropoulos, G.P.; Ferentinos, K.P. Determining the Use of Sentinel-2A MSI for Wildfire Burning & Severity Detection. *Int. J. Remote Sens.* **2019**, *40*, 905–930. [CrossRef]
44. Chou, Y.H. Management of Wildfires with a Geographical Information System. *Int. J. Geogr. Inf. Syst.* **1992**, *6*, 123–140. [CrossRef]
45. Epp, H.; Lanoville, R. Satellite Data and Geographic Information Systems for Fire and Resource Management in the Canadian Arctic. *Geocarto Int.* **1996**, *11*, 97–103. [CrossRef]
46. Ruiz-Gallardo, J.R.; Castaño, S.; Calera, A. Application of Remote Sensing and GIS to Locate Priority Intervention Areas after Wildland Fires in Mediterranean Systems: A Case Study from South-Eastern Spain. *Int. J. Wildland Fire* **2004**, *13*, 241. [CrossRef]
47. Schulz, J.J.; Schröder, B. Identifying Suitable Multifunctional Restoration Areas for Forest Landscape Restoration in Central Chile. *Ecosphere* **2017**, *8*, e01644. [CrossRef]
48. Xie, H.; Yao, G.; Liu, G. Spatial Evaluation of the Ecological Importance Based on GIS for Environmental Management: A Case Study in Xingguo County of China. *Ecol. Indic.* **2015**, *51*, 3–12. [CrossRef]
49. Fernández-Guisuraga, J.M.; Suárez-Seoane, S.; Calvo, L. Modeling Pinus Pinaster Forest Structure after a Large Wildfire Using Remote Sensing Data at High Spatial Resolution. *For. Ecol. Manag.* **2019**, *446*, 257–271. [CrossRef]
50. Brown, A.R.; Petropoulos, G.P.; Ferentinos, K.P. Appraisal of the Sentinel-1 & 2 Use in a Large-Scale Wildfire Assessment: A Case Study from Portugal's Fires of 2017. *Appl. Geogr.* **2018**, *100*, 78–89. [CrossRef]
51. Marjokorpi, A.; Otsamo, R. Prioritization of Target Areas for Rehabilitation: A Case Study from West Kalimantan, Indonesia. *Restor. Ecol.* **2006**, *14*, 662–673. [CrossRef]
52. Drobne, S.; Lisec, A. Multi-Attribute Decision Analysis in GIS: Weighted Linear Combination and Ordered Weighted Averaging. *Informatica* **2009**, *33*, 459–474.
53. Silva, V.A.M.; Mello, K.D.; Vettorazzi, C.A.; Costa, D.R.D.; Valente, R.A. Priority areas for forest conservation, aiming at the maintenance of water resources, through the multicriteria evaluation. *Rev. Árvore* **2017**, *41*, e410119. [CrossRef]
54. Ehrgott, M.; Figueira, J.R.; Greco, S. (Eds.) *Trends in Multiple Criteria Decision Analysis*; International Series in Operations Research & Management Science; Springer: Boston, MA, USA, 2010; Volume 142, ISBN 978-1-4419-5903-4.
55. Eimil-Fraga, C.; Rodríguez-Soalleiro, R.; Sánchez-Rodríguez, F.; Pérez-Cruzado, C.; Álvarez-Rodríguez, E. Significance of Bedrock as a Site Factor Determining Nutritional Status and Growth of Maritime Pine. *For. Ecol. Manag.* **2014**, *331*, 19–24. [CrossRef]
56. García, M.J.L.; Caselles, V. Mapping Burns and Natural Reforestation Using Thematic Mapper Data. *Geocarto Int.* **1991**, *6*, 31–37. [CrossRef]
57. Key, C.H.; Benson, N.C. *Landscape Assessment: Ground Measure of Severity, the Composite Burn Index; and Remote Sensing of Severity, the Normalized Burn Ratio*; USDA Forest Service, Rocky Mountain Research Station: Ogden, UT, USA, 2006.
58. Beven, K.J.; Kirkby, M.J. A Physically Based, Variable Contributing Area Model of Basin Hydrology/Un Modèle à Base Physique de Zone d'appel Variable de l'hydrologie Du Bassin Versant. *Hydrol. Sci. Bull.* **1979**, *24*, 43–69. [CrossRef]
59. Saaty, R.W. The Analytic Hierarchy Process—What It Is and How It Is Used. *Math. Model.* **1987**, *9*, 161–176. [CrossRef]
60. Nakos, G. *Land Classification, Mapping and Evaluation: Technical Specifications*; Ministry of Agriculture: Athens, Greece, 1991.
61. Long, J.W.; Quinn-Davidson, N.; Skinner, C.N. *Science Synthesis to Support Socioecological Resilience in the Sierra Nevada and Southern Cascade Range*; U.S. Department of Agriculture, Forest Service: Albany, CA, USA, 2014; p. 723.
62. Aguirre-Salado, C.; Miranda-Aragón, L.; Pompa-García, M.; Reyes-Hernández, H.; Soubervielle-Montalvo, C.; Flores-Cano, J.; Méndez-Cortés, H. Improving Identification of Areas for Ecological Restoration for Conservation by Integrating USLE and MCDA in a GIS-Environment: A Pilot Study in a Priority Region Northern Mexico. *ISPRS Int. J. Geo-Inf.* **2017**, *6*, 262. [CrossRef]
63. Saaty, T.L.; Vargas, L.G. Uncertainty and Rank Order in the Analytic Hierarchy Process. *Eur. J. Oper. Res.* **1987**, *32*, 107–117. [CrossRef]

 *land*

Article

Unraveling the European Agricultural Policy Sustainable Development Trajectory

Yannis E. Doukas [1,*,†], Luca Salvati [2,†] and Ioannis Vardopoulos [3,4,*,†]

1   Department of Agricultural Development, Agri-Food and Natural Resources Management, School of Agricultural Development, Nutrition and Sustainability, National and Kapodistrian University of Athens (UoA), 34400 Psachna, Greece
2   Department of Methods and Models for Territory, Economics and Finance (MEMOTEF), Faculty of Economics, Sapienza University of Rome, Via del Castro Laurenziano 9, 00161 Rome, Italy
3   School of Environment, Geography and Applied Economics, Harokopio University (HUA), 17676 Kallithea, Greece
4   Department of Regional and Economic Development, School of Applied Economics and Social Sciences, Agricultural University of Athens (AUA), 33100 Amfissa, Greece
*   Correspondence: jodoukas@agro.uoa.gr (Y.E.D.); ivardopoulos@post.com (I.V.)
†   These authors contributed equally to this work.

Abstract: Amidst growing concerns about the impact of agriculture on the environment, the Common Agricultural Policy (CAP) has been overhauled to prioritize sustainable rural development in European agriculture. Based on this line of thought, the present contribution delves into the details of the CAP's shift, focusing on the main environmental concerns faced in the policy-making framework. Grounded in a political science perspective, the current study looks at how environmental and climate change concerns were gradually elevated inside the CAP's policy-making framework and how they helped create the "green architecture" for European agriculture. Examining the process of policy change under the lens of historical institutionalism and neo-institutionalism within the multilevel governance framework of the European Union (EU), the key role played by the gradual introduction of measures aimed at promoting measurable environmental criteria and climatic targets is highlighted. For instance, measures aimed at preserving carbon-rich soils and enhancing water resources can have positive impacts on the environment. However, these measures were also recognized to increase the cost of production for the European farmers, who faced serious difficulties in adjusting to the new framework. Within this context, this research delves into the roles played by two additional fundamental entities: the consumer and environmental activism. Additionally, the study underscores the EU's commitment to addressing climate change and sustainable development challenges and how conditionality is being used to link funding to results. Upon analyzing the CAP's shift, the reflection of a more flexible and rational approach is argued to be embodied by the new policy architecture. By incorporating both CAP pillars, encouraging collaboration with compatible policies, and allowing for greater adaptability in response to the unique circumstances and objectives of each member state, the CAP is taking significant steps towards sustainability and climate action. These insights into the significance and implications of the CAP's shift towards sustainability offer valuable recommendations for future policy developments, emphasizing the need to balance environmental concerns with the needs of farmers and other stakeholders.

Keywords: sustainability; environmental impact; circular economy; climate targets; policy-making; policy change; stakeholder engagement; agriculture; common agricultural policy (CAP)

Citation: Doukas, Y.E.; Salvati, L.; Vardopoulos, I. Unraveling the European Agricultural Policy Sustainable Development Trajectory. *Land* **2023**, *12*, 1749. https://doi.org/10.3390/land12091749

Academic Editors: Kleomenis Kalogeropoulos, Andreas Tsatsaris, Nikolaos Stathopoulos, Demetrios E. Tsesmelis, Nilanchal Patel and Xiao Huang

Received: 12 July 2023
Revised: 2 September 2023
Accepted: 7 September 2023
Published: 8 September 2023

1. Introduction

The Common Agricultural Policy (CAP) in the European Union (EU) has had associated costs and benefits that are largely variable over time and country, and based on such differences, EU member states (MSs) have consolidated diverging agendas and expectations regarding it [1]. The CAP involves multiple actors, from bureaucrats to sectoral interests' stakeholders, from governmental institutions to pressure groups, such as consumer and environmental activism, interested in agriculture [2]. In addition to economic benefits, the CAP integrates social and environmental aspects, promoting a resilient and sustainability-oriented agricultural structure throughout the EU [3]. This strategy preserves favorable environmental conditions that allow farmers to benefit from soil resources and maintain financial stability through continuous agro-food production [4], with broader implications (refer to [5]). Agricultural income not only supports farming households and communities in peri-urban and rural areas [6] but also contributes to society's overall gains from agricultural production [7]. In addition, agricultural activities are susceptible to climate change [8] while contributing to mitigation of global warming through greenhouse gas emissions reduction and carbon storage [9].

Especially during the last three decades, the CAP policy-making process has been evolving towards the creation of a sustainable and eco-friendly framework for European agriculture [10], promoting environmental consciousness at all levels of decision-making within the EU's complex multilevel governance framework [11,12]. However, despite the evident evolution of the CAP towards sustainability, there is a need to comprehensively examine the mechanisms driving this shift, especially considering the institutional characteristics of the CAP operation. Therefore, this study seeks to address the following research questions:

1. What are the key drivers of institutional shifts in the EU CAP, particularly in relation to environmental and climate change considerations?
2. How have these institutional shifts affected the sustainability and environmental impact of agricultural practices within the EU?

Framed within this context, our contribution delves into the theoretical underpinnings of policy change, specifically examining the concepts of neo-institutionalism and historical institutionalism to better understand the mechanisms driving this shift. These theoretical approaches emphasize the role of institutional factors in shaping policy outcomes and the importance of historical contexts in understanding policy development. Examining the process of policy change under the lens of historical institutionalism and neo-institutionalism, we discuss the role played by the gradual introduction of measures aimed at promoting measurable environmental criteria and climatic targets. Additionally, our study unravels the EU's commitment to climate change and sustainable development challenges, such as the Agreements on Agriculture (AoA) stated at the Uruguay Round (1986–1994) [13,14] and the UN-Paris Agreement in 2015 [12,15], and how conditionality is being used to link funding to results [16]. These insights into the significance and implications of the CAP's shift towards sustainability offer valuable recommendations for future policy developments, emphasizing the need to balance environmental concerns with the needs of farmers and other stakeholders in line with the international context in which European agriculture operates. Therefore, our study, based on an extended literature review, proposes a thorough analysis of the challenging long-term "greening" transformation of the CAP through an integrated methodology, using instruments from 'historical institutionalism' and 'neo-institutionalism' and, at the same time, taking into consideration the multilevel governance framework of the decision-making process within the EU.

The above procedure seems to be the most appropriate to explain the delay in implementing environmentally effective reforms due to the institutional characteristics of CAP operation, as illustrated below. Moreover, for example, decisions on agricultural funding are now made not on an annual basis, as before 1992, but on a multi-year basis. During these multi-year periods, changes occur in the international environment, developments in the market for agricultural products, foreign trade, etc. These influence the decisions of the European Commission (EC) and the MSs as far as their choices for agricultural policy are concerned. In addition, different pressure groups, in different phases, dominate inside the EU, while the composition of the EC is also changing. These shifts often contribute to policy changes and drive wider CAP reforms [17].

This paper is structured into eight main sections, each working towards providing answers to the research questions and objectives. First, we delve into 'historical institutionalism' and 'neo-institutionalism' as a logical framework for interpreting policy change. Then, we proceed to unveil the forces of change by examining institutional shifts in public policy. Following this, we continue with exploring the dynamics and forces driving institutional changes within the EU CAP, with a particular focus on environmental considerations. Subsequently, we examine the CAP's journey towards sustainability and its role in safeguarding agriculture in the face of climate change. We also analyze the competing goals in CAP implementation and emphasize the latent shift towards new environmental targets. We chart the course for a greener CAP while exploring new territories. Finally, in the conclusions section, we summarize key findings and insights.

2. Methodology

In this section, the methodological approach employed for synthesizing and analyzing the existing literature related to the CAP within the EU is elucidated. The chosen methodology ensures a systematic and comprehensive scholarly synthesis, encompassing the process of systematically gathering, analyzing, and summarizing existing knowledge, research, and academic literature on a particular topic. A range of sources, including peer-reviewed academic articles, reports from EU authorities, and other relevant publications, has been selected based on their significance and contribution to the understanding of CAP policy evolution, institutional shifts, environmental considerations, and their impacts. The analytical approach comprises three distinct steps. Firstly, a thematic analysis is conducted, dividing the literature into key themes and concepts related to CAP policy evolution, institutional shifts, and environmental concerns, allowing for the identification of common trends and recurring patterns in the literature. Secondly, historical contextualization is employed, organizing the literature chronologically to understand the historical aspects of CAP transformation, trace the evolution of CAP policies over time, and assess the factors driving these changes. Lastly, a comparative analysis of the different studies is undertaken to identify variations, consensus, or contradictions in the literature. This approach, visually represented in Figure 1, aids in synthesizing diverse perspectives and viewpoints, ultimately contributing to a more comprehensive understanding of the CAP's evolution.

From a delimitation perspective, it should be noted that the quality and comprehensiveness of the available literature may vary, potentially introducing bias into the analysis. However, through the selection of a diverse range of sources and the application of rigorous analytical methods, an effort has been made to mitigate this potential bias and provide a robust synthesis of the existing literature.

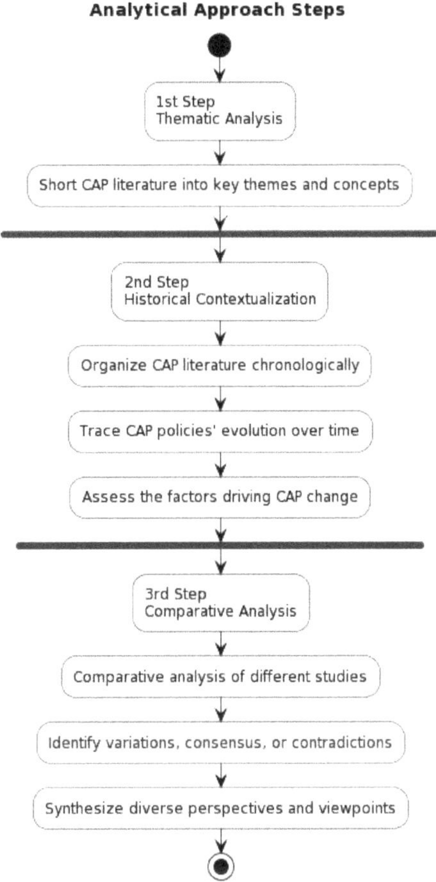

Figure 1. Analytical Approach Steps: Flowchart illustrating the three key steps in the methodology for synthesizing and analyzing existing literature related to the CAP within the EU.

3. 'Historical Institutionalism' and 'Neo-Institutionalism' as a Logical Framework Interpreting Policy Change

During the latter half of the nineteenth century and the early portion of the twentieth century, social scientists created a body of literature focusing on organizations and their interconnections and concentrating research efforts on how bureaucratic structures of organizations as well as the organizational structure within societies fostered 'institutionalization' processes [18]. Until the 1950s, political science was concerned with the analysis of governmental structures (from local to state), mainly in the United States and the United Kingdom. This earlier approach, known as 'old institutionalism', focused on the formal structures and institutions constituting such political entities [19]. However, a behavioral movement that popularized new theories on how policies are formulated and altered, including behaviorism, positivism, and rational choice theory, emerged to complement it [20,21]. This led to a narrowed emphasis on institutions, which was progressively dropped in favor of a thorough evaluation of people as the main policy influencers instead of the institutions administering (and/or surrounding) them [22]. In 1977, Meyer and Rowan's work completely revised the 'institutionalism' approach [19], leading to a resurgence of the issue in the ensuing decade, with important contributions from various fields outside social science.

The 'neo-institutionalism' term was coined in 1983 by March and Olsen, distinguishing it from the 'old institutionalism' notion [23]. Neo-institutionalism focused on comparative examinations of the autonomous impact of institutions on political behavior and outcomes [24] as a means of expressing disagreement with the dominant 'behaviorist' mainstream [25]. Behaviorist viewpoints overestimate the role of institutions exerting a direct impact on politics, disregarding that political institutions are more than passive platforms for political conduct but also elements that may influence, at least in some ways, political behaviors [19]. Historical institutionalism scholars have emphasized the idea that every political action takes place within a given temporal context and that history influences decisions, actions, and occurrences in future times [26]. This perspective considers history not as a collection of specific events but as a factor determining policy change [18]. The terms 'new institutionalism' and 'historical institutionalism', addressing progressive changes to existing institutions or new and innovative policies and their connection to policy reforms [27], are both highlighted and discussed widely in this study.

Historical institutionalism faced the challenge of institutional change despite its emphasis on institutional constancy [28]. A specific body of work on path dependence provided a useful perspective for examining the persistence (or absence) of policy change. Path dependency refers to the fact that institutions, once established, tend to follow historically set, particular trajectories, making it costly to change the pre-existing regime [29]. Institutions become locked into and evolve along the trajectories that govern their dependency, encompassing unintended outcomes and inefficiencies [19]. As a consequence, even when the existing model is inadequate, key players tend to uphold it since it matches the needs of its founders. Altering policies is rather difficult due to institutions' resilience [30]. Owing to the reinforcement of policy continuity by earlier decisions, public policies, and formal frameworks often turn out to be difficult to modify [28].

As strategic actors integrate into the institutional environment, their effects grow, and systemic factors increasingly constrain and delineate the strategic options available to them [31]. Therefore, institutional change may transpire within a particular setting whose breadth and attributes were influenced by earlier political and institutional decisions [32]. The idea of 'point equilibrium' was extensively studied in historical institutionalist analyses of institutional evolution [33]. Institutions tend to be in a state of equilibrium throughout the majority of their history, functioning based on the decisions made at the time of their establishment or the most recent circumstances.

The 'point equilibrium' emphasizes how crucial the institutional atmosphere is in influencing policy dynamics and the outcomes of future reforms [34]. Prolonged periods of institutional stability and adherence to historical patterns can lead to critical junctures, where pivotal decisions are made that can greatly impact the trajectory of a policy system or a policy strategy [35]. When opportunities for significant institutional transformations are both evident and practical, a critical juncture is characterized as a brief span of time whereby uncertainties regarding the prospects of an institution create the groundwork for policymakers to place the institution on an alternative development path [18]. However, a critical juncture point does not always happen at a moment when its effects may be seen retrospectively [36]. Thelen and Steinmo (1992) [24] contend that the critical juncture actually happens much sooner in the process, prior to its impacts becoming apparent. A 'short-period' time interval pertains to how briefly institutions may alter their direction before reverting to their prior dependencies. To deal with new issues, actors can select how an institution should evolve over time and what new policies should be put in place [37].

In their systematic approach to institutional transformation, Streeck and Thelen (2005) [38] added supplementary concepts to the conceptual framework of the historical institution. As per their research, policies that establish norms, allocate entitlements and obligations to players that are backed by norms, and enable third-party enforcement can also be thought of as institutions. Based on these premises, 'institutional transformation' can be considered a theory of policy change. Thus, policies may constitute rules for players other than policymakers, and, if required, they will be enforced by agents acting in the interests of society, legitimizing institutions as policymakers. Streeck and Thelen (2005) [38] identified five categories of institutional change: displacement, layering, conversion, drift, and shock.

Displacement refers to the gradual modification of existing regulatory structures. This represents the most basic form of institutional change and occurs when pre-existing arrangements are contested (or ignored) in favor of new institutions and related behavioral paradigms. Layering involves actively supporting changes introduced to an established set of institutions by appending new regulations and/or institutions alongside or atop the older ones. Andreou (2018) [19] proposes that in highly partisan states, layering is a prevalent practice where new frameworks and/or laws are created to 'control' party supporters without directly affecting state institutions. This creates tensions between institutions and policies, potentially giving rise to conflicts and institutional change. Moreover, recent studies have shown that the COVID-19 pandemic [39] has created a shock to institutional structures, culminating in significant policy changes—and likely more—across the globe [39–42]. This unforeseen and abrupt disruption has compelled governments to enact new policies and regulations in order to address the health and economic repercussions of the pandemic [43].

Streeck and Thelen (2005) [38] highlighted the possibility of institutional drift, gradual deterioration, or atrophy of institutions due to the failure to adapt to shifting political and economic conditions. This drift can arise from gaps in regulations, and political sophistication is essential for effecting the needed changes. The neglect may or may not be intentional, but institutions may be diverted to new goals, functions, or purposes due to emerging environmental concerns, changes in the balance of power, or political struggles [44]. As a result, unexpected outcomes are to be expected, and change requires compromise, which may take time. Additionally, in contrast to the other four shift cycles, Streeck and Thelen (2005) [38] identified exhaustion as a process that causes failure, occurring when an institution's normal activities degrade its external environment and available resources. It usually happens gradually rather than suddenly. When an institution experiences exhaustion, activities within the organization degrade its functioning, contrary to drift, where the formal integrity of an organization is retained despite becoming progressively dysfunctional.

Recent studies have emphasized the importance of institutional resilience, which refers to the intrinsic ability of institutions to adapt to changing circumstances and maintain their functions and goals [45,46]. Institutional resilience can be attained through proactive strategies, such as scenario planning and regular assessments of the institution's performance [47,48]. Furthermore, institutional resilience can be strengthened through the establishment of flexible structures and decision-making processes that allow for quick adaptation to change [49,50]. Finally, studies have shown that successful institutional adaptation requires the involvement of multiple stakeholders, including policymakers, private actors, and civil society [51,52]. The engagement of diverse perspectives can augment the institution's capacity to identify (and respond to) emerging challenges [53].

4. Unveiling the Forces of Change: Examining Institutional Shifts in Public Policy

Extensive research in public policy has thoroughly examined the significance of concepts and knowledge in systemic change [17,54,55]. Built on this, a significant part of the political debate is considered to be in a continuous process of societal advancement stage manifested through public policy. With the previous policy exerting the most significant cognitive influence and the current policy reacting to the effects of earlier initiatives, public policy functions as an educational endeavor or as a means of learning. Hall (1993) [56] refers to this mechanism as "the purposeful endeavor to adjust the goals or tactics of public policy in order to conform with old knowledge and new facts", termed as 'social learning'. The majority of those involved in this learning process are professionals in public policy who advise or serve the public sector from high-ranking positions at the intellectual subcultures of society and the bureaucracy nexus.

Three stages make up the process of changing public policy: the broad objectives that guide policy in a given area, the methods (or tools) employed to accomplish those objectives, and the real costs associated with those methods and instruments. Each stage is composed of an equal number of variables. Historical institutionalists contend that the creation of institutions and policies frequently leads to conflicts between groups with different spheres of influence because they understand that institutions reflect, organize, and reproduce uneven power relations. This conflict often culminates in changes in the institution or policy under consideration [19].

5. From Policy Reform to Environmental Stewardship: Examining the CAP's Journey towards Sustainability

A supportive environment is required for agricultural operations in order to use natural resources, generate agro-food, and maintain farmers' financial stability. Agriculture yields benefits to the whole society, far beyond specific targets, such as sustaining farming households and communities in rural or peri-urban areas [57]. However, the primary sector has a two-fold impact on the environment. First, it directly impacts agricultural practices. Second, it contributes to climate change by releasing greenhouse gases into the atmosphere. To create a sustainable agricultural system throughout the EU, the CAP has integrated social, economic, and environmental concerns and was more recently committed to additional international agreements addressing climate change and sustainable development challenges (see also [58]). Building upon a more innovative, impactful, and comprehensive approach to tackle the issues of climate change and sustainable development, ambitious and forward-thinking frameworks for environmental-friendly actions were proposed. The original principles of the CAP did not prioritize environmental conservation, but this perception shifted as environmental issues became more politicized and pressing in the early 1970s [59]. During the 1980s, several policy guidelines were released that highlighted the significance of safeguarding the environment. These include the "Green Paper" on the future of the CAP, a 1988 Communication on "Environment and Agriculture", and the guidebook "The Future of Rural Society", among others. This stream of literature emphasized the immediacy of minimizing environmental degradation [60]. The Green Paper report acknowledged the importance of establishing institutionalized measures to mitigate and prevent environmental degradation caused by intensive farming practices [61]. Techniques that prioritize greenhouse gas emissions reduction, carbon storage, and maintenance/stabilization of food production have the potential to alleviate the repercussions of climate change [62].

In the late 1980s and the early 1990s, there was a surge in consumer and environmental activism that advocated for policy reform. This movement was prompted by several food-related scandals and the adverse ecological consequences of the agricultural practices supported by the CAP. Moreover, the EU intensified its international efforts to address environmental issues with global consequences [63], particularly after the United Nations (UN) Conference on the Environment and Development in 1992. These internal and external forces led to a major reform of the CAP in 1992, whereby environmental concerns became increasingly critical in the subsequent policy revisions. The 1992 agri-environmental measures were novel and reflected the initial notable attempt to back a new type of agriculture providing commodities and services that enhance the environment [64]. The notion of the 'second pillar' for rural development, initially introduced in Agenda 2000, continued to develop and expand this operational concept [64,65]. The Rural Development Pillar of the CAP was included in the Agenda 2000 reform package, which emphasized the importance of safe agri-food products and environmental outcomes. This reform package aimed at striking a balance between the need for environmental conservation and the provision of direct incentives to producers in order to adopt more environmentally friendly practices for the production of safe and healthy food.

Furthermore, the EU was required to change a number of domestic support policies in order to be compliant with the Green Box criteria and non- or minimally trade distorting following the General Agreement on Tariffs and Trade (GATT), the AoA, and the current negotiations within the World Trade Organization (WTO) [13,66]. In light of the General Food Law (REGULATION (EC) No 178/2002 [67]) and traceability, which is defined as "the ability to trace and follow food, feed, and ingredients through all stages of production, processing, and distribution", the CAP encouraged farmers to improve their agricultural practices, taking into consideration food safety and environmental issues [13,68].

In accordance to the CAP regulations, MSs were required to take adequate environmental safeguards while being granted flexibility in how they supported farmers alongside environmental measures [69]. When MSs failed to comply with regulations, their funding from pillar 1 (mainly the direct payments scheme) was reduced or revoked, and the unpaid amounts were redirected to their respective rural development programs [69,70]. Generally, the policy for rural development tries to integrate the local/territorial element, taking advantage of particular characteristics of lagging regions by financing measures of a structural nature [71] (Figure 2).

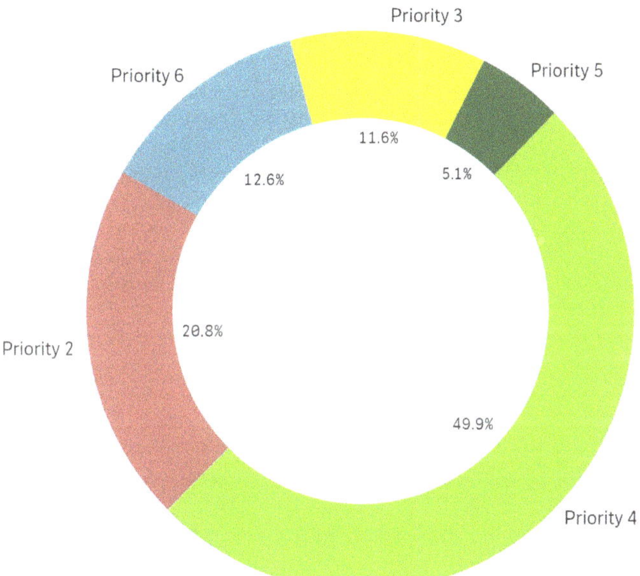

Figure 2. Distribution of rural development expenditure by priority area. Priority 2: Farm Viability and Competitiveness. Priority 3: Food Chain Organization and Risk Management. Priority 4: Restoring, Preserving, and Enhancing Ecosystems. Priority 5: Resource-efficient, Climate-resilient Economy. Priority 6: Social Inclusion and Economic Development. (Source: EC, Director-General for Agriculture and Rural Development).

Additionally, MSs were incentivized to allocate a portion of these funds for developing more environmentally friendly production techniques in the dairy and cattle industries and training farmers in ecologically friendly practices to assist forests with high ecological value and underserved areas (e.g., [72,73]) (see Tables 1 and 2). These measures were geared towards enhancing the efforts toward environmental preservation and climate action by requiring MSs to formulate complete national or regional programs that included environmental conservation, among other rural activities [74,75].

Moreover, other than environmental measures are included within the Rural Development Pillar, improving the competitiveness of agriculture and forestry, the quality of life of the residents in the countryside and diversifying the rural economy to meet the new challenges in a changing economic, social and climatic context within the EU and globally [17]. Lastly, special attention is given to actions to promote research and innovation, food safety, and the containment of populations in the European countryside. All of the above intend to formulate an effective and sustainable framework of operation for rural areas, contributing to its economic and social cohesion.

In the 2003 Mid-Term Review (MTR) of the CAP, cross-compliance became a mandatory requirement for all direct payments. The cross-compliance criteria were crafted to ensure that farmers meet environmental and other standards before becoming eligible for funding. The regulations for statutory management under Union law and the requirements for maintaining excellent agricultural and environmental conditions were included in the cross-compliance norms [76]. The MTR also introduced a 'one farm payment' system, which departs from relying on output while being related to compliance with environmental, food safety, and animal welfare criteria. The obligation to maintain all farms in excellent agricultural and environmental condition was preeminent to this operational perspective [77]. To promote the environment, quality, or animal welfare, direct payments were reduced for larger farms, freeing up more funds for programs that meet these goals [77,78].

To maximize the use of natural resources, the CAP has urged farmers to adopt eco-friendly procedures for growing plants and rearing animals over the last two decades, incorporating new technology into their production processes. Farmers that satisfy three environmental criteria are eligible for the green direct payment, which makes up to 30% of the direct payment program budget under the CAP system (Figure 3).

Table 1. CAP expenditure and CAP reform path (by financial year). (Source: EC, Directore-General for Agriculture and Rural Development).

Year	Export Refunds (€ bn)	Other Market Support (€ bn)	Coupled Direct Payments (€ bn)	Decoupled Direct Payments. Excl: (€ bn)	Greening (€ bn)	Total Rural Development. Excl: (€ bn)	Rural Development Environment/Climate (€ bn)	CAP as % EU GDP (%)
2006	2.22	5.27	18.072	12.488	0.00	11.02	0.00	0.48
2007	1.34	3.41	6.617	26.613	0.00	8.025	2.675	0.46
2008	0.91	3.18	6.087	28.06	0.00	7.53	2.51	0.44
2009	0.62	3.27	6.256	29.538	0.00	6.307	2.102	0.45
2010	0.37	3.54	5.811	30.559	0.00	8.212	2.737	0.47
2011	0.18	3.16	3.314	33.567	0.00	8.775	2.925	0.46
2012	0.15	3.20	3.183	34.412	0.00	9.397	3.132	0.47
2013	0.06	3.09	2.79	35.583	0.00	9.292	3.097	0.47
2014	0.00	2.43	2.683	35.781	0.00	7.35	3.15	0.44
2015	0.00	2.63	2.994	35.295	0.00	7.819	3.351	0.43
2016	0.00	3.07	4.46	22.324	10.803	8.064	3.456	0.42
2017	0.00	2.86	4.524	22.676	10.840	7.371	3.159	0.40
2018	0.00	2.61	4.679	22.439	10.840	8.316	3.564	0.39
2019	0.00	2.37	4.632	22.311	10.806	9.394	4.026	0.39
2020	0.002	2.59	4.692	22.262	10.858	9.676	4.146	0.38
2021	0.00	2.56	4.696	21.978	10.775	10.255	4.395	0.38
2022	0.00	2.953	4.695	21.834	10.763	10.670	4.573	0.36

Table 2. The integration of environmental concerns into the CAP through fundamental agri-environmental indicators (EU-28, 2017–2018) (Source: European Parliament Fact Sheets on the EU—The CAP).

	Share of UAA Classed as Natura 2000 Area Including Natural Grassland (%)	Share of Energy Used in Agroforestry of Total Energy Consumption (%)	Share of UAA Managed by Farms by Low Input Intensity (%)	Share of UAA under Organic Farming (%)	Share of Agriculture in Production of Renewable Energy (%)	Share of Forestry in Production of Renewable Energy (%)
	2018	2018	2017	2018	2018	2018
BE	7.1	2.4	63.0	6.6	18.8	36.3
BG	22.5	1.9	10.3	2.6	6.6	59.5
CZ	6.6	2.6	27.8	14.8	16.6	67.2
DK	4.0	4.3	42.7	9.8	7.1	42.9
DE	10.7	1.7	34.4	7.3	24.1	27.2
EE	5.8	4.3	29.2	20.6	0.3	94.5
EL	18.7	1.7	40.2	9.3	6.0	25.9
ES	16.6	3.0	27.0	9.3	9.8	29.0
FR	8.4	2.9	31.6	7.0	13.3	37.0
IE	3.6	2.0	31.1	2.6	2.6	18.6
IT	9.5	2.4	35.0	15.2	8.3	26.5
CY	6.3	2.7	61.4	4.6	5.9	11.8
HR	25.7	3.2	26.9	6.9	2.7	62.6
LV	6.4	4.5	28.4	14.5	6.0	85.8
LT	4.5	2.0	35.2	8.1	10.3	79.1
LU	21.1	0.6	30.6	4.4	11.5	52.2
HU	14.8	3.6	24.6	3.9	17.2	71.1
MT	8.1	0.9	38.2	0.4	4.8	0.0
NL	4.2	8.1	23.6	3.2	37.0	23.7
AT	11.8	2.0	21.0	24.1	5.4	47.0
PL	11.5	5.6	33.5	3.3	11.8	69.3
PT	17.8	2.4	30.6	5.9	5.2	41.8
RO	12.7	2.4	16.1	2.4	3.2	58.3
SI	23.5	1.5	35.5	10.0	1.9	51.8
SK	16.2	1.3	20.2	9.9	17.8	56.3
FI	1.1	2.7	37.5	13.1	2.7	74.1
SE	4.0	1.9	30.4	20.3	2.4	48.5
UK	2.6	1.0	32.3	2.6	11.3	26.9
EU-27	11.2	2.9	27.0	8.0	12.1	41.4
EU-28	10.9	2.7	27.2	7.5	12.1	40.3

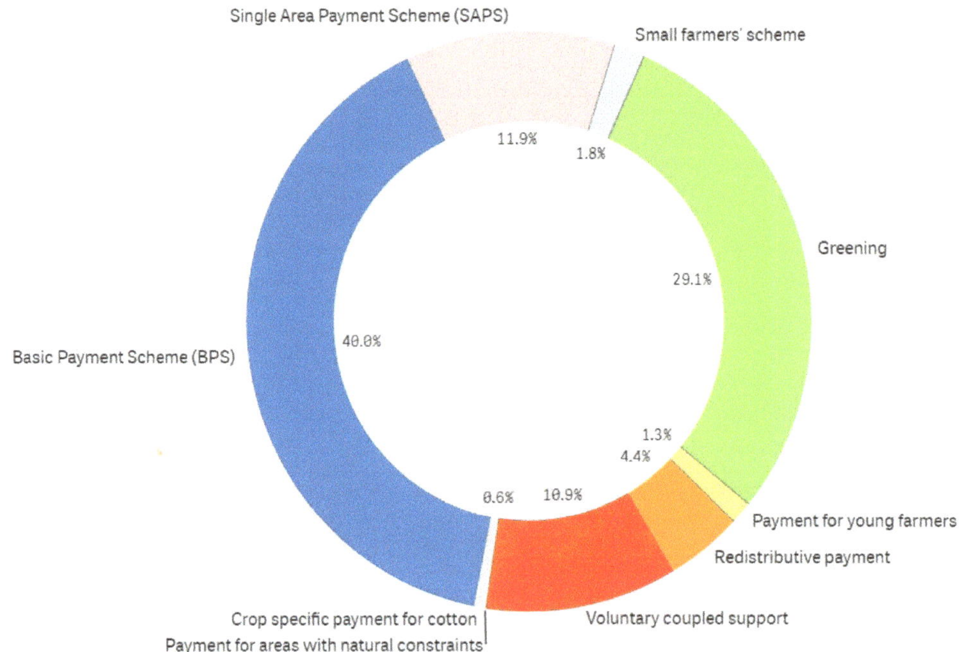

Figure 3. Distribution of direct payment expenditure by scheme (Claim Year 2021). (Source: EC, Directore-General for Agriculture and Rural Development).

Variegating their crops, preserving permanent grassland, upholding biodiversity, and allocating 5% of their arable land to ecologically beneficial areas (Ecological Focus Areas) are the relevant intervention areas. In EU countries, the ratio of permanent grassland to agricultural land was decided by national or regional authorities, with a 5% leeway [79]. Advanced digital applications and innovations, such as Blockchain, Internet-of-Things, and Artificial Intelligence, and Immerse Reality, are expected to provide effective tools to the farmers in order to achieve the above goals [80]. As an illustration, "Smart Farming" applications for intelligent-edge computing will make it easier to deploy edge capacity linked to agricultural equipment, enabling the real-time collection of agricultural data, improved services for farmers, such as harvest prediction and farm management, and the improvement of the food supply chain [80].

Several studies have suggested that cross-compliance is an efficacious tool for promoting environmental and agricultural sustainability [81–84]. However, some scholars have argued that cross-compliance alone may not be sufficient to meet the environmental goals of the CAP [85], and a thorough combination with other policy instruments, such as agri-environmental measures and payments for ecosystem services, could be indispensable [81,85]. Additionally, recent research has shown that the efficacy of the CAP's green direct payment scheme may vary depending on the farm characteristics and the local context [86], suggesting the need for more targeted and flexible policies. Therefore, policymakers and scholars should delve deeper into evaluating the effectiveness of the CAP's environmental policies and identifying ways to improve them. The integration of agri-environmental measures, payments for ecosystem services, and the use of targeted policies contingent upon the farm and local context seems to be promising strategies bolstering environmental sustainability while supporting agricultural production [87].

In the EU, regulations are in place to protect designated sections of permanent grassland, which cannot be transformed or cultivated by farmers [88,89]. However, certain farmers, such as those enrolled in the small farmer's program, are exempt from the greening regulations, and organic farmers are rewarded for their environmentally friendly practices [8,90]. Failure to comply with greening regulations can result in reduced direct payments [77,85,91]. The green direct payment system holds crucial significance in promoting eco-friendly environmental practices and alleviating the impact of climate change on agriculture [77,92]. The Cross-Compliance regime, although relatively lenient, provides a framework for regulation and control mechanisms [83,93,94]. The agricultural industry worldwide faces escalating pressure stemming from an expanding populace [95–97], urbanization [98–101], resource depletion [102–106], and climate change [107–109]. In the EU, the repercussions of climate change are detrimentally impacting agricultural production and environments [110,111]. European agriculture is at risk of extreme weather events, rising temperatures, changing rainfall patterns, river flooding, coastal flooding, droughts, and other impacts due to climate change [12,112]. While some regions may benefit from certain climatic changes [113,114], most of them will experience negative effects, exacerbating the existing environmental issues [107,108].

6. Safeguarding Agriculture in the Face of Climate Change: Exploring the Role of CAP

Studies have shown that climate change is anticipated to amplify the frequency of extreme weather events, such as droughts, floods, or extreme temperatures, with a potentially serious impact on agricultural production and the quality of food products [16,115]. The risks posed by climate change are particularly profound in the Mediterranean zone, where agricultural output is expected to be negatively impacted by water shortage and a sudden process of local warming [116,117]. Furthermore, these effects on agriculture will likely exacerbate existing environmental issues, such as land resource depletion and soil degradation, all while heightening the susceptibility of local regions [118–120]. Therefore, implementing measures to mitigate and counterbalance the impact of climate change, such as the green direct payment system, is of crucial importance in promoting sustainable agricultural practices and preserving the environment [121].

Additionally, research has highlighted the importance of establishing quantifiable standards for the Cross-Compliance regime, considering its comparatively lenient structure and limitations in oversight mechanisms [79]. Moreover, farmers who disregard greening regulations are penalized through reduced direct payments, indicating the importance of complying with these regulations. Finally, the exemptions granted to certain farmers, such as those enrolled in the small farmer's program and organic farmers, ought to be carefully evaluated for reasons of administration and proportionality [90].

Overall, protecting the environment and mitigating the impact of climate change emerge as imperative objectives for upholding sustainable agricultural production in the EU. The implementation of regulations, such as the green direct payment system and the Cross-Compliance regime, hold crucial significance in promoting good environmental practices ensuring compliance with environmental standards. To address the challenges facing agriculture from climate change and land resource depletion, it is therefore essential to implement sustainable production techniques and promote climate change and natural resource management. The CAP endeavors to attain these aspirations, casting European farmers in the role of primary environmental stewards, spending money, and engaging in producing within (economically disadvantaged) rural areas [122]. The new CAP framework for 2021–2027 champions a competitive and sustainable agricultural sector, supporting farmers' livelihoods and providing food to society while fostering vibrant rural communities. In this perspective, the European Green Deal (EGD), with a focus on agriculture and rural areas, is a crucial tool in achieving the Farm to Fork (F2F) and biodiversity objectives. The EGD represents a collection of policy initiatives strategically aimed at guiding the EU towards a greener trajectory, ultimately striving to attain climate neutrality by the year 2050. Its core mission revolves around reshaping the EU into an equitable and flourishing society equipped with a contemporary and competitive economy. By advocating for a comprehensive and cross-disciplinary strategy, the deal aligns various pertinent policy domains to contribute cohesively to the overarching climate-oriented goal. This comprehensive program encompasses a diverse array of actions spanning climate, environment, energy, transportation, industry, agriculture, and sustainable finance—all interconnected as they pursue the common objective of fostering sustainability (see [123–125]). Embedded within the EGD, the F2F strategy assumes a pivotal role, centering its attention squarely on the food system. Its central objective entails enhancing the sustainability of the EU's food system across its entire continuum—from production to consumption. This is accomplished through the promotion of healthful and sustainable dietary habits, curbing food wastage, enhancing food labeling standards, and endorsing ecologically conscious agricultural methods. The overarching aim of this strategy is to ensure the availability of secure, nourishing, and eco-friendly sustenance for all while concurrently mitigating the environmental footprint of the food industry (see [126,127]). However, further noting the holistic impact and salience of these initiatives lies beyond the scope of this present analysis.

The new CAP requires measurable environmental and climatic requirements to be met in the context of obtaining direct payments [128]. This involves implementing agricultural practices, such as the rotation of crops rather than diversification, safeguarding wetlands to uphold carbon-abundant soils, and managing water resources in a sustainable manner. Each MS is enjoined to formulate programs or incentives for farmers aimed at fostering judicious agricultural practices. The new CAP also includes an escalated allocation of resources from pillar 1 to pillar 2 for environmental and climatic policies, reflecting a 15% financial upgrade of environmental issues [80,129].

7. Competing Goals in CAP Implementation: Actors, Interests, and Bargaining

Historical institutionalism and rational choice institutionalism represent two research methods used to elucidate the actions of MS governments and their impact on the EU institutions. Leading scholars in the field of historical institutionalism, such as Pierson (1996) [130] and Bulmer (2009) [131], have emphasized, on the one hand, the importance of scrutinizing political and policy-related maneuvers in the context of multi-tier government. To this effect, Bulmer introduced the concept of a 'governance regime' to analyze assorted subsystem policies within the EU. On the other hand, rational choice institutionalism finds its foundation in the realm of economics, revolving around the theory of rational behavior. This approach was mainly harnessed to delineate the goals of MS governments within the context of EU integration. According to this theory, governments willingly engage in (and cede power to) the EU because they believe it affords them several advantages. These benefits may include diminished transaction expenses, heightened efficacy of policy formulation, improved operational efficiency, and increased adherence to regulatory frameworks [132].

However, it is worth noting that while rational choice institutionalism furnishes a useful framework for dissecting the EU integration process, it has also been criticized for oversimplifying the intricate interplays between actors and institutions [133]. Moreover, recent developments, such as the Brexit referendum [134] and the rise of populist movements in Europe [135], including the surge of right-wing ethnonationalism and the emergence of authoritarianist governments –largely attributed to the austerity-driven economic strategies of the preceding decade across various European nations [136], as well as the ongoing migration predicament [137]– have cast a spotlight on the underlying assumptions of rational choice institutionalism. These events highlight the importance of examining the roles of emotions, identity, and culture in the realm of EU politics [138].

The implementation of the CAP engenders varying costs and benefits across different MSs, giving rise to competing goals and anticipations among actors involved in its development. These entities may comprise various stakeholders, such as government officials with specific agendas, committee bureaucrats, interest groups representing various industries, and groups advocating for agricultural issues, among others [17,139]. By leveraging future decisions regarding CAP reforms as a bargaining tool [139], the EC and the Council of Agriculture and Rural Development Ministers historically sought to champion European farmers' interests through a leveling act involving major international partners (such as the UN and WTO members). However, these determinations have often disregarded the opinions of consumer and environmentalist factions, especially in regard to environmental issues [140].

Interestingly, a notable transformation in the annals of the CAP, especially in the late 1990s, should also be ascribed to a number of unwanted episodes, or better, incidents, including the advent of bovine spongiform encephalopathy [141], the detection of dioxin contamination in Belgian food products [142], and the spread of foot and mouth diseases [143]. These events engendered an elevated degree of awareness among consumers, propelling them to clamor for heightened vigilance over matters of food safety and quality. This change in consumer demand also brought to light apprehensions regarding the influence of the CAP on animal welfare and the environment, spurring the formulation of policies that address such issues [144]. The priority placed on food safety and quality has been significantly accentuated among EU residents, largely due to the demand from consumer groups. Consequently, these issues have been thrust to the forefront of the CAP agenda. While the CAP's position on these issues may vary, its political response has been to address them in policy reform efforts [140].

8. The Latent Shift toward New Environmental Targets

More recent studies have emphasized the importance of harmonizing CAP policies with the EU's climate and biodiversity aspirations [145,146] while accentuating the imperative to integrate environmental and climate objectives into the CAP to ensure its contribution to the EGD's objectives [147,148]. Academic research has also emphasized the necessity of dovetailing the CAP with the Sustainable Development Goals established by the UN [147]. Specifically, the pertinence of Goal 2, which aims at reducing hunger, attaining food security, enhancing nutrition, and fostering sustainable agriculture, is of critical significance from this perspective [149,150].

The CAP, at least in its original formulation established in 1962 and lasting until 1992, saw little opposition from pressure groups representing farmers' professional associations [64]. However, in the following years, the European agricultural model's detrimental environmental impacts diminished the influence of consumer and environmental advocacy movements [151]. These movements subsequently garnered greater sway within EU institutions, propelled by food-related scandals and escalating environmental effects [70].

Although politically justified, the economic philosophy behind cross-compliance, which involved the addition of new conditions to existing income support, remains unclear. The query emerges as to whether this constitutes a valuable policy tool. Several studies have intimated that direct payments were allocated in accordance with agricultural policy objectives rather than environmental objectives [152–154]. As a result, farmers who heavily rely on direct payments may not necessarily align with those who inflict substantial harm to the environment [155].

It is crucial to contemplate the decision-making process regarding the allocation of direct payment savings, particularly those stemming from farmers' non-compliance [156]. MSs could only withhold 25% of funds obtained through the implementation of cross-compliance (see [157]), thereby offering little incentive for states to establish a reliable control system. This deficiency in follow-through by the MSs indicates a discrepancy between political rhetoric and reality [13]. Consequently, new 'players', such as consumers and environmental movements, gained importance in the reform process, as well as the gradual wane of producer pressure organizations resulting from the contraction in rural populations. The decline in rural populations, both in terms of absolute numbers and as a share of the overall workforce in the EU [13,158], adds a pivotal contextual layer. This decline prompts an exploration into the underlying causes behind the dwindling appeal of agriculture as a viable livelihood, even in the presence of substantial funding. This necessitates a broader examination of the political economy driving the agrarian crisis, considering its far-reaching consequences and implications. Moreover, it is essential to recognize the role of disparities and social injustices in this landscape. While the shift in dynamics has paved the way for new participants, such as consumers and environmental movements, to influence reform processes, the gradual decrease in rural populations has impacted producer pressure organizations. These organizations have historically played a significant role but have diminished in influence due to the changing demographic composition in rural areas. It is worth acknowledging that between 1960 and 1990, there was an overarching reduction of the active rural populace by 54.9% [17]. This shift occurred as the strain on rural incomes gradually compelled a significant portion of the workforce engaged in the primary sector to transition into the secondary and tertiary sectors of the economy, often necessitating a move from rural areas to urban centers [17]. Recent literature suggests that CAP's current system perpetuates socioeconomic disparities, particularly in peripheral (rural) areas while favoring large agricultural enterprises and consolidating land ownership [159]. This consolidation of land ownership contributes to the displacement of small-scale farmers and the exclusion of marginalized groups, thereby sustaining patterns of social injustice. Moreover, studies show that CAP has faltered to deliver on its environmental objectives despite earmarked budget allocations for these specific purposes [146], indicating the urgent need for a reform ensuring sustainable agricultural practices.

The Green Payment Scheme represents a critical juncture in the progression of the CAP in terms of both its configuration and implementation. With the inception of this scheme, the compliance of producers with environmental regulations became quantifiable, and explicit criteria were established for the allocation of direct payments. Furthermore, the most recent iteration of the CAP, covering the period from 2021 to 2027, places a greater emphasis on the targeted allocation of resources in order to align with the EU's climate goals. This change is propelled by the recognition of the severe and widespread impacts of climate change on agriculture, exerting its impact across the entirety of the food production continuum. The new framework of the CAP addresses these issues by implementing measures that laser-focus on climate action and sustainability within the realm of agriculture.

As set out earlier, the focus on cultivating green architecture within the CAP is clear, aiming to create a robust yet adaptable framework for 'greening' the policy. The CAP has been under scrutiny for a considerable period due to the detrimental effect it has on the environment. The escalation in agricultural production intensity and the depletion of land resources rank among the factors catalyzing this apprehension. Under the new CAP paradigm, rational decision-making is paramount within a structured framework of shared commitments and goals. The policy's diminished budget places considerable emphasis on conditionality, accentuating its significance. At the same time, the policy raises awareness of (and addresses) environmental and climate change-related issues through specific funding mechanisms. Realizing these goals requires the harmonious integration of both CAP pillars, coupled with cooperation with allied policies while providing flexibility based on national priorities [17]. As a result, the EU27's proportion of global emissions witnessed a decline from 16.8% in 1990 to 7.3% in 2021 [160,161].

In this perspective, the Green Payment Scheme represents a new system of incentives aimed at promoting environmentally friendly agricultural practices as well as rewarding farmers who have consistently upheld certain environmental standards. This is in line with the 'public money for public goods' principle, which advocates that public funding allocated to agriculture should be reciprocated through the provision of public goods, such as the protection of the environment and the preservation of biodiversity. The scheme is also strategically crafted to encourage farmers to take advantage of new technologies and management practices that curtail the environmental impact of agricultural production.

9. Charting the Course for a Greener Common Agricultural Policy

Ten main objectives for the years 2023 to 2027 shape the operating framework of the new CAP. These goals, which placed a focus on social, environmental, and economic issues, served as a guide for how EU countries built their CAP Strategic Plans. It is noteworthy that, along with the promotion of knowledge and innovation, the top goals for the upcoming years include combating climate change, protecting the environment, and preserving landscapes and biodiversity, as the EU aims to be climate-neutral by 2050 [148,162,163].

Additionally, the new CAP builds on the previous policy framework, incorporating changes that reflect the EU's increasing emphasis on environmental quality, sustainable agronomic practices, and climate action [164]. Sustainable farming practices include precision agriculture, organic farming, agro-ecology, agro-forestry, and more stringent animal welfare standards [80]. By shifting the emphasis from compliance to performance (e.g., [165]), eco-schemes, for instance, are poised to remunerate farmers for improved environmental and climatic performance, such as managing and storing soil carbon and improving fertilizer management to improve water quality and reduce emissions [166,167]. The EC has recommended that climate action should secure a minimum of 40% of the total funding for the CAP from 2021 to 2027 [145]. Therefore, the F2F strategy [168] will bolster European farmers' efforts to combat climate change, protect the environment, and preserve biodiversity.

In line with this strategy, the EGD is centered on agriculture and rural areas, and the new CAP aims to be a crucial tool in achieving the F2F and biodiversity aspirations, as well as facilitating the transition to a climate-neutral, circular, and resilient economic paradigm. Employing an innovative approach, the policy's implementation hinges on the combination of mandatory and voluntary measures, which vary according to the country's environmental and agricultural exigencies. The flexibility of the policy enables MSs to tailor their implementation strategies, contingent upon their individual circumstances, while the obligatory nature of direct payments serves as an incentive compelling farmers to attain environmental and climate-related objectives.

In summary, the new CAP's framework alongside the Green Payment Scheme represents a significant shift towards a more sustainable and environmentally attuned approach to agriculture in the EU. The policy's focus on incentivizing environmentally friendly practices, combined with flexibility and conditionality, furnishes a scaffold for amalgamating environmental concerns into the broader agricultural sector. However, the success of the new framework will depend on its implementation and the extent to which MSs and farmers embrace the new spectrum of incentives and adapt to the policy's changing priorities [17].

10. Considering Research Scope and Charting New Territories

This work contributes to the study of sustainable development by specifically influencing the level and spatial direction of local and regional development in the field of the primary sector. The current study adopts a predominantly theoretical and descriptive approach rooted in the synthesis of existing literature. The theoretical underpinnings are illuminated, and a comprehensive overview of pertinent literature is presented, establishing the foundation for deeper insights into the subject matter.

The study highlights a few important figures and basic indicators derived from official records that aim to reflect the differentiated mechanisms of rural development and the intrinsic complexity of the quantitative information required to evaluate processes of (economic) growth. In this forward-looking perspective, the evolving nature of the field is underscored, along with the continued contributions that can shape its trajectory. This invites further scrutiny and empirical examination, potentially involving empirical methodologies and primary data collection.

The utilization of econometric modeling for these indicators can further capture the complexity of the study subject, thereby advocating for a multi-disciplinary research strategy that integrates socio-demographic and economic approaches. These approaches necessitate continually updated, spatially relevant, and comparable indicators that are capable of describing possible changes occurring over short time windows.

Official statistics and data at both country and supra-national scales (e.g., Eurostat, OECD, FAO, UN) should thus build up the quality of socioeconomic indicators reflecting rural development and agrarian change. This task is considered particularly urgent, contributing to informing regional science, urban studies, and spatial planning in both advanced and emerging economies.

Concurrently, the development of a set of figures and indicators of exurban development based on official statistics data may eventually solidify the use of mixed econometric techniques and exploratory, multivariate approaches. These approaches can be applied to differentiated socioeconomic contexts and territorial conditions, aiming to disentangle the inherent complexity underlying rural development. Notably, future research avenues could encompass detailed case studies, cross-sectional analyses, and longitudinal assessments to provide empirical substantiation for the theoretical constructs discussed herein.

11. Conclusions

Despite the fact that environmental issues became more politicized and pressing in the early 1970s, only over the last three decades has there been a greater emphasis on policy change and adaptation to address the impact of economic activity and agricultural output on the environment and climate change. This change has been driven by internal and external factors as well as informal and formal changes in the institutional framework of the CAP and the power of the main groups of interest.

Among the factors that pushed the "greening" process are the growing awareness of the issue among consumers, environmentalists, and the public in the EU, along with the gradual decline of farmer's power. Other evolutions include the recognition of agriculture's vulnerability to climate change due to the direct impact of weather on farming activities and its contribution to greenhouse gas emissions. Moreover, the EU's international commitments to climate change and sustainable development challenges, such as the AoA [13,14] and the 2015 UN Paris Agreement [12,15], played a significant role in gradually including "green" measures in several CAP reforms since 1992 [169,170].

Consequently, the new CAP (2023–2027) reflects a significant shift towards a more sustainable and environmentally friendly approach. The CAP's new green architecture integrates environmental and climate criteria that are quantifiable. The inclusion of wetland preservation for carbon-rich soils, responsible management of water resources, and crop rotation in lieu of diversification is some of the measures taken to ensure sustainability. In addition, the CAP allocates at least 30% of funding for environmental and climate change activities under the second pillar, amounting to roughly 23 billion euros.

The new CAP resonates with a collective dedication to confronting climate change and the trials of sustainable development, constituting a pivotal stride toward their realization. It encourages collaboration with other similar policies, incorporates both CAP pillars, and amplifies malleability contingent upon national priorities. The new architecture stands as a robust yet supple framework designed to imbue the CAP with ecological hues and tackle the persistent environmental and climate change apprehensions that have endured over numerous years.

Author Contributions: All authors have contributed equally to this work. All authors have read and agreed to the published version of the manuscript.

Funding: This research received no external funding.

Data Availability Statement: No new data were created.

Acknowledgments: Thanks are due to Napoleon Maravegias, for the insightful and constructive comments from which the paper greatly improved.

Conflicts of Interest: The authors declare no conflict of interest.

References

1. Guth, M.; Smędzik-Ambroży, K.; Czyżewski, B.; Stępień, S. The Economic Sustainability of Farms under Common Agricultural Policy in the European Union Countries. *Agriculture* **2020**, *10*, 34. [CrossRef]
2. Cagliero, R.; Bellini, F.; Marcatto, F.; Novelli, S.; Monteleone, A.; Mazzocchi, G. Prioritising CAP Intervention Needs: An Improved Cumulative Voting Approach. *Sustainability* **2021**, *13*, 3997. [CrossRef]
3. Smędzik-Ambroży, K.; Guth, M.; Stępień, S.; Brelik, A. The Influence of the European Union's Common Agricultural Policy on the Socio-Economic Sustainability of Farms (the Case of Poland). *Sustainability* **2019**, *11*, 7173. [CrossRef]
4. Volkov, A.; Balezentis, T.; Morkunas, M.; Streimikiene, D. Who Benefits from CAP? The Way the Direct Payments System Impacts Socioeconomic Sustainability of Small Farms. *Sustainability* **2019**, *11*, 2112. [CrossRef]
5. Manola, M.; Koufadakis, S.X. The Gastronomy as an Art and Its Role in the Local Economic Development of a Tourism Destination: A Literature Review. *SPOUDAI J. Econ. Bus.* **2020**, *70*, 81–92.
6. Kaldis, P. The Greek Fresh-Fruit Market in the Framework of the Common Agricultural Policy. Ph.D. Thesis, Coventry University, Coventry, UK, 2002.
7. *European Commission CAP Post-2020: Environmental Benefits and Simplification*; European Commission: Brussels, Belgium, 2021.
8. Doukas, Y.E. Η Πολιτική Οικονομία Της Μεταρρύθμισης Της Κοινής Αγροτικής Πολιτικής. Ph.D. Thesis, National and Kapodistrian University of Athens, Greece, 2010.
9. Malhi, G.S.; Kaur, M.; Kaushik, P. Impact of Climate Change on Agriculture and Its Mitigation Strategies: A Review. *Sustainability* **2021**, *13*, 1318. [CrossRef]
10. Kyriakopoulos, G.L.; Sebos, I.; Triantafyllou, E.; Stamopoulos, D.; Dimas, P. Benefits and Synergies in Addressing Climate Change via the Implementation of the Common Agricultural Policy in Greece. *Appl. Sci.* **2023**, *13*, 2216. [CrossRef]
11. Melchior, I.C.; Newig, J. Governing Transitions towards Sustainable Agriculture—Taking Stock of an Emerging Field of Research. *Sustainability* **2021**, *13*, 528. [CrossRef]
12. Doukas, Y.E.; Petides, P. The Common Agricultural Policy's Green Architecture and the United Nation's Development Goal for Climate Action: Policy Change and Adaptation. *Reg. Peripher.* **2021**, *11*, 107–128.
13. Maravegias, N.; Doukas, Y.E. Traceability and the New CAP. In *A Resilient European Food Industry in a Challenging World*; European Political, Economic, and Security Issues; Agriculture Issues and Policies; Baourakis, G., Mattas, K., Zopounidis, C., van Dijk, G., Eds.; Nova Science Publishers: Hauppauge, NY, USA, 2011; pp. 245–253.
14. Castellano-Álvarez, F.J.; Parejo-Moruno, F.M.; Rangel-Preciado, J.F.; Cruz-Hidalgo, E. Regulation of Agricultural Trade and Its Implications in the Reform of the CAP. The Continental Products Case Study. *Agriculture* **2021**, *11*, 633. [CrossRef]
15. Vrolijk, H.; Poppe, K. Cost of Extending the Farm Accountancy Data Network to the Farm Sustainability Data Network: Empirical Evidence. *Sustainability* **2021**, *13*, 8181. [CrossRef]
16. Doukas, Y.E. The Common Agricultural Policy under the Pressure of the New Financial Framework (2021–2027): Nationalization and Adaptation. *Reg. Peripher.* **2019**, *8*, 133–142. [CrossRef]
17. Doukas, Y.E.; Maravegias, N. Ευρωπαϊκή Αγροτική Οικονομία και Πολιτική: Μετασχηματισμοί και Προκλήσεις Προσαρμογής; Kritiki: Athens, Greece, 2021; ISBN 9789605863753.
18. Hall, P.A.; Taylor, R.C.R. Political Science and the Three New Institutionalisms. *Polit. Stud.* **1996**, *44*, 936–957. [CrossRef]
19. Andreou, G. Η Νεοθεσμική Προσέγγιση της Πολιτικής; Kritiki: Athens, Greece, 2018; ISBN 9789605862565.
20. Ostrom, E. A Behavioral Approach to the Rational Choice Theory of Collective Action: Presidential Address, American Political Science Association, 1997. *Am. Polit. Sci. Rev.* **1998**, *92*, 1–22. [CrossRef]
21. Strom, K. A Behavioral Theory of Competitive Political Parties. *Am. J. Pol. Sci.* **1990**, *34*, 565. [CrossRef]
22. Tina Dacin, M.; Goodstein, J.; Richard Scott, W. Institutional Theory and Institutional Change: Introduction to the Special Research Forum. *Acad. Manag. J.* **2002**, *45*, 45–56. [CrossRef]
23. March, J.G.; Olsen, J.P. The New Institutionalism: Organizational Factors in Political Life. *Am. Polit. Sci. Rev.* **1983**, *78*, 734–749. [CrossRef]
24. Thelen, K.; Steinmo, S. Historical Institutionalism in Contemporary Politics. In *Structuring Politics: Historical Institutionalism in Comparative Analysis*; Steinmo, S., Thelen, K., Longstreth, F., Eds.; Cambridge University Press: Cambridge, MA, USA, 1992; pp. 1–32.
25. Kraatz, M.S.; Zajac, E.J. Exploring the Limits of the New Institutionalism: The Causes and Consequences of Illegitimate Organizational Change. *Am. Sociol. Rev.* **1996**, *61*, 812. [CrossRef]
26. Christiansen, T.; Verdun, A. Historical Institutionalism in the Study of European Integration. In *Oxford Research Encyclopedia of Politics*; Oxford University Press: Oxford, UK, 2020.
27. Bennett, C.J.; Howlett, M. The Lessons of Learning: Reconciling Theories of Policy Learning and Policy Change. *Policy Sci.* **1992**, *25*, 275–294. [CrossRef]
28. Pierson, P. Increasing Returns, Path Dependence, and the Study of Politics. *Am. Polit. Sci. Rev.* **2000**, *94*, 251–267. [CrossRef]
29. Levi, M. A Model, a Method, and a Map: Rational Choice in Comparative and Historical Analysis. In *Comparative Politics: Rationality, Culture and Structure*; Lichbach, M., Zuckerman, A., Eds.; Cambridge University Press: Cambridge, MA, USA, 1997; pp. 19–41.
30. Greener, I. Understanding NHS Reform: The Policy-Transfer, Social Learning, and Path-Dependency Perspectives. *Governance* **2002**, *15*, 161–183. [CrossRef]

31. Beckert, J. Agency, Entrepreneurs, and Institutional Change. The Role of Strategic Choice and Institutionalized Practices in Organizations. *Organ. Stud.* **1999**, *20*, 777–799. [CrossRef]
32. Clemens, E.S.; Cook, J.M. Politics and Institutionalism: Explaining Durability and Change. *Annu. Rev. Sociol.* **1999**, *25*, 441–466. [CrossRef]
33. Thelen, K. Historical Institutionalism in Comparative Politics. *Annu. Rev. Polit. Sci.* **1999**, *2*, 369–404. [CrossRef]
34. Romanelli, E.; Tushman, M.L. Organizational Transformation as Punctuated Equilibrium: An Empirical Test. *Acad. Manag. J.* **1994**, *37*, 1141–1166. [CrossRef]
35. Collier, R.; Collier, D. Critical Junctures and Historical Legacies. In *Shaping the Political Arena: Critical Junctures, the Labor Movement and Regime Dynamics in Latin America*; Princeton University Press: Princeton, NJ, USA, 1991; pp. 27–39.
36. Capoccia, G. Critical Junctures and Institutional Change. In *Advances in Comparative-Historical Analysis*; Cambridge University Press: Cambridge, UK, 2015; pp. 147–179.
37. Di Bartolomeo, G.; Saltari, E.; Semmler, W. The Effects of Political Short-Termism on Transitions Induced by Pollution Regulations. In *Dynamic Analysis in Complex Economic Environments*; Dynamic Modeling and Econometrics in Economics and Finance; Dawid, H., Arifovic, J., Eds.; Springer: Cham, Switzerland, 2021; Volume 26, pp. 109–122.
38. Streeck, W.; Thelen, K. Introduction: Institutional Change in Advanced Political Economies. In *Beyond Continuity: Institutional Change in Advanced Political Economies*; Streeck, W., Thelen, K., Eds.; Oxford University Press: Oxford, UK, 2005; pp. 1–39.
39. Alaimo, L.S.; Ciommi, M.; Vardopoulos, I.; Nosova, B.; Salvati, L. The Medium-Term Impact of the COVID-19 Pandemic on Population Dynamics: The Case of Italy. *Sustainability* **2022**, *14*, 13995. [CrossRef]
40. Trebesch, C.; Konradt, M.; Ordoñez, G.; Herrera, H. The Political Consequences of the Covid Pandemic: Lessons from Cross-Country Polling Data. VoxEU.org. 2020. Available online: https://cepr.org/voxeu/columns/political-consequences-covid-pandemic-lessons-cross-country-polling-data (accessed on 6 September 2023).
41. Bol, D.; Giani, M.; Blais, A.; Loewen, P.J. The Effect of COVID-19 Lockdowns on Political Support: Some Good News for Democracy? *Eur. J. Polit. Res.* **2021**, *60*, 497–505. [CrossRef]
42. Chantzaras, A.; Yfantopoulos, J. The Impact of COVID-19 Pandemic and Its Associations with Government Responses in Europe. *Reg. Peripher.* **2022**, *13*, 23–40. [CrossRef]
43. Katsikas, D. Health and Economic Responses to the COVID-19 Crisis in the EU. *Reg. Peripher.* **2022**, *13*, 9–21. [CrossRef]
44. Polishchuk, L. Misuse of Institutions: Lessons from Transition. In *Economies in Transition*; Palgrave Macmillan: London, UK, 2012; pp. 172–193.
45. Barasa, E.; Mbau, R.; Gilson, L. What Is Resilience and How Can It Be Nurtured? A Systematic Review of Empirical Literature on Organizational Resilience. *Int. J. Health Policy Manag.* **2018**, *7*, 491–503. [CrossRef]
46. Jung, K. Sources of Organizational Resilience for Sustainable Communities: An Institutional Collective Action Perspective. *Sustainability* **2017**, *9*, 1141. [CrossRef]
47. Lengnick-Hall, C.A.; Beck, T.E.; Lengnick-Hall, M.L. Developing a Capacity for Organizational Resilience through Strategic Human Resource Management. *Hum. Resour. Manag. Rev.* **2011**, *21*, 243–255. [CrossRef]
48. Kantabutra, S.; Ketprapakorn, N. Toward an Organizational Theory of Resilience: An Interim Struggle. *Sustainability* **2021**, *13*, 13137. [CrossRef]
49. Carmichael, D.G. Incorporating Resilience through Adaptability and Flexibility. *Civ. Eng. Environ. Syst.* **2015**, *32*, 31–43. [CrossRef]
50. Miceli, A.; Hagen, B.; Riccardi, M.P.; Sotti, F.; Settembre-Blundo, D. Thriving, Not Just Surviving in Changing Times: How Sustainability, Agility and Digitalization Intertwine with Organizational Resilience. *Sustainability* **2021**, *13*, 2052. [CrossRef]
51. Wamsler, C. Stakeholder Involvement in Strategic Adaptation Planning: Transdisciplinarity and Co-Production at Stake? *Environ. Sci. Policy* **2017**, *75*, 148–157. [CrossRef]
52. Rivera-Ferre, M.; Ortega-Cerdà, M.; Baumgärtner, J. Rethinking Study and Management of Agricultural Systems for Policy Design. *Sustainability* **2013**, *5*, 3858–3875. [CrossRef]
53. Holley, K.A. Interdisciplinary Strategies as Transformative Change in Higher Education. *Innov. High. Educ.* **2009**, *34*, 331–344. [CrossRef]
54. Pereira, L.M.; Drimie, S.; Maciejewski, K.; Tonissen, P.B.; Biggs, R. (Oonsie) Food System Transformation: Integrating a Political–Economy and Social–Ecological Approach to Regime Shifts. *Int. J. Environ. Res. Public Health* **2020**, *17*, 1313. [CrossRef]
55. Wolfram, M.; Frantzeskaki, N. Cities and Systemic Change for Sustainability: Prevailing Epistemologies and an Emerging Research Agenda. *Sustainability* **2016**, *8*, 144. [CrossRef]
56. Hall, P.A. Policy Paradigms, Social Learning, and the State: The Case of Economic Policymaking in Britain. *Comp. Polit.* **1993**, *25*, 275. [CrossRef]
57. Birkhaeuser, D.; Evenson, R.E.; Feder, G. The Economic Impact of Agricultural Extension: A Review. *Econ. Dev. Cult. Change* **1991**, *39*, 607–650. [CrossRef]
58. Vardopoulos, I.; Karytsas, S. An Exploratory Path Analysis of Climate Change Effects on Tourism. *Sustain. Dev. Cult. Tradit. J.* **2019**, 132–152. [CrossRef]
59. Islam, M. Sustainability through the Lens of Environmental Sociology: An Introduction. *Sustainability* **2017**, *9*, 474. [CrossRef]
60. Louloudis, L.; Beopoulos, N.; Vlachos, G. Policy for the Protection of the Rural Environment in Greece with a Horizon in 2010. In *Η Ελληνική Γεωργία προς το 2010*; Maravegias, N., Ed.; Papazisis: Athens, Greece, 1999.

61. Vardopoulos, I.; Falireas, S.; Konstantopoulos, I.; Kaliora, E.; Theodoropoulou, E. Sustainability Assessment of the Agri-Environmental Practices in Greece. Indicators' Comparative Study. *Int. J. Agric. Resour. Gov. Ecol.* **2018**, *14*, 368. [CrossRef]
62. *European Court of Auditors The Common Agricultural Policy and Climate Change: Ambitious or Inadequate?* Special Report 16/2021. 2021. Available online: https://www.eca.europa.eu/Lists/ECADocuments/SR21_16/SR_CAP-and-Climate_EN.pdf (accessed on 6 September 2023).
63. Carpenter, A. Introduction: The Role of the European Union as a Global Player in Environmental Governance. *J. Contemp. Eur. Res.* **2012**, *8*, 167–172. [CrossRef]
64. Maravegias, N. *Croissance Economique et Espace Rural. Agriculture et Agro-Industrie Dans Le Development Regional de La Grece*; Universite des Sciences Sociales de Grenoble, Faculte des Sciences Economiques: Grenoble, France, 1983.
65. Maravegias, N.; Doukas, Y.E.; Petides, P. The Political Economy of the Common Agricultural Policy's Green Architecture. *Sustain. Dev. Cult. Tradit. J.* **2023**, *1*, 73–84. [CrossRef]
66. Borsellino, V.; Schimmenti, E.; El Bilali, H. Agri-Food Markets towards Sustainable Patterns. *Sustainability* **2020**, *12*, 2193. [CrossRef]
67. Regulation (EC) No 178/2002 of the European Parliament and of the Council of 28 January 2002 Laying down the General Principles and Requirements of Food Law, Establishing the European Food Safety Authority and Laying down Procedures in Matters of Food safety. *Official Journal L*, 1 February 2002.
68. Linares Quero, A.; Iragui Yoldi, U.; Gava, O.; Schwarz, G.; Povellato, A.; Astrain, C. Assessment of the Common Agricultural Policy 2014–2020 in Supporting Agroecological Transitions: A Comparative Study of 15 Cases across Europe. *Sustainability* **2022**, *14*, 9261. [CrossRef]
69. Louloudis, L.; Maravegias, N. Farmers and Agricultural Policy in Greece since the Accession to the European Union. *Sociol. Ruralis* **1997**, *37*, 270–286. [CrossRef]
70. Doukas, Y.E. Η Κοινή Αγροτική Πολιτική και ο Ελληνικός Αγροτικός Τομέας. In *Ελλάδα και Ευρωπαϊκή Ενοποίηση: Η Ιστορία μιας Πολυκύμαντης Σχέσης 1962–2018*; Maravegias, N., Sakellaropoulos, T., Eds.; Dionicos: Athens, Greece, 2018.
71. Karaveli, E.; Doukas, Y.E. Rural Development Policy: An Alternative Development Strategy for The Case of Greece. *Reg. Peripher.* **2012**, *1*, 21–36. [CrossRef]
72. Cecchini, M.; Zambon, I.; Pontrandolfi, A.; Turco, R.; Colantoni, A.; Mavrakis, A.; Salvati, L. Urban Sprawl and the 'Olive' Landscape: Sustainable Land Management for 'Crisis' Cities. *GeoJournal* **2019**, *84*, 237–255. [CrossRef]
73. Xeni, F.; Gavriel, I.; Tsangas, M.; Doula, M.K.; Zorpas, A.A. A Preliminary Assessment of Sustainable Agricultural Practices in Cyprus. In Proceedings of the 7th International Conference on Sustainable Solid Waste Management, Heraklion, Greece, 26–29 June 2019.
74. Tal, A. Making Conventional Agriculture Environmentally Friendly: Moving beyond the Glorification of Organic Agriculture and the Demonization of Conventional Agriculture. *Sustainability* **2018**, *10*, 1078. [CrossRef]
75. Leinonen, I. Achieving Environmentally Sustainable Livestock Production. *Sustainability* **2019**, *11*, 246. [CrossRef]
76. OECD. *Environmental Cross-Compliance in Agriculture*; OECD: Paris, France, 2010.
77. Cortignani, R.; Severini, S.; Dono, G. Complying with Greening Practices in the New CAP Direct Payments: An Application on Italian Specialized Arable Farms. *Land Use Policy* **2017**, *61*, 265–275. [CrossRef]
78. Maravegias, N.; Martinos, N. L'agriculture Grecque Face à l'union Économique et Monétaire de La Communauté Européenne. In *Agricultures Familiales et Politiques Agricoles en Méditerranée: Enjeux et Perspectives*; Abaab, A., Campagne, P., Elloumi, M., Fragata, A., Zagdouni, L., Eds.; CIHEAM: Montpellier, France, 1997; pp. 277–284.
79. Doukas, Y.E. Ο Αγροτικός Τομέας την Περίοδο 2014–2020. In *Οι Πολιτικές που Χρηματοδοτούνται από τον Κοινοτικό Προυπολογισμό και η Ελληνική Οικονομία*; Mitsos, A., Ed.; Hellenic Foundation for European and Foreign Policy: Athens, Greece, 2014.
80. Doukas, Y.E.; Maravegias, N.; Chrysomallidis, C. Digitalization in the EU Agricultural Sector: Seeking a European Policy Response. In *Food Policy Modelling Responses to Current Issues*; Mattas, K., Baourakis, G., Zopounidis, C., Staboulis, C., Eds.; Springer: Cham, Swijtzerland, 2022; pp. 83–98.
81. Meyer, C.; Matzdorf, B.; Müller, K.; Schleyer, C. Cross Compliance as Payment for Public Goods? Understanding EU and US Agricultural Policies. *Ecol. Econ.* **2014**, *107*, 185–194. [CrossRef]
82. Bennett, H.; Osterburg, B.; Nitsch, H.; Kristensen, L.; Primdahl, J.; Verschuur, G. Strengths and Weaknesses of Crosscompliance in the CAP. *EuroChoices* **2006**, *5*, 50–57. [CrossRef]
83. Juntti, M. Implementing Cross Compliance for Agriculture in the EU: Relational Agency, Power and Action in Different Socio-Material Contexts. *Sociol. Ruralis* **2012**, *52*, 294–310. [CrossRef]
84. Ragazou, K.; Garefalakis, A.; Zafeiriou, E.; Passas, I. Agriculture 5.0: A New Strategic Management Mode for a Cut Cost and an Energy Efficient Agriculture Sector. *Energies* **2022**, *15*, 3113. [CrossRef]
85. Matthews, A. Greening Agricultural Payments in the EU's Common Agricultural Policy. *Bio-Based Appl. Econ.* **2013**, *2*, 1–27. [CrossRef]
86. Hristov, J.; Clough, Y.; Sahlin, U.; Smith, H.G.; Stjernman, M.; Olsson, O.; Sahrbacher, A.; Brady, M.V. Impacts of the EU's Common Agricultural Policy "Greening" Reform on Agricultural Development, Biodiversity, and Ecosystem Services. *Appl. Econ. Perspect. Policy* **2020**, *42*, 716–738. [CrossRef]

87. Johansson, R.C.; Kaplan, J.D. A Carrot-and-Stick Approach to Environmental Improvement: Marrying Agri-Environmental Payments and Water Quality Regulations. *Agric. Resour. Econ. Rev.* **2004**, *33*, 91–104. [CrossRef]
88. Nitsch, H.; Osterburg, B.; Roggendorf, W.; Laggner, B. Cross Compliance and the Protection of Grassland–Illustrative Analyses of Land Use Transitions between Permanent Grassland and Arable Land in German Regions. *Land Use Policy* **2012**, *29*, 440–448. [CrossRef]
89. Viira, A.; Kall, K.; Ariva, J.; Oper, L.; Jürgenson, E.; Maasikamäe, S.; Põldaru, R. Restricting the Eligible Maintenance Practices of Permanent Grassland—A Realistic Way towards More Active Farming? *Agron. Res.* **2020**, *18*, 1556–1572. [CrossRef]
90. Rydén, R. Smallholders, Organic Farmers, and Agricultural Policy. *Scand. J. Hist.* **2007**, *32*, 63–85. [CrossRef]
91. Cimino, O.; Henke, R.; Vanni, F. The Effects of CAP Greening on Specialised Arable Farms in Italy. *New Medit* **2015**, *2*, 22–31.
92. Wu, J.; Babcock, B.A. Optimal Design of a Voluntary Green Payment Program under Asymmetric Information. *J. Agric. Resour. Econ.* **1995**, *20*, 316–327.
93. Nitsch, H.; Osterburg, B. Criteria for an Efficient Enforcement of Standards in Relation to Cross Compliance. In Proceedings of the European Association of Agricultural Economists (EAAE) 12th International Congress, Ghent, Belgium, 26–29 August 2008.
94. Webster, P.; Williams, N. Environmental Cross Compliance—Panacea or Placebo? In Proceedings of the The 13th International Farm Management Congress, Wageningen, The Netherlands, 7–12 July 2002.
95. Karim, A.H.M.Z. Impact of a Growing Population in Agricultural Resource Management: Exploring the Global Situation with a Micro-Level Example. *Asian Soc. Sci.* **2013**, *9*, 14. [CrossRef]
96. Bongaarts, J. Population Pressure and the Food Supply System in the Developing World. *Popul. Dev. Rev.* **1996**, *22*, 483. [CrossRef]
97. Balakuntala, M.V.; Ayad, M.; Voyles, R.M.; White, R.; Nawrocki, R.; Sundaram, S.; Priya, S.; Chiu, G.; Donkin, S.; Min, B.-C.; et al. Global Sustainability through Closed-Loop Precision Animal Agriculture. *Mech. Eng.* **2018**, *140*, S19–S23. [CrossRef]
98. Satterthwaite, D.; McGranahan, G.; Tacoli, C. Urbanization and Its Implications for Food and Farming. *Philos. Trans. R. Soc. B Biol. Sci.* **2010**, *365*, 2809–2820. [CrossRef] [PubMed]
99. Beckers, V.; Poelmans, L.; Van Rompaey, A.; Dendoncker, N. The Impact of Urbanization on Agricultural Dynamics: A Case Study in Belgium. *J. Land Use Sci.* **2020**, *15*, 626–643. [CrossRef]
100. Atta-ur-Rahman; Surjan, A.; Parvin, G.A.; Shaw, R. Impact of Urban Expansion on Farmlands. In *Urban Disasters and Resilience in Asia*; Shaw, R., Surjan, A., Rahman, A., Parvin, G., Eds.; Elsevier: Oxford, UK, 2016; pp. 91–112.
101. Masters, W.A.; Djurfeldt, A.A.; De Haan, C.; Hazell, P.; Jayne, T.; Jirström, M.; Reardon, T. Urbanization and Farm Size in Asia and Africa: Implications for Food Security and Agricultural Research. *Glob. Food Secur.* **2013**, *2*, 156–165. [CrossRef]
102. Turner, S.W.D.; Hejazi, M.; Calvin, K.; Kyle, P.; Kim, S. A Pathway of Global Food Supply Adaptation in a World with Increasingly Constrained Groundwater. *Sci. Total Environ.* **2019**, *673*, 165–176. [CrossRef]
103. Foley, J.A.; Ramankutty, N.; Brauman, K.A.; Cassidy, E.S.; Gerber, J.S.; Johnston, M.; Mueller, N.D.; O'Connell, C.; Ray, D.K.; West, P.C.; et al. Solutions for a Cultivated Planet. *Nature* **2011**, *478*, 337–342. [CrossRef]
104. Kahn, B.M.; LeZaks, D. Investing in Agriculture: Far-Reaching Challenge, Significant Opportunity: An Asset Management Perspective. *SSRN Electron. J.* **2009**. [CrossRef]
105. Hazell, P.; Wood, S. Drivers of Change in Global Agriculture. *Philos. Trans. R. Soc. B Biol. Sci.* **2008**, *363*, 495–515. [CrossRef]
106. Mancosu, N.; Snyder, R.; Kyriakakis, G.; Spano, D. Water Scarcity and Future Challenges for Food Production. *Water* **2015**, *7*, 975–992. [CrossRef]
107. Arora, N.K. Impact of Climate Change on Agriculture Production and Its Sustainable Solutions. *Environ. Sustain.* **2019**, *2*, 95–96. [CrossRef]
108. Baba, S.A.; Adamu, I.A. Implications of Climate Change on Agricultural Productivity: A Review. *J. Food Technol. Nutr. Sci.* **2021**, *365*, 1–5. [CrossRef]
109. Rai, R. Heat Stress in Crops: Driver of Climate Change Impacting Global Food Supply. In *Contemporary Environmental Issues and Challenges in Era of Climate Change*; Springer: Singapore, 2020; pp. 99–117.
110. Shrestha, S.; Himics, M.; Van Doorslaer, B.; Ciaian, P. Impacts of Climate Change on EU Agriculture. *Rev. Agric. Appl. Econ.* **2013**, *16*, 24–39. [CrossRef]
111. Olesen, J.E.E.; Trnka, M.; Kersebaum, K.C.C.; Skjelvåg, A.O.O.; Seguin, B.; Peltonen-Sainio, P.; Rossi, F.; Kozyra, J.; Micale, F. Impacts and Adaptation of European Crop Production Systems to Climate Change. *Eur. J. Agron.* **2011**, *34*, 96–112. [CrossRef]
112. Iglesias, A.; Garrote, L.; Quiroga, S.; Moneo, M. A Regional Comparison of the Effects of Climate Change on Agricultural Crops in Europe. *Clim. Change* **2012**, *112*, 29–46. [CrossRef]
113. Santos, J.A.; Fraga, H.; Malheiro, A.C.; Moutinho-Pereira, J.; Dinis, L.-T.; Correia, C.; Moriondo, M.; Leolini, L.; Dibari, C.; Costafreda-Aumedes, S.; et al. A Review of the Potential Climate Change Impacts and Adaptation Options for European Viticulture. *Appl. Sci.* **2020**, *10*, 3092. [CrossRef]
114. Maharjan, K.L.; Joshi, N.P. Effects of Climate Change on Regional Agriculture Production, Food Price and Food Insecurity. In *Climate Change, Agriculture and Rural Livelihoods in Developing Countries*; Springer: Tokyo, Japan, 2013; pp. 93–103.
115. Skendžić, S.; Zovko, M.; Živković, I.P.; Lešić, V.; Lemić, D. The Impact of Climate Change on Agricultural Insect Pests. *Insects* **2021**, *12*, 440. [CrossRef]
116. Fraga, H.; Moriondo, M.; Leolini, L.; Santos, J.A. Mediterranean Olive Orchards under Climate Change: A Review of Future Impacts and Adaptation Strategies. *Agronomy* **2020**, *11*, 56. [CrossRef]

117. del Pozo, A.; Brunel-Saldias, N.; Engler, A.; Ortega-Farias, S.; Acevedo-Opazo, C.; Lobos, G.A.; Jara-Rojas, R.; Molina-Montenegro, M.A. Climate Change Impacts and Adaptation Strategies of Agriculture in Mediterranean-Climate Regions (MCRs). *Sustainability* **2019**, *11*, 2769. [CrossRef]

118. Salvati, L.; Zitti, M. Territorial Disparities, Natural Resource Distribution, and Land Degradation: A Case Study in Southern Europe. *GeoJournal* **2007**, *70*, 185–194. [CrossRef]

119. Salvati, L.; Gemmiti, R.; Perini, L. Land Degradation in Mediterranean Urban Areas: An Unexplored Link with Planning? *Area* **2012**, *44*, 317–325. [CrossRef]

120. Zambon, I.; Colantoni, A.; Carlucci, M.; Morrow, N.; Sateriano, A.; Salvati, L. Land Quality, Sustainable Development and Environmental Degradation in Agricultural Districts: A Computational Approach Based on Entropy Indexes. *Environ. Impact Assess. Rev.* **2017**, *64*, 37–46. [CrossRef]

121. Zambon, I.; Benedetti, A.; Ferrara, C.; Salvati, L. Soil Matters? A Multivariate Analysis of Socioeconomic Constraints to Urban Expansion in Mediterranean Europe. *Ecol. Econ.* **2018**, *146*, 173–183. [CrossRef]

122. Delfanti, L.; Colantoni, A.; Recanatesi, F.; Bencardino, M.; Sateriano, A.; Zambon, I.; Salvati, L. Solar Plants, Environmental Degradation and Local Socioeconomic Contexts: A Case Study in a Mediterranean Country. *Environ. Impact Assess. Rev.* **2016**, *61*, 88–93. [CrossRef]

123. Kakabouki, I.; Tataridas, A.; Mavroeidis, A.; Kousta, A.; Roussis, I.; Katsenios, N.; Efthimiadou, A.; Papastylianou, P. Introduction of Alternative Crops in the Mediterranean to Satisfy EU Green Deal Goals. A Review. *Agron. Sustain. Dev.* **2021**, *41*, 71. [CrossRef]

124. Tsangas, M.; Zorpas, A.A.; Jeguirim, M. Sustainable Renewable Energy Policies and Regulations, Recent Advances, and Challenges. In *Renewable Energy Production and Distribution*; Jeguirim, M., Ed.; Elsevier: Cambridge, MA, USA, 2022; pp. 449–465.

125. Loizia, P.; Voukkali, I.; Zorpas, A.A.; Navarro Pedreño, J.; Chatziparaskeva, G.; Inglezakis, V.J.; Vardopoulos, I.; Doula, M. Measuring the Level of Environmental Performance in Insular Areas, through Key Performed Indicators, in the Framework of Waste Strategy Development. *Sci. Total Environ.* **2021**, *753*, 141974. [CrossRef]

126. Kritikou, T.; Panagiotakos, D.; Abeliotis, K.; Lasaridi, K. Investigating the Determinants of Greek Households Food Waste Prevention Behaviour. *Sustainability* **2021**, *13*, 11451. [CrossRef]

127. Byrne, K.A.; Zabetakis, I. Farm to Fork: Balancing the Needs for Sustainable Food Production and Food for Health Promotion. *Sustainability* **2023**, *15*, 3396. [CrossRef]

128. Volkov, A.; Melnikienė, R. CAP Direct Payments System's Linkage with Environmental Sustainability Indicators. *Public Policy Adm.* **2017**, *16*, 231–244. [CrossRef]

129. Bielik, P.; Turčeková, N.; Adamičková, I.; Belinská, S.; Bajusová, Z. Will Changes in the Common Agricultural Policy Bring a Respectful Approach to Environment in EU Countries? *Visegr. J. Bioecon. Sustain. Dev.* **2022**, *11*, 21–25. [CrossRef]

130. Pierson, P. The Path to European Integration. *Comp. Polit. Stud.* **1996**, *29*, 123–163. [CrossRef]

131. Bulmer, S. Politics in Time Meets the Politics of Time: Historical Institutionalism and the EU Timescape. *J. Eur. Public Policy* **2009**, *16*, 307–324. [CrossRef]

132. Moravcsik, A. Preferences and Power in the European Ommunity: A Liberal Intergovernmentalist Approach. *JCMS J. Common Mark. Stud.* **1993**, *31*, 473–524. [CrossRef]

133. Schneider, G.; Ershova, A. Rational Choice Institutionalism and European Integration. In *Oxford Research Encyclopedia of Politics*; Oxford University Press: Oxford, UK, 2018.

134. Harrison, S. What Is Electoral Psychology?—Scope, Concepts, and Methodological Challenges for Studying Conscious and Subconscious Patterns of Electoral Behavior, Experience, and Ergonomics. *Societies* **2020**, *10*, 20. [CrossRef]

135. Yilmaz, I.; Morieson, N. A Systematic Literature Review of Populism, Religion and Emotions. *Religions* **2021**, *12*, 272. [CrossRef]

136. Baker, S.; Quinn, M.J. Populism, Austerity and Governance for Sustainable Development in Troubled Times: Introduction to Special Issue. *Sustainability* **2022**, *14*, 3271. [CrossRef]

137. Ciommi, M.; Egidi, G.; Vardopoulos, I.; Chelli, F.M.; Salvati, L. Toward a 'Migrant Trap'? Local Development, Urban Sustainability, Sociodemographic Inequalities, and the Economic Decline in a Mediterranean Metropolis. *Soc. Sci.* **2022**, *12*, 26. [CrossRef]

138. Manners, I. Political Psychology of European Integration: The (Re)Production of Identity and Difference in the Brexit Debate. *Polit. Psychol.* **2018**, *39*, 1213–1232. [CrossRef]

139. Maravegias, N. Agriculture Méditerranéenne et Politique Agricole Commune: L'expérience de La Grèce. In *Choix Technologiques, Risques et Sécurité Dans les Agricultures Méditerranéennes*; Bedrani, S., Campagne, P., Eds.; CIHEAM: Montpellier, France, 1991; pp. 159–165.

140. Swinnen, J.F.M. *A Fischler Reform of the Common Agricultural Policy?* Centre for European Policy Studies: Brussels, Belgium, 2001.

141. Mizik, T. Agri-Food Trade Competitiveness: A Review of the Literature. *Sustainability* **2021**, *13*, 11235. [CrossRef]

142. Kaczorowska, J.; Prandota, A.; Rejman, K.; Halicka, E.; Tul-Krzyszczuk, A. Certification Labels in Shaping Perception of Food Quality—Insights from Polish and Belgian Urban Consumers. *Sustainability* **2021**, *13*, 702. [CrossRef]

143. Woldemariyam, F.T.; De Vleeschauwer, A.; Hundessa, N.; Muluneh, A.; Gizaw, D.; Tinel, S.; De Clercq, K.; Lefebvre, D.; Paeshuyse, J. Risk Factor Assessment, Sero-Prevalence, and Genotyping of the Virus That Causes Foot-and-Mouth Disease on Commercial Farms in Ethiopia from October 2018 to February 2020. *Agriculture* **2021**, *12*, 49. [CrossRef]

144. Buller, H.; Blokhuis, H.; Jensen, P.; Keeling, L. Towards Farm Animal Welfare and Sustainability. *Animals* **2018**, *8*, 81. [CrossRef] [PubMed]

145. Pe'er, G.; Bonn, A.; Bruelheide, H.; Dieker, P.; Eisenhauer, N.; Feindt, P.H.; Hagedorn, G.; Hansjürgens, B.; Herzon, I.; Lomba, Â.; et al. Action Needed for the EU Common Agricultural Policy to Address Sustainability Challenges. *People Nat.* **2020**, *2*, 305–316. [CrossRef] [PubMed]
146. Dupraz, P.; Guyomard, H. Environment and Climate in the Common Agricultural Policy. *EuroChoices* **2019**, *18*, 18–25. [CrossRef]
147. Cuadros-Casanova, I.; Cristiano, A.; Biancolini, D.; Cimatti, M.; Sessa, A.A.; Mendez Angarita, V.Y.; Dragonetti, C.; Pacifici, M.; Rondinini, C.; Di Marco, M. Opportunities and Challenges for Common Agricultural Policy Reform to Support the European Green Deal. *Conserv. Biol.* **2023**, *37*, e14052. [CrossRef]
148. Pe'er, G.; Zinngrebe, Y.; Moreira, F.; Sirami, C.; Schindler, S.; Müller, R.; Bontzorlos, V.; Clough, D.; Bezák, P.; Bonn, A.; et al. A Greener Path for the EU Common Agricultural Policy. *Science* **2019**, *365*, 449–451. [CrossRef] [PubMed]
149. Matthews, A. The New CAP Must Be Linked More Closely to the UN Sustainable Development Goals. *Agric. Food Econ.* **2020**, *8*, 19. [CrossRef]
150. Swaminathan, M.S.; Kesavan, P.C. Science for Sustainable Agriculture to Achieve UN SDG Goal 2. *Curr. Sci.* **2018**, *114*, 1585. [CrossRef]
151. Recanatesi, F.; Clemente, M.; Grigoriadis, E.; Ranalli, F.; Zitti, M.; Salvati, L. A Fifty-Year Sustainability Assessment of Italian Agro-Forest Districts. *Sustainability* **2015**, *8*, 32. [CrossRef]
152. Sadłowski, A.; Wrzaszcz, W.; Smędzik-Ambroży, K.; Matras-Bolibok, A.; Budzyńska, A.; Angowski, M.; Mann, S. Direct Payments and Sustainable Agricultural Development—The Example of Poland. *Sustainability* **2021**, *13*, 13090. [CrossRef]
153. Volkov, A.; Balezentis, T.; Morkunas, M.; Streimikiene, D. In a Search for Equity: Do Direct Payments under the Common Agricultural Policy Induce Convergence in the European Union? *Sustainability* **2019**, *11*, 3462. [CrossRef]
154. Morkunas, M.; Labukas, P. The Evaluation of Negative Factors of Direct Payments under Common Agricultural Policy from a Viewpoint of Sustainability of Rural Regions of the New EU Member States: Evidence from Lithuania. *Agriculture* **2020**, *10*, 228. [CrossRef]
155. Brown, C.; Kovács, E.; Herzon, I.; Villamayor-Tomas, S.; Albizua, A.; Galanaki, A.; Grammatikopoulou, I.; McCracken, D.; Olsson, J.A.; Zinngrebe, Y. Simplistic Understandings of Farmer Motivations Could Undermine the Environmental Potential of the Common Agricultural Policy. *Land Use Policy* **2021**, *101*, 105136. [CrossRef]
156. Sadłowski, A.; Beluhova-Uzunova, R.; Popp, J.; Atanasov, D.; Ivanova, B.; Shishkova, M.; Hristov, K. Direct Payments Distribution between Farmers in Selected New EU Member States. *Agris Online Pap. Econ. Inform.* **2022**, *14*, 97–107. [CrossRef]
157. Regulation (EU) No 1306/2013 Regulation (EU) No 1306/2013 of the European Parliament and of the Council on the Financing, Management and Monitoring of the Common Agricultural Policy and Repealing Council Regulations (EEC) No 352/78, (EC) No 165/94, (EC) No 2799/98, (EC) No 814/2000. 2013. Available online: https://eur-lex.europa.eu/legal-content/EN/TXT/?uri=celex:32013R1306 (accessed on 6 September 2023).
158. Brady, M.; Hristov, J.; Höjgård, S.; Jansson, T.; Johansson, H.; Larsson, C.; Nordin, I.; Rabinowicz, E. *Impacts of Direct Payments—Lessons for CAP Post-2020 from a Quantitative Analysis*; Rapport/AgriFood Economics Centre: Lund, Sweden, 2017.
159. Milczarek-Andrzejewska, D.; Zawalińska, K.; Czarnecki, A. Land-Use Conflicts and the Common Agricultural Policy: Evidence from Poland. *Land Use Policy* **2018**, *73*, 423–433. [CrossRef]
160. Crippa, M.; Guizzardi, D.; Banja, M.; Solazzo, E.; Muntean, M.; Schaaf, E.; Pagani, F.; Monforti-Ferrario, F.; Olivier, J.; Quadrelli, R.; et al. CO_2 Emissions of All World Countries. In *JRC Science for Policy Report*; European Commission: Brussels, Belgium, 2022.
161. Vardopoulos, I.; Konstantinou, Z. Study of the Possible Links between CO_2 Emissions and Employment Status. *Sustain. Dev. Cult. Tradit. J.* **2017**, *1*, 100–112. [CrossRef]
162. Đokić, D.; Matkovski, B.; Jeremić, M.; Đurić, I. Land Productivity and Agri-Environmental Indicators: A Case Study of Western Balkans. *Land* **2022**, *11*, 2216. [CrossRef]
163. Platis, D.P.; Papoui, E.; Bantis, F.; Katsiotis, A.; Koukounaras, A.; Mamolos, A.P.; Mattas, K. Underutilized Vegetable Crops in the Mediterranean Region: A Literature Review of Their Requirements and the Ecosystem Services Provided. *Sustainability* **2023**, *15*, 4921. [CrossRef]
164. Matthews, A. The CAP in the 2021–2027 MFF Negotiations. *Intereconomics* **2018**, *53*, 306–311. [CrossRef]
165. Kosmas, C.; Karamesouti, M.; Kounalaki, K.; Detsis, V.; Vassiliou, P.; Salvati, L. Land Degradation and Long-Term Changes in Agro-Pastoral Systems: An Empirical Analysis of Ecological Resilience in Asteroussia–Crete (Greece). *CATENA* **2016**, *147*, 196–204. [CrossRef]
166. Arata, L.; Sckokai, P. The Impact of Agri-Environmental Schemes on Farm Performance in Five E.U. Member States: A DID-Matching Approach. *Land Econ.* **2016**, *92*, 167–186. [CrossRef]
167. Zafeiriou, E.; Azam, M.; Garefalakis, A. Exploring Environmental—Economic Performance Linkages in EU Agriculture: Evidence from a Panel Cointegration Framework. *Manag. Environ. Qual. An Int. J.* **2023**, *34*, 469–491. [CrossRef]
168. European Commission Communication from the Commission to the European Parliament, the Council, the European Economic and Social Committee and the Committee of the Regions. A Farm to Fork Strategy for a Fair, Healthy and Environmentally-Friendly Food System. COM(2020)381 Fina. 2020. Available online: https://eur-lex.europa.eu/legal-content/EN/TXT/?uri=celex%3A52020DC0381 (accessed on 6 September 2023).

169. Garske, B.; Heyl, K.; Ekardt, F.; Weber, L.; Gradzka, W. Challenges of Food Waste Governance: An Assessment of European Legislation on Food Waste and Recommendations for Improvement by Economic Instruments. *Land* **2020**, *9*, 231. [CrossRef]
170. Tataridas, A.; Kanatas, P.; Chatzigeorgiou, A.; Zannopoulos, S.; Travlos, I. Sustainable Crop and Weed Management in the Era of the EU Green Deal: A Survival Guide. *Agronomy* **2022**, *12*, 589. [CrossRef]

 *land*

Article

Straw Mulch Application Enhanced Soil Properties and Reduced Diffuse Pollution at a Steep Vineyard in Istria (Croatia)

Ivan Dugan [1], Paulo Pereira [2], Jasmina Defterdarovic [1], Lana Filipovic [1], Vilim Filipovic [1,3] and Igor Bogunovic [1,*]

[1] Faculty of Agriculture, University of Zagreb, Svetosimunska 25, 10000 Zagreb, Croatia; idugan@agr.hr (I.D.); jdefterdarovic@agr.hr (J.D.); lfilipovic@agr.hr (L.F.); vfilipovic@agr.hr (V.F.)

[2] Environmental Management Laboratory, Mykolas Romeris University, LT-08303 Vilnius, Lithuania; pereiraub@gmail.com

[3] Future Regions Research Centre, Geotechnical and Hydrogeological Engineering Research Group, Federation University, Gippsland, VIC 3841, Australia

[*] Correspondence: ibogunovic@agr.hr

Abstract: Straw mulching is a sustainable practice used to control soil erosion. However, different doses of mulch affect the efficiency of straw conservation. This study presents detailed research on how soil physicochemical properties and the hydrological response react to different types of vineyard soil management (Tilled, Grass, Low Straw, High Straw) and seasons (spring, summer, autumn) under conventional management on Anthrosols in Mediterranean conditions. To assess soil properties, core samples and disturbed samples were taken from the topsoil layer (0–10 cm). To evaluate erosion rates, a rainfall simulation experiment was conducted (58 mm h^{-1} for 30 min) with 10 replicates per treatment and season (120 in total). The results show higher water-stable aggregates (WSA) and soil organic matter (SOM) and lower bulk density (BD) in the mulch and grass treatment groups compared with the Tilled treatment group. High Straw treatment successfully mitigated runoff, while other treatments had significantly higher runoff that triggered sediment loss (SL) and translocation of P, K, Zn and Ni down the slope. There were 254% and 520% higher K losses with Tilled treatment in autumn compared with Low Straw and Grass treatments, respectively. Statistical analysis showed a strong association between element loss and SL, which indicates an ecological threat in degraded and endangered vineyards. Mulch application and grass cover reduce the vulnerability of vineyards, reduce evaporation, act as insulation against high temperatures, reduce erosion and suppress weed growth. The mulch dosage varies depending on the goals and conditions of the vineyard; thus, lower mulch dosage (2 t/ha) is appropriate when soil conditions are favourable and there is no significant need for moisture retention, while higher mulch dosage is necessary in dry regions to maintain soil moisture during high-temperature periods, as well as in sloped areas subjected to erosion.

Keywords: sustainable management; mulch application; element behaviour; soil erosion

Citation: Dugan, I.; Pereira, P.; Defterdarovic, J.; Filipovic, L.; Filipovic, V.; Bogunovic, I. Straw Mulch Application Enhanced Soil Properties and Reduced Diffuse Pollution at a Steep Vineyard in Istria (Croatia). *Land* **2023**, *12*, 1691. https://doi.org/10.3390/land12091691

Academic Editors: Andreas Tsatsaris, Nikolaos Stathopoulos, Xiao Huang, Demetrios E. Tsesmelis, Kleomenis Kalogeropoulos and Nilanchal Patel

Received: 14 July 2023
Revised: 15 August 2023
Accepted: 21 August 2023
Published: 29 August 2023

1. Introduction

The rising global population has certain demands (e.g., food, fuel and fibre), but at what cost? Agriculture is facing many changes. On one hand, there is a need to ensure sufficient quantities of food for the growing population, while on the other hand, this needs to be achieved sustainably, without endangering the soil and water [1,2]. Applying sustainable agricultural practices can be challenging, particularly bearing in mind that the environmental, social and economic impacts of agriculture are multifarious and interact with each other [3,4].

Soil and water conservation are necessary for mitigating and reversing land degradation. This can be achieved by incorporating conservation tillage management of highly jeopardized soils [5,6]. These soils are well represented in areas with intensive crop production such as croplands, fruit plantations and vineyards [7–9]. Vineyards are known for their high vulnerability in terms of soil and plant health. Soil health is widely affected by intensive management with tractor traffic, pesticide and fertilizer use, and frequent tillage [10–12]. All these activities have serious effects on soil degradation; they can disturb the soil pore system, increase bulk density, decrease soil organic matter and water-stable aggregates and increase diffuse pollution in steep areas [13–17]. To mitigate these consequences, soil conservation methods such as conservation tillage, permanent grass cover, mulching and the use of cover crops are needed [18–20]. Even though tillage has shown some positive effects, including reduced compaction and enhanced soil air-to-water ratio, these effects are short-lived, while in the long term, negative effects such as crust formation, reduced soil organic matter, deteriorated structure and increased soil compaction and erosion appear [21–24]. More sustainable practices are currently being implemented, particularly the use of mulch, which can take different forms such as harvest residues [25–29], sawmill byproducts [30–33], agricultural industry byproducts [34,35], pruning residues [36,37] and synthetic materials [38–40]. Mulching is one of the main features of conservation practice, along with reduced tillage. Conservation practice has certain advantages, including improving soil structure [31,41,42], reducing compaction [43,44], improving the air-to-water ratio [45,46], increasing soil micro and macro elements [47,48] and preventing further soil particle transport in sloped areas [49–51]. These measures are well-recognized and can be implemented in endangered agricultural areas such as vineyards. Croatia has recorded an increase in vineyards [52], and great concern has risen with regard to increasing soil erosion events, which often result in high rates of soil loss and diffuse pollution.

Diffuse pollution in vineyards can be very dangerous if the concentrations of elements exceed their normal rates. This is more pronounced in intensively managed vineyards with high pesticide use. In these vineyards, vines are exposed to phytotoxicity, and soil and water pollution often occurs, especially in lower-sloped areas. Additionally, higher amounts of trace elements affect yield quality, thus endangering human health [53,54]. This study presents a comprehensive evaluation of the impacts of soil management practices and seasonal variations on soil physicochemical properties, and the hydrological response and element transport in an intensively managed vineyard in Istria, Croatia.

By investigating the effects of different soil management techniques and seasonal variations on diffuse pollution and soil properties during simulated rainfall, we aimed to address the pressing concern of increasing soil erosion events and the associated soil loss and diffuse pollution in vineyards. This study contributes to the existing knowledge by providing valuable insights into the specific challenges faced by vineyards in terms of soil degradation and highlights the importance of implementing sustainable soil management practices to mitigate these issues in the context of the increasing number of vineyards in Croatia. The research was conducted to advance our knowledge of conservation management and land degradation in vineyards. The present research takes us a step forward by evaluating two dosages of mulch as well as bare and grass-covered soils and testing them in different seasons (under different soil conditions). The results presented in this work confirm the need for repeated experiments in different seasons to obtain a solid and accurate conclusion.

2. Materials and Methods

2.1. Study Area

The investigated vineyard is located in Grimalda (Central Istria, Croatia; Figure 1) and is situated on an average slope of 6.65° (3° min to 9° max) with NE–SW exposure at an average elevation of 285 m a.s.l. The vineyard is mostly surrounded by forest and olive orchards. The climate in the investigated area is classified as Cfb (temperate humid climate with warm summer) according to the Köppen climate classification [55]. The parent material is flysch, and the soil is classified as Anthrosols derived from Stagnosols by deep ploughing and ameliorative fertilisation [56]. The mean annual temperature (2021) is 12.7 °C and ranges from a minimum average of 3.9 °C in January to a maximum average of

23.7 °C in July. The mean annual precipitation is 829.9 mm, ranging from a minimum of 10.8 mm in October to a maximum of 113.7 mm in May (Figure 2).

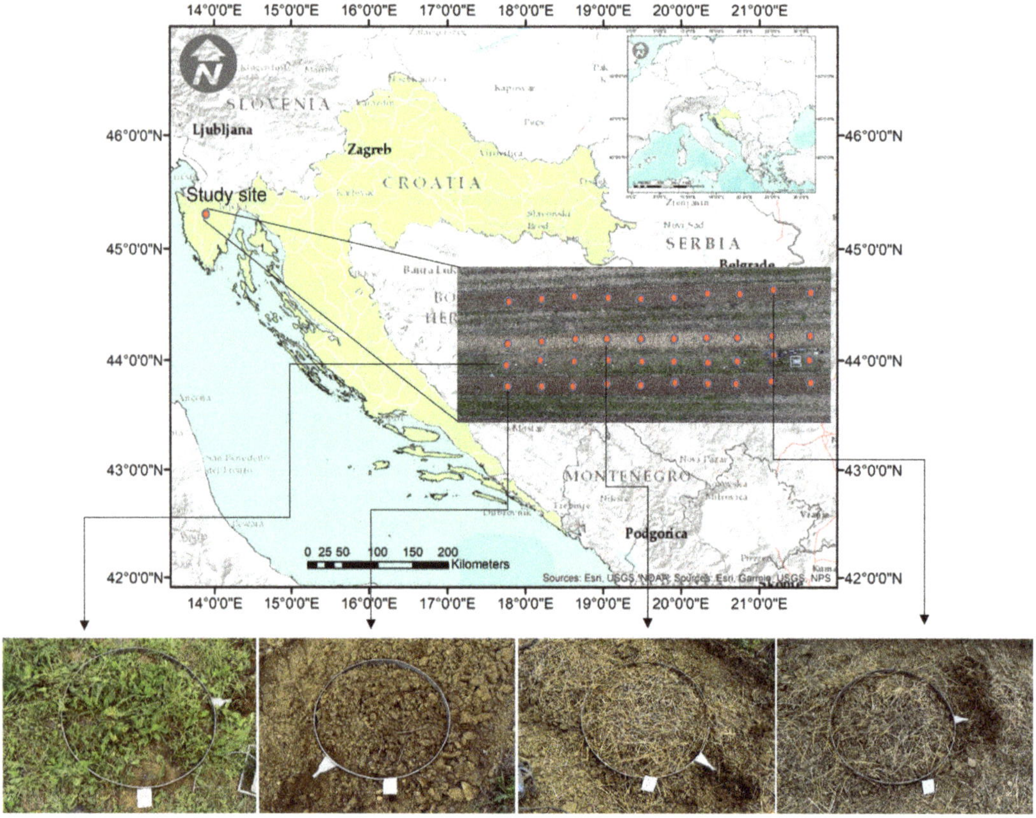

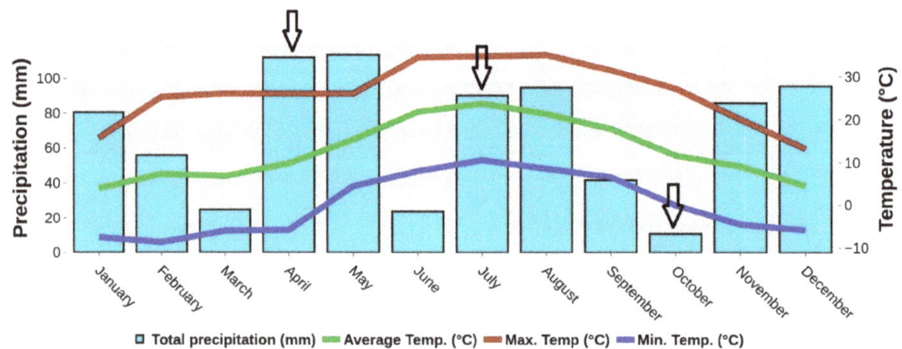

Figure 1. Study area and rainfall simulation experimental plots.

Figure 2. Monthly precipitation and temperature throughout the experimental period. Arrows indicate when measurements were taken.

2.2. Experimental Design

The experimental vineyard consisted of 4 treatment sections: grass-covered (Grass): grass mowed in June and right before the harvest; tillage (Tilled): conducted once a year (in spring); straw mulch: application of 2 t/ha (Low Straw) or a double dose of 4 t/ha (High Straw). Inter-row treatments were applied on the same soil type, slope, aspect and exposure using a paired-plot strategy (Figure 1). In each treatment, 10 plots were separated by 5 m. This is a 4-year-old vineyard where cv. Istrian Teran was planted on Kober 5BB (*Vitis berlandieri* × *Vitis riparia*) rootstock with 2 m between rows and 0.8 m between vines. Over the year, a tractor passed 25 times to apply pesticide and fertiliser, mow the grass, manage the canopy and conduct tillage. Pesticide and fertiliser were applied frequently (10–12 times during the growing period). Rainfall simulation experiments were conducted throughout the vegetation season to determine the impact of seasonal viticultural practices; 40 experiments were conducted per season (spring, summer and autumn 2021) for a total of 120 experiments. The simulations were conducted for 30 min with pressurized rainfall simulators (UGT Rainmaker, Müncheberg, Germany), with a rainfall intensity of 58 mm h^{-1}. The time and intensity of rainfall simulations were selected because 93% of annual soil loss was measured in a single rainstorm at that intensity [57]. Before the rainfall simulation experiments, the rainfall simulator was calibrated using a plastic vessel. The catchment area was enclosed by a metal ring with a small faucet (1 m diameter, 0.785 m^2) pressed 5–7 cm into the soil and extending 10 cm above the soil surface. The faucet on the ring was connected to a plastic canister to collect the total overland flow during the simulations. A Casio HS-6-1EF chronometer (Tokyo, Japan) was used during the rainfall experiment to determine the time to ponding (TP) and time to runoff (TR).

2.3. Laboratory Work

Undisturbed soil samples (4 treatments × 10 replicates × 3 seasons) and core samples (4 treatments × 10 replicates × 3 seasons) were taken from 0 to 10 cm depths close to the rainfall simulation plots and used to determine soil water content (SWC), bulk density (BD) and water-stable aggregates (WSA). SWC and BD were determined by wetting, weighing and drying soil core samples in an oven at 105 °C for 48 h, and all undisturbed samples were prepared by hand and broken down into aggregates [58] and left to dry for 1 week at room temperature (25 °C). The wet sieving method described in [59] was used to determine WSA. Eijkelkamp's wet sieving apparatus was utilized for the wet sieving of all samples that had previously undergone dry sieving. The percentage of WSA was obtained using the equation:

$$\text{WSA} = \frac{Wds}{Wds + Wdw},\tag{1}$$

where WSA is the percentage of water-stable aggregates, *Wds* is the weight of aggregates dispersed in dispersing solution (g) and *Wdw* is the weight of aggregates dispersed in distilled water (g). The rest of the undisturbed soil samples (aggregates not used for WSA determination) were milled and sieved through a 2 mm mesh, in preparation for further chemical analysis. Soil pH was measured using the electrometric method in a 1:5 (*w/v*) ratio using a Beckman Φ72 pH meter in H$_2$O. The digestion method by Walkley and Black [60] was used to measure the soil organic matter (SOM) content, utilizing the previously dried, milled and sieved samples.

Using the method described in [49], sediment concentration (SC) was calculated by dividing the sediment mass by the overland flow mass. Sediment yield was determined after air-drying at room temperature (25 °C) for 1 week to obtain the mass of the filter paper and sediment. Sediment loss (SL) was determined by calculating the mass of overland flow to obtain the runoff.

After drying, milling and sieving, the soil samples were placed in a plastic cylinder with a protective foil on the bottom and a plastic lid on the top. The cylinder was later placed in the workstation and set for pXRF analysis (Olympus, Waltham, MA, USA) after exposing the samples to X-ray beams for 2 min [61,62] to obtain the concentrations of elements (phosphorous (P), potassium (K), nickel (Ni), zinc (Zn) and lead (Pb)) in the soil and sediment. These elements were chosen due to their overall potentially harmful effects on agroecosystems.

2.4. Statistical Analysis

Before data analysis, the normality and homogeneity of variance were assessed using the Kolmogorov–Smirnov test. The data normality of variance was observed at $p > 0.05$. Many of the variables did not follow the Gaussian distribution and heteroscedasticity. Consequently, the dataset was normalized using natural logarithm and Box–Cox transformations. Two-way ANOVA and the Kruskal–Wallis test were performed to assess the impact of treatment and seasonality on soil properties, hydrological response and element transport (SOM, pH, TP, TR, runoff, SC, SL, P loss, K loss, Ni loss, Zn loss, Pb loss). When significant differences were observed at $p < 0.05$, Tukey's LSD test was applied post hoc. Using the Box–Cox transformed data, principal component analysis (PCA) was performed based on the correlation matrix to identify interrelations between variables. All data analyses were performed using Statistica 12.0 [63] and graphs were created using Plotly [64]. All the data in tables and graphs are presented in their original states.

3. Results

3.1. Soil Physical Properties

Statistical analysis showed a significant difference following Tilled treatment, with higher BD values observed in autumn (1.59 g cm^{-3}) compared with spring (1.34 g cm^{-3}) and summer (1.47 g cm^{-3}). Under the Low Straw treatment, lower BD values were recorded in spring (1.31 g cm^{-3}) while higher values were recorded in summer (1.47 g cm^{-3}) and autumn (1.52 g cm^{-3}). Significantly higher values were recorded following the High Straw treatment in autumn (1.48 g cm^{-3}) compared with spring (1.31 g cm^{-3}) and summer (1.27 g cm^{-3}). Grass treatment showed no significant differences between seasons. There were no significant differences between treatments in spring, while significantly lower values were observed in summer and autumn following the Grass treatment compared with the other treatments (Table 1). For SWC, statistical differences were noted only for treatments through seasons. For the Tilled treatment, SWC was significantly higher in spring (40.86%) than in summer (25.07%), while in autumn, the SWC of 30.93% was not statistically different compared with the spring or summer. For the Low Straw treatment, there were no significant differences in SWC between the seasons, with a minimum of 26.21% in autumn, 32.05% in summer and a maximum of 32.69% in spring. For the High Straw treatment, there were no significant differences in SWC through the seasons, with a minimum of 28.66% in autumn, 29.71% in summer and a maximum of 32.36% in autumn. For the Grass treatment, SWCF was significantly lower in summer (20.21%) than in spring (37.47%) and autumn (38.38%). Statistical analysis of WSA showed significant differences between treatments and seasons. For the Tilled treatment, WSA was significantly lower in summer (64.18%) and autumn (62.05%) than in spring (75.12%). For the Low Straw treatment, there were no significant differences through the seasons—values ranged from 73.26% in summer to 74.08% in autumn. For the High Straw treatment, significantly lower WSA values were recorded in spring (70.73%) compared with summer (81.11%) and autumn (80.77%). Grass treatment followed the same pattern as Low Straw treatment, with no significant differences through the seasons—values ranged from 82.33% in spring to 85.86% in summer. In spring, significantly higher WSA values were observed in the Grass treatment group than in the other treatment groups; the same pattern was observed in summer and autumn (Table 1).

Table 1. Results of two-way ANOVA based on soil properties. Different letters after mean values in columns represent significant differences at $p < 0.05$. Capital letters indicate statistical differences between treatments; lowercase letters indicate statistical differences between seasons. BD, bulk density; SWC, soil water content; WSA, water-stable aggregates.

Season	Treatment	BD (g cm^{-3})	SWC (%)	WSA (%)
Spring	Tilled	1.34 Ca	40.86 Aa	75.12 Ab
	Low Straw	1.31 Ba	32.69 Aa	73.80 Ab
	High Straw	1.31 Ba	29.71 Aa	70.73 Bb
	Grass	1.38 Aa	37.47 Aa	82.33 Aa
Summer	Tilled	1.47 Ba	25.07 Ba	64.18 Bd
	Low Straw	1.47 Ba	32.05 Aa	73. 26 Ac
	High Straw	1.27 Bb	28.66 Aa	81.11 Ab
	Grass	1.38 Aab	20.21 Ba	85.86 Aa
Autumn	Tilled	1.59 Aa	30.93 ABa	62.05 Bc
	Low Straw	1.52 Aab	26.21 Aa	74.08 Ab
	High Straw	1.48 Aab	32.36 Aa	80.77 Aab
	Grass	1.45 Ab	38.38 Aa	84.04 Aa

3.2. Soil Chemical Properties

Regarding the chemical properties, significant differences were noted for treatments and seasons. There were no significant differences in soil organic matter in the Tilled, Low Straw and Grass treatment groups through the seasons, with values ranging from a minimum of 1.38% for Tilled treatment in spring to a maximum of 2.51% for Grass treatment in autumn. For the High Straw treatment, there were significantly lower SOM values in spring (1.37%) compared with summer (1.56%) and autumn (1.73%). In every season, SOM values were significantly higher following Grass treatment (Table 2). There were no significant differences in soil pH based on treatments through the seasons. In spring, soil pH values were significantly higher (8.04) in the High Straw treatment group compared with the Tilled (7.54), Low Straw (7.93) and Grass (7.99) treatment groups. In summer, there were significantly lower pH values in the Tilled (7.52) and Low Straw (7.97) treatment groups compared with the Grass (8.16) treatment group, while pH values did not significantly differ between the High Straw (8.09) and the Low Straw and Grass treatment groups. The same pattern in soil pH values was observed in autumn, as presented in Table 2.

Soil P concentrations showed significant differences based on treatment and season. With Tilled treatment, there were no significant differences in soil P through the seasons, with a minimum of 390.66 kg ha^{-1} in summer, 425.29 kg ha^{-1} in spring and a maximum of 515.45 kg ha^{-1} in autumn. The same pattern was observed for p values in the High Straw treatment group (Table 2). For Low Straw treatment, p values were significantly lower in spring (380.58 kg ha^{-1}) than in summer (545.89 kg ha^{-1}) and autumn (533.79 kg ha^{-1}). For Grass treatment, there were significantly higher p values in spring (561.57 kg ha^{-1}) than in summer (377.2 kg ha^{-1}) and autumn (376.54 kg ha^{-1}). In spring, significantly higher P values were observed in the Grass treatment group compared with the other treatment groups, while in summer, significantly higher P values were recorded for the Low Straw treatment group compared with the other treatment groups, and in autumn, significantly lower values were recorded in the Tilled, Low Straw and High Straw treatment groups compared with the Grass treatment group (Table 2).

Table 2. Results of two-way ANOVA based on element concentrations. Different letters after mean values in columns represent significant differences at $p < 0.05$. Capital letters indicate statistical differences between treatments; lowercase letters indicate statistical differences between seasons. SOM, soil organic matter; pH; P, phosphorous; K, potassium; Ni, nickel; Zn, zinc; Pb; lead. Soil P, K, Ni, Zn and Pb data are given in kg ha^{-1}.

Season	Treatment	SOM (%)	pH	Soil P	Soil K	Soil Ni	Soil Zn	Soil Pb
Spring	Tilled	1.38 Ab	7.54 Ab	425.29 Aab	27,297.09 Ba	210.78 Ba	184.80 Ba	43.68 Aa
	Low Straw	1.40 Ab	7.93 Aab	380.58 Bb	26,282.59 Ba	206.08 ABa	180.54 ABa	34.72 Aa
	High Straw	1.37 Bb	8.04 Aa	418.21 Aab	26,460.90 Ba	200.48 Ba	184.58 Aa	40.32 Aa
	Grass	2.11 Aa	7.99 Aab	561.57 Aa	27,149.47 Aa	209.22 Aa	183.90 Aa	42.34 Aa
Summer	Tilled	1.52 Ab	7.52 Ac	390.66 Ab	31,471.10 Aa	249.76 Aa	219.97 Aa	41.44 Aa
	Low Straw	1.58 Aab	7.97 Ab	545.89 Aa	25,781.50 Bc	192.42 Bb	169.12 Bb	40.77 Aab
	High Straw	1.56 Ab	8.09 Aab	500.64 Aab	28,148.51 ABb	201.82 ABb	175.17 Ab	37.18 Ab
	Grass	2.29 Aa	8.16 Aa	377.22 Bb	27,588.51 Ab	211.23 Ab	174.72 Ab	32.26 Bb
Autumn	Tilled	1.59 Ab	7.50 Ac	515.45 Aab	28,293.22 Ba	218.40 Ba	193.06 Ba	44.38 Aa
	Low Straw	1.64 Aab	7.87 Ab	533.79 Aa	27,647.20 Aa	224.90 Aa	183.90 Aa	40.32 Aab
	High Straw	1.73 Aab	8.00 Aab	392.22 Ab	28,429.18 Aa	224.67 Aa	183.90 Aa	33.82 Ab
	Grass	2.51 Aa	8.13 Aa	376.54 Bb	27,327.78 Aa	219.30 Aa	180.32 Aa	34.50 Bb

Potassium values in the Tilled treatment group were significantly higher in summer (31,471.10 kg ha^{-1}) than in spring (27,297.09 kg ha^{-1}) and autumn (28,293.22 kg ha^{-1}). For the Low Straw treatment group, lower K values were recorded in spring (26,282.59 kg ha^{-1}) and summer (25,781.50 kg ha^{-1}) than in autumn (27,647.20 kg ha^{-1}). For the High Straw treatment group, significantly higher K values were observed in autumn (28,429.18 kg ha^{-1}) than in spring (26,460.90 kg ha^{-1}), while the value in summer (28,148.51 kg ha^{-1}) was not statistically different from those recorded in spring or autumn. For Grass treatment, there were no significant differences in soil K concentrations between seasons, with a minimum of 27,149.47 kg ha^{-1} in spring, 27,327.78 kg ha^{-1} in autumn and a maximum of 27,588.51 kg ha^{-1} in summer. There were no significant differences in soil K values between the treatments in spring and autumn, while in summer, the values were significantly lower for Low Straw, High Straw and Grass treatments compared with Tilled treatment (Table 2).

Soil Ni values differed significantly based on treatment and season. Significantly higher Ni values were recorded for the Tilled treatment in summer (249.76 kg ha^{-1}) compared with spring (210.78 kg ha^{-1}) and autumn (218.40 kg ha^{-1}). Significantly higher Ni values were observed for the Low Straw treatment in autumn (224.90 kg ha^{-1}) than in summer (192.42 kg ha^{-1}), but there were no significant differences between spring (206.08 kg ha^{-1}) and the other two seasons. Significantly lower Ni values were recorded for the High Straw treatment group in spring (200.48 kg ha^{-1}) than in autumn (224.67 kg ha^{-1}), but the value in summer (201.82 kg ha^{-1}) was not statistically different from those of the other seasons. There were no statistical differences in soil Ni values in the Grass treatment group in all seasons, with a minimum of 209.22 kg ha^{-1} in spring, 211.23 kg ha^{-1} in summer and a maximum of 219.30 kg ha^{-1} in autumn. Significantly higher Ni values were observed in the Tilled treatment group in summer, while no differences were observed in spring and autumn (Table 2).

Zinc values differed significantly based on treatment and season. Significantly lower Zn values were recorded for the Tilled treatment group in spring (184.80 kg ha^{-1}) and autumn (193.06 kg ha^{-1}) compared with summer (219.97 kg ha^{-1}). Significantly higher Zn values were recorded for the Low Straw treatment group in autumn (183.90 kg ha^{-1}) compared with summer (169.12 kg ha^{-1}), while the value did not differ statistically from other treatments in spring (180.54 kg ha^{-1}). There were no significant differences in Zn values in the High Straw treatment group across seasons, with a minimum of 175.17 kg ha^{-1} in summer, 183.90 kg ha^{-1} in autumn and a maximum of 184.58 kg ha^{-1} in spring. The same pattern was observed for Grass treatment, with lower values recorded in summer (174.72 kg ha^{-1}) and autumn (180.32 kg ha^{-1}) and higher Zn values recorded in spring (183.90 kg ha^{-1}).

Soil Pb values showed significant differences based on treatment and season. There were no significant differences in soil Pb in the Tilled treatment group through the seasons, with a minimum of 41.44 kg ha^{-1} in summer and a maximum of 43.68 kg ha^{-1} in spring and 44.38 kg ha^{-1} in autumn. The same pattern was observed in the Low Straw and High Straw treatment groups, as presented in Table 2. Significantly higher Pb values were recorded for Grass treatment in spring (42.34 kg ha^{-1}) compared with summer (32.26 kg ha^{-1}) and autumn (34.50 kg ha^{-1}). No significant differences between treatments were observed in spring. Significantly higher values were recorded in summer in the Tilled treatment group compared with the High Straw and Grass treatment groups, but the Pb value for the Low Straw treatment group was not significantly different from the Pb values for the other three treatment groups; the same pattern was observed in autumn (Table 2).

3.3. Hydrological Response

Statistical analysis of the hydrological response showed significant differences based on treatment and season (Table 3). Significantly higher values of time to ponding were recorded for Tilled treatment in spring (1326 s) compared with autumn (75.5 s), while the value recorded in summer (444 s) was not statistically different from the values recorded in the other two seasons. Significantly higher TP values were also recorded in the Low Straw treatment group in spring (1800 s) compared with autumn (96.5 s), while the value recorded in summer (1326 s) was not statistically different from the values recorded in spring and autumn. Significantly lower TP values were recorded in the High Straw treatment group in autumn (149 s) compared with spring (1800 s) and summer (1800 s). No significant differences were observed in the Grass treatment group across seasons, with a minimum of 49.5 s in spring and a maximum of 372 s in summer and 108.5 s in autumn. Regarding the differences between treatments by season, differences in TP were noted only in spring, with significantly lower values in the Grass treatment group compared with the other treatment groups. There were no significant differences in TP values between treatments in summer and autumn.

Time to runoff was significantly different in the Tilled treatment group, with higher values recorded in spring (1800 s) than in autumn (294 s), while the value recorded in summer (1164 s) was not significantly different from the values recorded in the other two seasons. Significantly lower TR values were recorded in the Low Straw treatment group in autumn (444 s) than in spring (1800 s) and summer (1800 s). No significant differences in TR were recorded between seasons in the High Straw treatment group since the value was the same in each season (1800 s). The same pattern was observed for Grass treatment, with TR values ranging from a minimum of 453 s and 460 s in autumn and spring to a maximum of 1416 s in summer. Significantly higher TR values were recorded for Tilled, Low Straw and High Straw treatments compared with Grass treatment in spring; there were no significant differences between treatments in summer, while significantly higher values were recorded for High Straw treatment compared with the other treatments in autumn (Table 3).

Regarding runoff, Tilled treatment resulted in significantly lower values in spring (0.00 m^3 ha^{-1}) than in summer (7.11 m^3 ha^{-1}), while the runoff value recorded in autumn (3.40 m^3 ha^{-1}) was not significantly different from the values recorded in spring and summer. Significantly higher runoff values were recorded for the Low Straw treatment in autumn (6.17 m^3 ha^{-1}) compared with spring (0.00 m^3 ha^{-1}) and summer (0.00 m^3 ha^{-1}). No significant differences in runoff were observed for High Straw treatment across seasons, with a value of 0.00 m^3 ha^{-1} for every season. Significantly lower runoff values were observed with Grass treatment in summer (1.20 m^3 ha^{-1}) compared with autumn (6.53 m^3 ha^{-1}), while the value in spring (4.48 m^3 ha^{-1}) was not statistically different from the values recorded in the other two seasons. Significantly lower values were recorded for Tilled, Low Straw and High Straw treatments compared with Grass treatment in spring. Significantly lower runoff values were recorded for Low Straw and High Straw treatments compared with Tilled treatment in summer, while the value recorded in the Grass treatment group was not statistically different from the values recorded in the other three treatment groups. Significantly lower runoff values were reported for High Straw treatment compared with Grass treatment in autumn, while there were no significant differences between Tilled and Low Straw treatments and High Straw and Grass treatments.

Table 3. Results of two-way ANOVA based on overland flow properties. Different letters after mean values in columns indicate significant differences at $p < 0.05$. Capital letters indicate statistical differences between treatments; lowercase letters indicate statistical differences between seasons. TP, time to ponding; TR, time to runoff; SC, sediment concentration; SL, sediment loss.

Season	Treatment	TP (s)	TR (s)	Runoff (m^3 ha^{-1})	SC (g kg^{-1})	SL (kg ha^{-1})
Spring	Tilled	1326 Aa	1800 Aa	0.00 Bb	0.00 Ba	0.00 Ba
	Low Straw	1800 Aa	1800 Aa	0.00 Bb	0.00 Ba	0.00 Ba
	High Straw	1800 Aa	1800 Aa	0.00 Ab	0.00 Aa	0.00 Aa
	Grass	49.5 Ab	460 Ab	4.48 ABa	1.31 Aa	6.15 Aa
Summer	Tilled	444 ABa	1164 ABa	7.11 Aa	18.19 Aa	138.34 Aa
	Low Straw	1326 ABa	1800 Aa	0.00 Bb	0.00 Bb	0.00 Bb
	High Straw	1800 Aa	1800 Aa	0.00 Ab	0.00 Ab	0.00 Ab
	Grass	372 Aa	1416 Aa	1.20 Bab	3.85 Aab	14.05 Ab
Autumn	Tilled	75.5 Ba	294 Bb	3.40 ABab	15.03 Aa	50.99 Aa
	Low Straw	96.5 Ba	444 Bb	6.17 Aab	6.59 Aa	41.49 Aab
	High Straw	149 Ba	1800 Aa	0.00 Ab	0.00 Ab	0.00 Ac
	Grass	108.5 Aa	453 Ab	6.53 Aa	3.94 Aa	25.91 Ab

Sediment concentrations showed significant differences between treatments and seasons. Significantly lower SC values were observed in the Tilled treatment group in spring (0.00 g kg^{-1}) compared with summer (18.19 g kg^{-1}) and autumn (15.03 g kg^{-1}). Similar observations were made for Low Straw treatment, with significantly higher values recorded in autumn (6.59 g kg^{-1}) than in spring (0.00 g kg^{-1}) and summer (0.00 g kg^{-1}). There were no significant differences between seasons in the High Straw treatment group, with an SC value of 0.00 g kg^{-1} in all seasons. Significantly higher SC values were recorded for Grass treatment in autumn (3.94 g kg^{-1}) than in spring (1.31 g kg^{-1}), while the value recorded in summer (3.85 g kg^{-1}) was not statistically different from the values recorded in the other treatment groups. Significantly higher SC values were recorded for Grass treatment than compared with the other treatments in spring, while significantly lower values were recorded for Low Straw and High Straw treatments compared with Tilled treatment; the value recorded in the Grass treatment group was not statistically different from the values recorded in the other treatments. Significantly lower SC values were observed in autumn in the High Straw treatment group compared with the other treatment groups (Table 3).

Sediment loss differed significantly between treatments and seasons. Significantly lower SL values were noted in Tilled treatment in spring (0.00 kg ha^{-1}) compared with summer (138.34 kg ha^{-1}) and autumn (50.99 kg ha^{-1}); the same pattern was observed in the Low Straw treatment group (Table 3). No SL was reported for High Straw treatment; thus, no significant differences were noted between seasons. There were also no significant differences in Grass treatment between seasons, with a minimum of 6.15 kg ha^{-1} in spring, 14.05 kg ha^{-1} in summer and a maximum of 25.91 kg ha^{-1} in autumn. There were no significant differences in SL between all treatments in spring, while SL values were significantly lower in Low Straw, High Straw and Grass treatment groups compared with the Tilled treatment group in summer. Finally, significantly higher SL values were recorded for Tilled treatment in autumn compared with High Straw and Grass treatments, while the SL value in the Low Straw treatment group was not significantly different from SL values in the Tilled and Grass treatment groups (Table 3).

3.4. Element Losses

Statistical analysis revealed noteworthy disparities in element losses across treatments and seasons (Table 4). A significant reduction in P losses was found for Tilled treatment during spring (0.00 g ha^{-1}) compared with substantial losses recorded in autumn (2353.07 g ha^{-1}). However, there was a statistically significant difference in P losses between treatments in summer (1127.66 g ha^{-1}). In contrast, significantly higher P losses were noted in the Low Straw treatment group in autumn (1041.62 g ha^{-1}) compared with negligible losses in spring and summer. Surprisingly, no P losses were observed in the High Straw treatment group throughout the seasons. Consistent P losses were noted in Grass treatment across all seasons, with no statistically significant differences. Notably, there were significantly higher P losses in Grass treatment compared with the other treatments in spring. There were significantly lower P losses in the Low Straw and High Straw treatment groups compared with the Tilled treatment group in summer, while the value recorded in the Grass treatment group did not differ significantly from the values recorded in the other three treatment groups. Moreover, Tilled treatment resulted in significantly higher P losses in autumn compared with High Straw treatment, while no significant differences were observed between Low Straw and Grass treatments and the other two treatments.

Table 4. Results of two-way ANOVA based on element losses. Different letters after mean values in columns indicate significant differences at $p < 0.05$. Capital letters indicate statistical differences between treatments; lowercase letters indicate statistical differences between seasons. P, phosphorous; K, potassium; Ni, nickel; Zn, zinc; Pb, lead. All data are given in g ha^{-1}.

Season	Treatment	P Loss	K Loss	Ni Loss	Zn Loss	Pb Loss
Spring	Tilled	0.00 Bb	0.00 Bb	0.00 Bb	0.00 Bb	0.00 Bb
	Low Straw	0.00 Bb	0.00 Bb	0.00 Bb	0.00 Bb	0.00 Bb
	High Straw	0.00 Ab	0.00 Ab	0.00 Ab	0.00 Ab	0.00 Ab
	Grass	397.27 Aa	19,999.94 Aa	150.64 Aa	234.37 Aa	23.47 Aa
Summer	Tilled	1127.66 ABa	55,044.90 ABa	394.76 ABa	452.67 ABa	64.00 ABa
	Low Straw	0.00 Bb	0.00 Bb	0.00 Bb	0.00 Bb	0.00 Bb
	High Straw	0.00 Ab	0.00 Ab	0.00 Ab	0.00 Ab	0.00 Ab
	Grass	180.89 Aab	9106.56 Aab	68.59 Aab	106.72 Aab	10.69 Aab
Autumn	Tilled	2353.07 Aa	115,627.66 Aa	888.93 Aa	1010.08 Aa	131.63 Aa
	Low Straw	1041.62 Aab	45,436.13 Aab	0.00 Ab	488.99 Aab	37.37 Aab
	High Straw	0.00 Ab	0.00 Ab	0.00 Ab	0.00 Ab	0.00 Ab
	Grass	465.50 Aab	22,227.22 Aab	170.33 Aab	264.71 Aab	25.59 Aab

Regarding K losses, a significant reduction in K was found in the Tilled treatment group in spring (0.00 g ha^{-1}) compared with substantial losses in autumn (115,627.66 g ha^{-1}) and no statistically significant difference in summer (55,044.90 g ha^{-1}). Surprisingly, significantly higher K losses were recorded in the Low Straw treatment group in autumn (45,436.13 g ha^{-1}) compared with negligible losses in spring and summer. Conversely, no significant variations in K losses were observed in the High Straw treatment group across seasons, as no losses were recorded. Notably, consistent K losses were found for Grass treatment across all seasons, with no statistically significant differences. However, significantly higher K losses were found for Grass treatment in spring compared with the other treatments. In summer, significantly lower K losses were recorded in the Low Straw and High Straw treatment groups compared with the Tilled treatment group, while the value recorded in the Grass treatment group did not differ significantly from the values recorded in the other three treatment groups. Additionally, in autumn, Tilled treatment resulted in significantly higher K losses compared with High Straw treatment, while no significant differences were found between Low Straw and Grass treatments and the other two treatments.

Analysis of Ni losses indicated significant differences between seasons for the Tilled treatment. Significantly lower Ni losses were recorded in the Tilled treatment group in spring (0.00 g ha^{-1}) compared with substantial losses in autumn (888.93 g ha^{-1}) and no statistically significant differences in summer (394.76 g ha^{-1}). Intriguingly, significantly higher Ni losses were observed in the Low Straw treatment group in autumn (408.56 g ha^{-1}) compared with negligible losses in spring and summer. Conversely, significant seasonal variations in Ni losses were not reported for High Straw treatment, as no losses were recorded. Notably, Grass treatment showed consistent Ni losses across all seasons, with no statistically significant differences. However, Grass had significantly higher Ni losses in spring compared with the other treatments. In summer, Low Straw and High Straw treatments showed significantly lower Ni losses compared with Tilled treatment, while Grass treatment did not show significantly different values compared with the other three treatments. Furthermore, Tilled treatment resulted in significantly higher Ni losses in autumn compared with High Straw treatment, while no significant differences were observed between Low Straw and Grass treatments and the other two treatments.

In terms of Zn losses, there was a significant reduction in the Tilled treatment group in spring (0.00 g ha^{-1}) compared with considerable losses in autumn (1010.08 g ha^{-1}) and no statistically significant differences in summer (452.67 g ha^{-1}). Surprisingly, significantly higher Zn losses were found in the Low Straw treatment group in autumn (488.99 g ha^{-1}) compared with negligible losses in spring and summer. Conversely, significant seasonal variations in Zn losses were not observed in the High Straw treatment group, as no losses were recorded. Notably, consistent Zn losses were observed for Grass treatment across all seasons, with no statistically significant differences. However, significantly higher Zn losses were observed for Grass treatment in spring compared with the other treatments. In summer, significantly lower Zn losses were recorded for Low Straw and High Straw treatments compared with Tilled treatment, while the value recorded for Grass treatment did not differ significantly from the values recorded for the other three treatments. Additionally, in autumn, Tilled treatment resulted in significantly higher Zn losses compared with High Straw treatment, while no significant differences were observed between Low Straw and Grass treatments and the other two treatments.

Regarding Pb losses, significantly lower losses were found for Tilled treatment in spring (0.00 g ha^{-1}) compared with autumn (131.63 g ha^{-1}), with no statistically significant difference in summer (64.00 g ha^{-1}). Significantly higher Pb losses were observed for Low Straw treatment in autumn (37.37 g ha^{-1}) compared with negligible losses in spring and summer. Conversely, no significant seasonal variations in Pb losses were noted in the High Straw treatment group, as no losses were recorded. Notably, consistent Pb losses were recorded in the Grass treatment group across all seasons, with no statistically significant differences. However, significantly higher Pb losses were recorded in the Grass treatment group in spring compared with the other treatment groups. In summer, significantly lower Pb losses were found in the Low Straw and High Straw treatment groups compared with the Tilled treatment group, while the value recorded in the Grass treatment group did not differ significantly from the values recorded in the other three treatment groups. Moreover, in autumn, Tilled treatment resulted in significantly higher Pb losses compared with High Straw treatment, while no significant differences were observed between Low Straw and Grass treatments and the other two treatments.

3.5. Principal Component Analysis

Principal component analysis identified four factors that explained at least one variable. Factor 1 explained 51.46%, factor 2 explained 14.47%, factor 3 explained 8.24% and factor 4 explained 6.00% of the total variance (80.18% in total). Factor 1 explained most of the variables and their correlations; TP and TR had a positive loading relationship with each other; while runoff, SC and SL had a strong positive impact on the loss of all elements (P, K, Ni, Zn, Pb). In factor 2, soil pH had a strong negative correlation with soil Zn and Pb concentrations, and a strong correlation was noted between WSA and SOM. Factor 3 revealed a negative correlation between soil P and Ni, and factor 4 revealed a negative correlation between BD and SWC. The correlation between factors 1 and 2 are presented in Figure 3A,B, while the correlations between seasons and treatments are presented in Figure 3C.

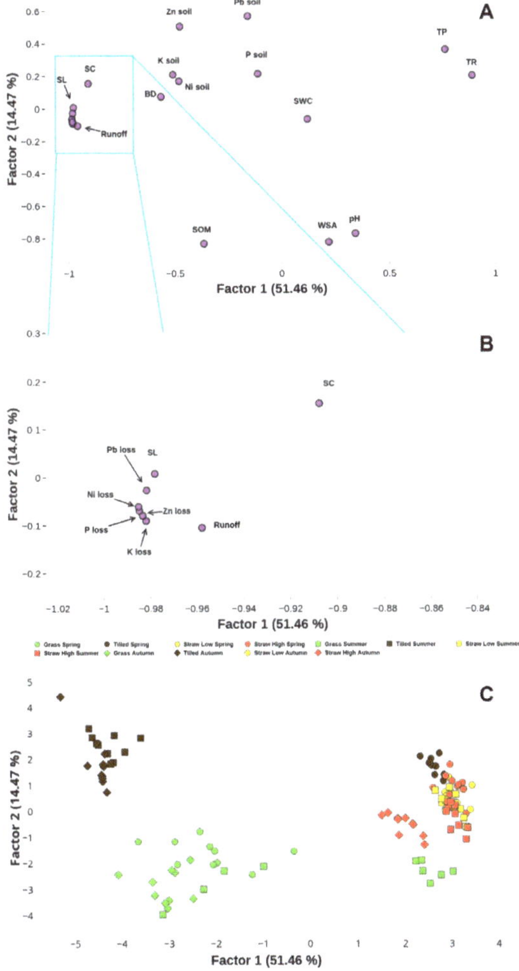

Figure 3. (**A**) Correlation between factors 1 and 2 (variables). BD, bulk density; SWC, soil water content; WSA, water-stable aggregates; SOM, soil organic matter; TP, time to ponding; TR, time to runoff; SC, sediment concentration; SL, sediment loss; P, phosphorous; K, potassium; Ni, nickel; Zn, zinc; Pb, lead. (**B**) Enlarged area, (**C**) Correlation between cases.

4. Discussion

Bulk density was higher in the Tilled, Low Straw and High Straw treatment groups in autumn compared with summer and spring, while the values in the Grass treatment group did not differ between seasons. Lower BD values in spring can be explained by (1) lower machinery traffic during that time of year, (2) freeze–thaw cycles that occurred in the previous winter [65] or (3) channels created by a growing and expanding grass root system, effectively loosening the soil and increasing pore spaces [66]. No differences were observed between treatments in spring. Higher values were noted with Tilled treatment in summer and autumn; this trend is expected since tillage is a well-known contributor to increased BD [67], and larger aboveground mass (such as in High Straw and Grass treatments) serves as a physical barrier between the soil surface and external forces, helping to disperse raindrop energy and reduce their impact on the soil, thus minimizing compaction [68].

Lower SWC values were recorded for Tilled and Grass treatments. Tillage instantly alters the soil air-to-water ratio and increases the soil's water infiltration capability [69], while grass favours soil structure via SOM accumulation, which directly improves the soil air-to-water ratio [70,71]. This is the main reason why WAS was significantly lower in all seasons in the Tilled treatment group, as noted in other studies [72,73]. Grass treatment had the highest WSA due to higher SOM content and grass mowing residue that is left on the soil surface to decompose [74]. This ensures aggregate stability during heavy rainfall [75].

Certain changes in SOM occurred between treatments and seasons. Higher amounts of SOM were found in the High Straw treatment group in autumn compared with spring and summer; this happened for two main reasons: (1) putting organic matter on the soil surface (twice as much as with Low Straw treatment) helped reduce the evaporation of water from the soil surface and increased soil temperature, thus creating favourable conditions for microbial activity and organic matter decomposition [76–78], and (2) reduced erosion left the soil intact, which allowed organic matter to accumulate and contribute to the soil's overall organic content [79,80]. Higher SOM was recorded in the Grass treatment group in all seasons; this was expected due to (1) the absence of annual soil disturbance [81], (2) the decomposition of larger amounts of mowed grass [82] and (3) the richness of organic compounds in grass root exudates [83].

The only differences in soil pH were noted between treatments based on season. In spring, the values were lower in Tilled and Low Straw treatment groups compared with the other treatment groups. The same pattern was observed in summer and autumn. This can be explained by (1) seasonal fluctuations and geomorphological characteristics of the investigated area [84,85] and (2) mineralization in the Tilled treatment, which causes faster decomposition and mineralization, releasing organic acids and other acidic compounds and causing soil acidification [86].

Soil P content did not differ between seasons in the Tilled and High Straw treatment groups; however, higher values were recorded in the Low Straw treatment group in summer and autumn and higher values were noted in the Grass treatment group in spring. Grass is well known for accumulating higher amounts of P on the soil surface and upper soil layer [87,88]. Higher P concentrations were recorded in the Low Straw treatment group in summer and autumn compared with the other treatment groups. This occurred due to (1) higher amounts of pesticide residues that collected on the ingrown grass under this treatment (which was mowed regularly), (2) organic fertilizer input and (3) decomposition of straw residue and ingrown grass, which released organic compounds (including P) in the soil [89,90]. Lower values in the Tilled treatment group were due to erosion events that occurred between seasons when P bound to soil particles and were then translocated to the lower parts, as noted in other studies [91,92].

There were no differences in soil K concentrations between treatment groups in spring and autumn; however, in summer, lower concentrations were noted in treatments with surface cover, very likely due to the uptake of potassium by vines, grass and weeds [93]. For treatments with surface cover (grass and mulch), no significant differences in soil Ni and Zn concentrations were recorded in the Tilled treatment group in spring and autumn; in summer, these values were significantly lower in the other treatment groups. This may be due to several reasons: (1) periodic vegetation consumption, (2) lower ability to incorporate previously applied fertilizers and (3) larger amounts of compounds absorbed from burning fossil fuel and residual oils from tractor traffic in the Tilled treatment group [94]. Further on, these changes occur because of sediment loss due to soil erosion, as reported in other studies. For example, Mirás-Avalos et al. [95] found that Zn losses were high, up to 30,000 g ha^{-1}, while Jiao et al. [96] recorded lower Ni losses, averaging 21.27 g ha^{-1}. Additionally, Dugan et al. [62] recorded Zn and Ni losses ranging from 360 and 200 g ha^{-1} in spring to 900 and 470 g ha^{-1} in autumn in the tilled treatment group. Finally, for treatments with the most surface cover (Grass and High Straw), significantly lower values of Pb were recorded in summer and autumn, while no differences were recorded between the treatments in summer. These values reflect constant tractor traffic and soil overuse [97,98].

Season and treatment had strong effects on overland flow properties, as presented in Table 3. In every season, time to ponding and TR were longer with High Straw treatment followed by Low Straw treatment. For Grass treatment, the shortest TP and TR were recorded in spring and summer; however, lower values were observed in the Tilled treatment group in autumn due to lower vegetation cover and higher BD in the later seasons. There were no differences between treatments in summer, while in spring, higher values were recorded for Tilled, Low Straw and High Straw treatments. This occurred due to straw application and surface roughness due to tillage intervention, which had a great effect on reducing BD, while surface cover served as a barrier between the impact of raindrops and the soil surface [99,100], thus postponing TP and TR. There were no differences in TP between treatments in autumn, but significantly higher TR values were found in the High Straw treatment group compared with the other treatment groups.

In the Tilled treatment group, runoff values were significantly lower in spring than in summer and autumn. While there were no differences in runoff between High Straw and Low Straw treatment groups in spring and summer, High Straw treatment completely mitigated runoff later in the season, meaning that applying less straw was not enough to successfully mitigate soil erosion. There was no runoff in the Tilled, High Straw and Low Straw treatment groups in spring due to (1) surface roughness on the freshly tilled ground and (2) application of straw mulch on previously tilled surfaces. There was higher runoff with Grass treatment in autumn than in spring, while the values did not statistically differ between summer and the other seasons. Lower runoff rates in summer can be explained by (1) higher vegetation cover [101], (2) soil cracking, which alters the overland flow [102], and (3) higher evaporation, enabling the soil to absorb more water, thus reducing the overland flow [103]. Low runoff rates in the straw mulch treatment groups are a consequence not only of surface cover but also of higher SOM content [104,105] and higher WSA [106]. Sediment concentrations and SL are strongly dependent on the previously mentioned soil properties [107,108]. For instance, recorded SC and SL values in the Tilled treatment group were higher in summer than in autumn, up to 121% and 271%. Due to the application of more straw, High Straw treatment successfully mitigated SL, as the straw improved the soil's physical properties, as confirmed by other studies [109–111], making the soil more resilient to raindrop-induced degradation [26]. Besides High Straw treatment, there are better conservation possibilities for Low Straw and Grass treatments than Tilled treatment. Cover on topsoil and enhanced SOM and physical status intercept raindrops, thus reducing SL and SC.

The loss of soil elements in our study was highly correlated with SL [112]. As stated before, there were higher element losses with Tilled treatment, proving that tillage has a negative impact on soil properties and the environment [113–115]. While there were element losses in the Grass treatment group only in spring, P losses in summer were 622% lower for Grass than for Tilled treatment. Higher element losses were recorded for Grass treatment in autumn than in spring and summer, which is a consequence of a lower percentage of surface vegetation cover [116,117] and higher BD [118] in that period. It is important to mention that higher losses in the Tilled treatment group were also associated with SOM loss [119] due to a strong connection between elements and SOM. Furthermore, the concentrations of elements such as P, Cu and Zn are higher on the soil surface, such as in grass treatment, where the elements decompose and bind to the surface cover as a result of repeated management [120].

Principal component analysis revealed that different soil management techniques alter soil physiochemical properties, hydrological response and element transport. Factor 1 showed a positive correlation between TP and TR. Additionally, strong correlations were found between runoff, SC and SL and P, K, Ni, Zn and Pb loss. Higher runoff increases sediment and element translocation since the elements are bound to soil particles and surface cover residues that can be easily transferred when they move into water flows. Similar findings were reported in [62,121–123]. The adsorption of certain elements, such as phosphorus, to clay-sized particles in surface runoff can contribute to eutrophication in the receiving water bodies [124]. Factor 2 showed a strong correlation between WSA and SOM, as stated in many previous studies [125–130]. Rain disrupts soil aggregates via (1) slaking, (2) differential swelling of clay, (3) mechanical dispersion and (4) physiochemical dispersion. SOM is assumed to stabilize soil aggregates by increasing the cohesion of aggregates through the binding of mineral particles by organic polymers but can also decrease the wettability of aggregates by slowing their wetting rate and thus the extent of slaking [131–133].

Strong negative correlations were observed between soil pH and soil Zn and Pb concentrations. Soil pH has a strong effect on Zn and Pb availability and solubility, hence acidic soils have higher Zn and Pb concentrations; On the other hand, under alkaline or basic conditions, Zn and Pb may become less soluble and available [134–136]. In factor 3, a negative correlation was observed between soil P and Ni; even though they have no direct effect on each other, higher P concentrations can affect soil pH due to the buffering capacity of phosphates [137], which increases soil pH, thus potentially affecting Ni behaviour by influencing its solubility and availability [136,138]. Finally, factor 4 showed a negative correlation between BD and SWC. High BD indicates compacted soil, which can negatively affect SWC since compacted soils have reduced pore space, thus limiting water infiltration and movement. This leads to decreased porosity and increased density, which restrict water retention and drainage [139]. It is also important to mention that the correlation between BD and SWC is not linear and can be influenced by many other factors, including SOM, clay content and soil structure.

Even though the research results are clear and precise, certain shortcomings should be highlighted. This research was carried out for only one year, which is not long enough to provide a full assessment of soil erosion rates. This was a short-term study, as the rainfall simulation experiments were conducted for 30 min, which may not fully capture long-term effects or seasonal variations. It is important to mention the lack of long-term implications because this study focused on the immediate effects and short-term implications of different soil management practices, thus it would be valuable to investigate the long-term effects of straw mulch application on soil erosion control, soil fertility and overall vineyard sustainability. Additionally, the rainfall simulator has limitations. For instance, the methods and equipment are not standardized. Furthermore, the plots were rather small, thus the data collected are not suitable for large-scale modelling. It is important to point out that straw mulch is a natural product that enhances soil functions in the long term and has an immediate effect on soil and water loss. Straw mulch also allows higher water availability for plants due to higher infiltration, as explained in [140]. Furthermore, straw is a natural byproduct of grain harvesting and as such is affordable and available for farmers, who can start employing sustainable management on their permanent plantations. An even more important advantage of straw mulch used in intensively managed vineyards is that pesticide residues do not land on the grass or bare soil surface, thus reducing the chances of potential phytotoxicity. Finally, in terms of soil erosion, straw mulch ensures the conservation of SOM and maintains all elements in their locations, preventing them from transferring to lower slope locations where they can increase mass water pollution.

5. Conclusions

The results presented in this research show that soil management has a significant impact on soil erosion and several soil properties in different seasons. For Tilled treatment, there were slight reductions in SOM and WSA but increased runoff, SC, SL and element transfer. Grass treatment significantly lowered BD, TP and TR but mitigated runoff and diffused pollution. Additionally, mulch treatments were a good substitute for grass-covered treatment, particularly the double dose of straw, as there was no runoff in all three investigated seasons. It successfully increased SOM and WSA but completely stopped soil loss and element transfer. Furthermore, there were considerable differences between seasons in all treatment groups, mainly in terms of overland flow; runoff, SC and SL were higher in autumn, while TP and TR were higher in spring. Conventional agricultural systems in intensively managed vineyards are detrimental and dangerous for the environment due to soil and water pollution and degradation. Straw application is a cost-effective and sustainable practice for minimizing the negative impact of conventional tillage.

Author Contributions: Conceptualization, I.B. and P.P.; methodology, I.B., I.D. and P.P.; software, I.B. and I.D.; validation, I.D., P.P., J.D., L.F., V.F. and I.B.; formal analysis, I.B. and I.D.; investigation, I.B. and I.D.; resources, I.B.; data curation, I.D.; writing—original draft preparation, I.D.; writing—review and editing, I.D., P.P., J.D., L.F., V.F. and I.B.; visualization, I.D.; supervision, I.B. and P.P.; project administration, I.B.; funding acquisition, I.B. All authors have read and agreed to the published version of the manuscript.

Funding: This work was supported by the Croatian Science Foundation through the "Soil erosion and degradation in Croatia" project (UIP-2017-05-7834) (SEDCRO).

Data Availability Statement: The data presented in this study are available on reasonable request from the corresponding author.

Acknowledgments: The authors are grateful for the support of Vina Matošević.

Conflicts of Interest: The authors declare no conflict of interest. The funders had no role in the design of the study; in the collection, analyses, or interpretation of data; in the writing of the manuscript, or in the decision to publish the result.

References

1. Arora, N.K. Impact of climate change on agriculture production and its sustainable solutions. *Environ. Sustain.* **2019**, *2*, 95–96. [CrossRef]
2. Tian, Z.; Wang, J.W.; Li, J.; Han, B. Designing future crops: Challenges and strategies for sustainable agriculture. *Plant J.* **2021**, *105*, 1165–1178. [CrossRef]
3. Pugliese, P. Organic farming and sustainable rural development: A multifaceted and promising convergence. *Sociol. Rural.* **2001**, *41*, 112–130. [CrossRef]
4. Khan, N.; Ray, R.L.; Kassem, H.S.; Hussain, S.; Zhang, S.; Khayyam, M.; Ihtisham, M.; Asongu, S.A. Potential role of technology innovation in transformation of sustainable food systems: A review. *Agriculture* **2021**, *11*, 984. [CrossRef]
5. Sanaullah, M.; Usman, M.; Wakeel, A.; Cheema, S.A.; Ashraf, I.; Farooq, M. Terrestrial ecosystem functioning affected by agricultural management systems: A review. *Soil Till. Res.* **2020**, *196*, 104464. [CrossRef]
6. Zhang, H.; Shi, Y.; Dong, Y.; Lapen, D.R.; Liu, J.; Chen, W. Subsoiling and conversion to conservation tillage enriched nitrogen cycling bacterial communities in sandy soils under long-term maize monoculture. *Soil Till. Res.* **2022**, *215*, 105197. [CrossRef]
7. Blanco-Canqui, H.; Lal, R. Soil and Water Conservation. In *Principles of Soil Conservation and Management*; Springer: Dordrecht, The Netherlands, 2010; pp. 1–19.
8. Van Geel, M.; Verbruggen, E.; De Beenhouwer, M.; Van Rennes, G.; Lievens, B.; Honnay, O. High soil phosphorus levels overrule the potential benefits of organic farming on arbuscular mycorrhizal diversity in northern vineyards. *Agric. Ecosyst. Environ.* **2017**, *248*, 144–152. [CrossRef]
9. Daelemans, R.; Hulsmans, E.; Honnay, O. Both organic and integrated pest management of apple orchards maintain soil health as compared to a semi-natural reference system. *J. Environ. Manag.* **2022**, *303*, 114191. [CrossRef]
10. Komárek, M.; Čadková, E.; Chrastný, V.; Bordas, F.; Bollinger, J.C. Contamination of vineyard soils with fungicides: A review of environmental and toxicological aspects. *Environ. Int.* **2010**, *36*, 138–151. [CrossRef]
11. Bogunovic, I.; Bilandzija, D.; Andabaka, Z.; Stupic, D.; Rodrigo Comino, J.; Cacic, M.; Brezinscak, L.; Maletic, E.; Pereira, P. Soil compaction under different management practices in a Croatian vineyard. *Arab. J. Geosci.* **2017**, *10*, 340. [CrossRef]

12. Turhun, M.; Eziz, M. Identification of the distribution, contamination levels, sources, and ecological risks of heavy metals in vineyard soils in the main grape production area of China. *Environ. Earth Sci.* **2022**, *81*, 40. [CrossRef]

13. Ferrero, A.; Usowicz, B.; Lipiec, J. Effects of tractor traffic on spatial variability of soil strength and water content in grass covered and cultivated sloping vineyard. *Soil Till. Res.* **2005**, *84*, 127–138. [CrossRef]

14. Coulouma, G.; Boizard, H.; Trotoux, G.; Lagacherie, P.; Richard, G. Effect of deep tillage for vineyard establishment on soil structure: A case study in Southern France. *Soil Till. Res.* **2006**, *88*, 132–143. [CrossRef]

15. Ferrero, A.L.D.O.; Lipiec, J.E.R.Z.Y.; Turski, M.A.R.C.I.N.; Nosalewicz, A.R.T.U.R. Stability and sorptivity of soil aggregates in grassed and cultivated sloping vineyards. *Polish J. Soil Sci.* **2007**, *40*, 1–8.

16. Ferreira, C.S.S.; Keizer, J.J.; Santos, L.M.B.; Serpa, D.; Silva, V.; Cerqueira, M.; Ferreira, A.J.D.; Abrantes, N. Runoff, sediment and nutrient exports from a Mediterranean vineyard under integrated production: An experiment at plot scale. *Agric. Ecosyst. Environ.* **2018**, *256*, 184–193. [CrossRef]

17. Capello, G.; Biddoccu, M.; Ferraris, S.; Cavallo, E. Effects of tractor passes on hydrological and soil erosion processes in tilled and grassed vineyards. *Water* **2019**, *11*, 2118. [CrossRef]

18. Marques, M.J.; García-Muñoz, S.; Muñoz-Organero, G.; Bienes, R. Soil conservation beneath grass cover in hillside vineyards under Mediterranean climatic conditions (Madrid, Spain). *Land Degrad. Dev.* **2010**, *21*, 122–131. [CrossRef]

19. Ramos, M.C.; Benito, C.; Martínez-Casasnovas, J.A. Simulating soil conservation measures to control soil and nutrient losses in a small, vineyard dominated, basin. *Agric. Ecosyt. Environ.* **2015**, *213*, 194–208. [CrossRef]

20. Prosdocimi, M.; Jordán, A.; Tarolli, P.; Keesstra, S.; Novara, A.; Cerdà, A. The immediate effectiveness of barley straw mulch in reducing soil erodibility and surface runoff generation in Mediterranean vineyards. *Sci. Total Environ.* **2016**, *547*, 323–330. [CrossRef]

21. Usón, A.; Poch, R.M. Effects of tillage and management practices on soil crust morphology under a Mediterranean environment. *Soil Till. Res.* **2000**, *54*, 191–196. [CrossRef]

22. Poulenard, J.; Podwojewski, P.; Janeau, J.L.; Collinet, J. Runoff and soil erosion under rainfall simulation of Andisols from the Ecuadorian Páramo: Effect of tillage and burning. *Catena* **2001**, *45*, 185–207. [CrossRef]

23. Dam, R.F.; Mehdi, B.B.; Burgess, M.S.E.; Madramootoo, C.A.; Mehuys, G.R.; Callum, I.R. Soil bulk density and crop yield under eleven consecutive years of corn with different tillage and residue practices in a sandy loam soil in central Canada. *Soil Till. Res.* **2005**, *84*, 41–53. [CrossRef]

24. Kainiemi, V.; Arvidsson, J.; Kätterer, T. Short-term organic matter mineralisation following different types of tillage on a Swedish clay soil. *Biol. Fertil.* **2013**, *49*, 495–504. [CrossRef]

25. Erenstein, O. Crop residue mulching in tropical and semi-tropical countries: An evaluation of residue availability and other technological implications. *Soil Till. Res.* **2002**, *67*, 115–133. [CrossRef]

26. Gholami, L.; Sadeghi, S.H.; Homaee, M. Straw mulching effect on splash erosion, runoff, and sediment yield from eroded plots. *Soil Sci. Soc. Am. J.* **2013**, *77*, 268–278. [CrossRef]

27. Hellin, J.; Erenstein, O.; Beuchelt, T.; Camacho, C.; Flores, D. Maize stover use and sustainable crop production in mixed crop–livestock systems in Mexico. *Field Crops Res.* **2013**, *153*, 12–21. [CrossRef]

28. Siczek, A.; Horn, R.; Lipiec, J.; Usowicz, B.; Łukowski, M. Effects of soil deformation and surface mulching on soil physical properties and soybean response related to weather conditions. *Soil Till. Res.* **2015**, *153*, 175–184. [CrossRef]

29. Akhtar, K.; Wang, W.; Ren, G.; Khan, A.; Feng, Y.; Yang, G. Changes in soil enzymes, soil properties, and maize crop productivity under wheat straw mulching in Guanzhong, China. *Soil Till. Res.* **2018**, *182*, 94–102. [CrossRef]

30. Sinkevičienė, A.; Jodaugienė, D.; Pupalienė, R.; Urbonienė, M. The influence of organic mulches on soil properties and crop yield. *Agron. Res.* **2009**, *7*, 485–491.

31. Patil Shirish, S.; Kelkar Tushar, S.; Bhalerao Satish, A. Mulching: A soil and water conservation practice. *Res. J. Agric. For. Sci.* **2013**, *2320*, 6063.

32. Gholami, L.; Khaledi Darvishan, A.; Kavian, A. Wood chips as soil conservation in field conditions. *Arab. J. Geosci.* **2016**, *9*, 729. [CrossRef]

33. Jourgholami, M.; Abari, M.E. Effectiveness of sawdust and straw mulching on postharvest runoff and soil erosion of a skid trail in a mixed forest. *Ecol. Eng.* **2017**, *109*, 15–24. [CrossRef]

34. Camposeo, S.; Vivaldi, G.A. Short-term effects of de-oiled olive pomace mulching application on a young super high-density olive orchard. *Sci. Hortic.* **2011**, *129*, 613–621. [CrossRef]

35. Ferjani, A.I.; Jeguirim, M.; Jellali, S.; Limousy, L.; Courson, C.; Akrout, H.; Thevenin, N.; Ruidavets, L.; Muller, A.; Bennici, S. The use of exhausted grape marc to produce biofuels and biofertilizers: Effect of pyrolysis temperatures on biochars properties. *Renew. Sustain. Energ. Rev.* **2019**, *107*, 425–433. [CrossRef]

36. Zribi, W.; Aragüés, R.; Medina, E.; Faci, J.M. Efficiency of inorganic and organic mulching materials for soil evaporation control. *Soil Till. Res.* **2015**, *148*, 40–45. [CrossRef]

37. Den Boer, J.; Dyjakon, A.; Den Boer, E.; García-Galindo, D.; Bosona, T.; Gebresenbet, G. Life-cycle assessment of the use of peach pruning residues for electricity generation. *Energies* **2010**, *13*, 2734. [CrossRef]

38. Belel, M.D. Effects of grassed and synthetic mulching materials on growth and yield of sweet pepper (*Capsicum annuum*) in Mubi, Nigeria. *J. Agric. Soc. Sci.* **2012**, *8*, 97–99.

39. Ferrara, G.; Fracchiolla, M.; Al Chami, Z.; Camposeo, S.; Lasorella, C.; Pacifico, A.; Aly, A.; Montemurro, P. Effects of mulching materials on soil and performance of cv. Nero di Troia grapevines in the Puglia region, southeastern Italy. *Am. J. Enol. Vitic.* **2012**, *63*, 269–276. [CrossRef]

40. El-Beltagi, H.S.; Basit, A.; Mohamed, H.I.; Ali, I.; Ullah, S.; Kamel, E.A.; Shalaby, T.A.; Ramadan, K.M.; Alkhateeb, A.A.; Ghazzawy, H.S. Mulching as a Sustainable Water and Soil Saving Practice in Agriculture: A Review. *Agronomy* **2022**, *12*, 1881. [CrossRef]

41. Puerta, V.L.; Pereira, E.I.P.; Wittwer, R.; Van Der Heijden, M.; Six, J. Improvement of soil structure through organic crop management, conservation tillage and grass-clover ley. *Soil Till. Res.* **2018**, *180*, 1–9. [CrossRef]

42. Eze, S.; Dougill, A.J.; Banwart, S.A.; Hermans, T.D.; Ligowe, I.S.; Thierfelder, C. Impacts of conservation agriculture on soil structure and hydraulic properties of Malawian agricultural systems. *Soil Till. Res.* **2020**, *201*, 104639. [CrossRef]

43. Sidhu, D.; Duiker, S.W. Soil compaction in conservation tillage: Crop impacts. *Agron. J.* **2006**, *98*, 1257–1264. [CrossRef]

44. Afzalinia, S.; Zabihi, J. Soil compaction variation during corn growing season under conservation tillage. *Soil Till. Res.* **2014**, *137*, 1–6. [CrossRef]

45. Głąb, T.; Kulig, B. Effect of mulch and tillage system on soil porosity under wheat (*Triticum aestivum*). *Soil Till. Res.* **2008**, *99*, 169–178. [CrossRef]

46. Bandyopadhyay, K.K.; Hati, K.M.; Singh, R. Management options for improving soil physical environment for sustainable agricultural production: A brief review. *J. Agric. Phys.* **2009**, *9*, 1–8.

47. Gómez-Rey, M.X.; Couto-Vázquez, A.; González-Prieto, S.J. Nitrogen transformation rates and nutrient availability under conventional plough and conservation tillage. *Soil Till. Res.* **2012**, *124*, 144–152. [CrossRef]

48. García-Marco, S.; Gómez-Rey, M.X.; González-Prieto, S.J. Availability and uptake of trace elements in a forage rotation under conservation and plough tillage. *Soil Till. Res.* **2014**, *137*, 33–42. [CrossRef]

49. Bogunovic, I.; Telak, L.J.; Pereira, P.; Filipovic, V.; Filipovic, L.; Percin, A.; Durdevic, B.; Birkás, M.; Dekemati, I.; Comino, J.R. Land management impacts on soil properties and initial soil erosion processes in olives and vegetable crops. *J. Hydrol. Hydromech.* **2020**, *68*, 328–337. [CrossRef]

50. Panagos, P.; Ballabio, C.; Himics, M.; Scarpa, S.; Matthews, F.; Bogonos, M.; Poesen, J.; Borrelli, P. Projections of soil loss by water erosion in Europe by 2050. *Environ. Sci. Policy* **2021**, *124*, 380–392. [CrossRef]

51. Novara, A.; Cerda, A.; Barone, E.; Gristina, L. Cover crop management and water conservation in vineyard and olive orchards. *Soil Till. Res.* **2021**, *208*, 104896. [CrossRef]

52. Croatian Bureau of Statistics. Available online: https://dzs.gov.hr/en (accessed on 13 December 2022).

53. Milićević, T.; Aničić Urošević, M.; Relić, D.; Jovanović, G.; Nikolić, D.; Vergel, K.; Popović, A. Environmental pollution influence to soil–plant–air system in organic vineyard: Bioavailability, environmental, and health risk assessment. *Environ. Sci. Pollut. Res.* **2021**, *28*, 3361–3374. [CrossRef] [PubMed]

54. Ortega, P.; Sánchez, E.; Gil, E.; Matamoros, V. Use of cover crops in vineyards to prevent groundwater pollution by copper and organic fungicides. Soil column studies. *Chemosphere* **2022**, *303*, 134975. [CrossRef] [PubMed]

55. Kottek, M.; Grieser, J.; Beck, C.; Rudolf, B.; Rubel, F. World map of the Köppen-Geiger climate classification updated. *Meteorol. Z.* **2016**, *15*, 259–263. [CrossRef] [PubMed]

56. IUSS—WRB. *World Reference Base for Soil Resources 2014: International Soil Classification System for Naming Soils and Creating Legends for Soil Maps*; World Soil Resources Reports No. 106; FAO: Rome, Italy, 2014.

57. Biddoccu, M.; Ferraris, S.; Pitacco, A.; Cavallo, E. Temporal variability of soil management effects on soil hydrological properties, runoff and erosion at the field scale in a hillslope vineyard, North-West Italy. *Soil Till. Res.* **2017**, *165*, 46–58. [CrossRef]

58. Diaz-Zorita, M.; Perfect, E.; Grove, J.H. Disruptive methods for assessing soil structure. *Soil Till. Res.* **2002**, *64*, 3–22. [CrossRef]

59. Kemper, W.D.; Rosenau, R.C. Aggregate stability and size distribution. In *Methods of Soil Analysis*; Klute, A., Ed.; American Society of Agronomy, Inc.: Madison, WI, USA, 1986; pp. 425–442. [CrossRef]

60. Walkley, A.; Black, I.A. An examination of the Digestion method for determining soil organic matter, and a proposed modification of the chromic acid titration method. *Soil Sci.* **1934**, *37*, 29–38. [CrossRef]

61. Kikić, D.; Kisić, I.; Zgorelec, Ž.; Perčin, A. Portable X-ray fluorescence as a tool for characterization of nutrient status in soil. In Proceedings of the 55th Croatian & 15th International Symposium on Agriculture, Vodice, Croatia, 16–21 February 2020; pp. 91–95.

62. Dugan, I.; Pereira, P.; Barcelo, D.; Telak, L.J.; Filipovic, V.; Filipovic, L.; Kisic, I.; Bogunovic, I. Agriculture management and seasonal impact on soil properties, water, sediment and chemicals transport in hazelnut orchard (Croatia). *Sci. Total Environ.* **2022**, *839*, 156346. [CrossRef]

63. Statsoft. *Statistica 12.0 Software*; StatSoft Inc.: Hamburg, Germany, 2015.

64. Plotly Chart Studio. Available online: https://chart-studio.plotly.com/ (accessed on 21 March 2023).

65. Jabro, J.D.; Iversen, W.M.; Evans, R.G.; Allen, B.L.; Stevens, W.B. Repeated freeze-thaw cycle effects on soil compaction in a clay loam in Northeastern Montana. *Soil Sci. Soc. Am. J.* **2014**, *78*, 737–744. [CrossRef]

66. De Baets, S.; Poesen, J.; Gyssels, G.; Knapen, A. Effects of grass roots on the erodibility of topsoils during concentrated flow. *Geomorphology* **2006**, *76*, 54–67. [CrossRef]

67. Osunbitan, J.A.; Oyedele, D.J.; Adekalu, K.O. Tillage effects on bulk density, hydraulic conductivity and strength of a loamy sand soil in southwestern Nigeria. *Soil Till. Res.* **2005**, *82*, 57–64. [CrossRef]

68. Blanco-Canqui, H. Energy crops and their implications on soil and environment. *Agron. J.* **2010**, *102*, 403–419. [CrossRef]

69. Silva, S.G.C.; Silva, Á.P.D.; Giarola, N.F.B.; Tormena, C.A.; Sá, J.C.D.M. Temporary effect of chiseling on the compaction of a Rhodic Hapludox under no-tillage. *Rev. Bras. Cienc. Solo* **2012**, *36*, 547–555. [CrossRef]

70. Rawls, W.J.; Pachepsky, Y.A.; Ritchie, J.C.; Sobecki, T.M.; Bloodworth, H. Effect of soil organic carbon on soil water retention. *Geoderma* **2003**, *116*, 61–76. [CrossRef]

71. Paustian, K.; Collins, H.P.; Paul, E.A. Management controls on soil carbon. In *Soil Organic Matter in Temperate Agroecosystems*; CRC Press: Boca Raton, FL, USA, 2019; pp. 15–49.

72. Madejón, E.; Moreno, F.; Murillo, J.M.; Pelegrín, F. Soil biochemical response to long-term conservation tillage under semi-arid Mediterranean conditions. *Soil Till. Res.* **2007**, *94*, 346–352. [CrossRef]

73. Kabiri, V.; Raiesi, F.; Ghazavi, M.A. Six years of different tillage systems affected aggregate-associated SOM in a semi-arid loam soil from Central Iran. *Soil Till. Res.* **2016**, *154*, 114–125. [CrossRef]

74. Xiang, Y.; Li, Y.; Liu, Y.; Zhang, S.; Yue, X.; Yao, B.; Xue, J.; Lv, W.; Zhang, L.; Xu, X.; et al. Factors shaping soil organic carbon stocks in grass covered orchards across China: A meta-analysis. *Sci. Total Environ.* **2022**, *807*, 150632. [CrossRef]

75. Pikul, J.L.; Osborne, S.; Ellsbury, M.; Riedell, W. Particulate organic matter and water-stable aggregation of soil under contrasting management. *Soil Sci. Soc. Am. J.* **2007**, *71*, 766–776. [CrossRef]

76. Ramakrishna, A.; Tam, H.M.; Wani, S.P.; Long, T.D. Effect of mulch on soil temperature, moisture, weed infestation and yield of groundnut in northern Vietnam. *Field Crops Res.* **2006**, *95*, 115–125. [CrossRef]

77. Chen, S.Y.; Zhang, X.Y.; Pei, D.; Sun, H.Y.; Chen, S.L. Effects of straw mulching on soil temperature, evaporation and yield of winter wheat: Field experiments on the North China Plain. *Ann. Appl. Biol.* **2007**, *150*, 261–268. [CrossRef]

78. Siczek, A.; Frac, M. Soil microbial activity as influenced by compaction and straw mulching. *Int. Agrophys.* **2012**, *26*, 65. [CrossRef]

79. Dick, W.A. Organic carbon, nitrogen, and phosphorus concentrations and pH in soil profiles as affected by tillage intensity. *Soil Sci. Soc. Am. J.* **1983**, *47*, 102–107. [CrossRef]

80. Rasmussen, P.E.; Collins, H.P. Long-term impacts of tillage, fertilizer, and crop residue on soil organic matter in temperate semiarid regions. *Adv. Agron.* **1991**, *45*, 93–134. [CrossRef]

81. Pouyat, R.V.; Yesilonis, I.D.; Golubiewski, N.E. A comparison of soil organic carbon stocks between residential turf grass and native soil. *Urban Ecosyst.* **2009**, *12*, 45–62. [CrossRef]

82. Tagliavini, M.; Tonon, G.; Scandellari, F.; Quinones, A.; Palmieri, S.; Menarbin, G.; Gioacchini, P.; Masia, A. Nutrient recycling during the decomposition of apple leaves (*Malus domestica*) and mowed grasses in an orchard. *Agric. Ecosyst. Environ.* **2007**, *118*, 191–200. [CrossRef]

83. Panchal, P.; Preece, C.; Peñuelas, J.; Giri, J. Soil carbon sequestration by root exudates. *Trends Plant Sci.* **2022**, *27*, 749–757. [CrossRef]

84. Singh, S.N.; Kulshreshtha, K.; Agnihotri, S. Seasonal dynamics of methane emission from wetlands. *Chemosphere-Glob. Chang. Sci.* **2000**, *2*, 39–46. [CrossRef]

85. Szalai, Z. Spatial and temporal pattern of soil pH and Eh and their impact on solute iron content in a wetland (Transdanubia, Hungary). *Acta Geogr. Debr. Landsc. Environ. Ser.* **2021**, *2*, 34–45.

86. McCauley, A.; Jones, C.; Jacobsen, J. Soil pH and organic matter. *Nutr. Manag. Modul.* **2009**, *8*, 1–12.

87. Sharpley, A.N. Soil mixing to decrease surface stratification of phosphorus in manured soils. *J. Environ. Qual.* **2003**, *32*, 1375–1384. [CrossRef] [PubMed]

88. Dorioz, J.M.; Wang, D.; Poulenard, J.; Trevisan, D. The effect of grass buffer strips on phosphorus dynamics—A critical review and synthesis as a basis for application in agricultural landscapes in France. *Agric. Ecosyst. Environ.* **2006**, *117*, 4–21. [CrossRef]

89. Hart, M.R.; Quin, B.F.; Nguyen, M.L. Phosphorus runoff from agricultural land and direct fertilizer effects: A review. *J. Environ. Qual.* **2004**, *33*, 1954–1972. [CrossRef]

90. Li, B.; Chen, X.; Shi, X.; Liu, J.; Wei, Y.; Xiong, F. Effects of ridge tillage and straw mulching on cultivation the fresh faba beans. *Agronomy* **2021**, *11*, 1054. [CrossRef]

91. Stevens, C.J.; Quinton, J.N.; Bailey, A.P.; Deasy, C.; Silgram, M.; Jackson, D.R. The effects of minimal tillage, contour cultivation and in-field vegetative barriers on soil erosion and phosphorus loss. *Soil Till. Res.* **2009**, *106*, 145–151. [CrossRef]

92. Alewell, C.; Ringeval, B.; Ballabio, C.; Robinson, D.A.; Panagos, P.; Borrelli, P. Global phosphorus shortage will be aggravated by soil erosion. *Nat. Commun.* **2020**, *11*, 4546. [CrossRef] [PubMed]

93. Meena, V.S.; Bahadur, I.; Maurya, B.R.; Kumar, A.; Meena, R.K.; Meena, S.K.; Verma, J.P. Potassium-solubilizing microorganism in evergreen agriculture: An overview. In *Potassium Solubilizing Microorganisms for Sustainable Agriculture*; Meena, V.S., Maurya, B.R., Verma, J.P., Meena, R.S., Eds.; Springer: New Delhi, India, 2016; pp. 1–20. [CrossRef]

94. Hooda, P.S. (Ed.) *Trace Elements in Soils*; Wiley: Hoboken, NJ, USA, 2010.

95. Mirás-Avalos, J.M.; Bertol, I.; de Abreu, C.A.; Vidal Vazquez, E.; Paz Gonzalez, A. Crop residue effects on total and dissolved losses of Fe, Mn, Cu, and Zn by runoff. *Commun. Soil Science Plant Anal.* **2015**, *46*, 272–282. [CrossRef]

96. Jiao, W.; Ouyang, W.; Hao, F.; Huang, H.; Shan, Y.; Geng, X. Combine the soil water assessment tool (SWAT) with sediment geochemistry to evaluate diffuse heavy metal loadings at watershed scale. *J. Hazard. Mater.* **2014**, *280*, 252–259. [CrossRef]

97. Materechera, S.A. Tillage and tractor traffic effects on soil compaction in horticultural fields used for peri-urban agriculture in a semi-arid environment of the North West Province, South Africa. *Soil Till. Res.* **2009**, *103*, 11–15. [CrossRef]

98. Cheraghi, M.; Lorestani, B.; Merrikhpour, H.; Rouniasi, N. Heavy metal risk assessment for potatoes grown in overused phosphate-fertilized soils. *Environ. Monit. Assess.* **2013**, *185*, 1825–1831. [CrossRef]

99. Shi, Z.H.; Yue, B.J.; Wang, L.; Fang, N.F.; Wang, D.; Wu, F.Z. Effects of mulch cover rate on interrill erosion processes and the size selectivity of eroded sediment on steep slopes. *Soil Sci. Soc. Am. J.* **2013**, *77*, 257–267. [CrossRef]
100. da Rocha Junior, P.R.; Bhattarai, R.; Alves Fernandes, R.B.; Kalita, P.K.; Vaz Andrade, F. Soil surface roughness under tillage practices and its consequences for water and sediment losses. *J. Soil Sci. Plant Nutr.* **2016**, *16*, 1065–1074. [CrossRef]
101. Wang, B.; Zhang, G.H.; Shi, Y.Y.; Zhang, X.C. Soil detachment by overland flow under different vegetation restoration models in the Loess Plateau of China. *Catena* **2014**, *116*, 51–59. [CrossRef]
102. Ran, Q.; Su, D.; Li, P.; He, Z. Experimental study of the impact of rainfall characteristics on runoff generation and soil erosion. *J. Hydrol.* **2012**, *424*, 99–111. [CrossRef]
103. Belnap, J.; Welter, J.R.; Grimm, N.B.; Barger, N.; Ludwig, J.A. Linkages between microbial and hydrologic processes in arid and semiarid watersheds. *Ecology* **2005**, *86*, 298–307. [CrossRef]
104. Rhoton, F.E.; Shipitalo, M.J.; Lindbo, D.L. Runoff and soil loss from midwestern and southeastern US silt loam soils as affected by tillage practice and soil organic matter content. *Soil Till. Res.* **2002**, *66*, 1–11. [CrossRef]
105. Roose, E.; Barthes, B. Organic matter management for soil conservation and productivity restoration in Africa: A contribution from Francophone research. In *Managing Organic Matter in Tropical Soils: Scope and Limitations*; Springer: Dordrecht, The Netherlands, 2021; pp. 159–170. [CrossRef]
106. Zhang, G.S.; Chan, K.Y.; Oates, A.; Heenan, D.P.; Huang, G.B. Relationship between soil structure and runoff/soil loss after 24 years of conservation tillage. *Soil Till. Res.* **2007**, *92*, 122–128. [CrossRef]
107. Keesstra, S.; Pereira, P.; Novara, A.; Brevik, E.C.; Azorin-Molina, C.; Parras-Alcántara, L.; Jordán, A.; Cerdà, A. Effects of soil management techniques on soil water erosion in apricot orchards. *Sci. Total Environ.* **2016**, *551*, 357–366. [CrossRef]
108. Parras-Alcántara, L.; Lozano-García, B.; Keesstra, S.; Cerdà, A.; Brevik, E.C. Long-term effects of soil management on ecosystem services and soil loss estimation in olive grove top soils. *Sci. Total Environ.* **2016**, *571*, 498–506. [CrossRef]
109. Mahmoud, E.; Ibrahim, M.; Robin, P.; Akkal-Corfini, N.; El-Saka, M. Rice straw composting and its effect on soil properties. *Compost Sci. Util.* **2009**, *17*, 146–150. [CrossRef]
110. Wang, L.; Wang, C.; Feng, F.; Wu, Z.; Yan, H. Effect of straw application time on soil properties and microbial community in the Northeast China Plain. *J. Soils Sediments* **2021**, *21*, 3137–3149. [CrossRef]
111. Cui, H.; Luo, Y.; Chen, J.; Jin, M.; Li, Y.; Wang, Z. Straw return strategies to improve soil properties and crop productivity in a winter wheat-summer maize cropping system. *Eur. J. Agron.* **2022**, *133*, 126436. [CrossRef]
112. Shi, P.; Schulin, R. Erosion-induced losses of carbon, nitrogen, phosphorus and heavy metals from agricultural soils of contrasting organic matter management. *Sci. Total Environ.* **2018**, *618*, 210–218. [CrossRef]
113. Lal, R. Tillage effects on soil degradation, soil resilience, soil quality, and sustainability. *Soil Till. Res.* **1993**, *27*, 298–307. [CrossRef]
114. Ishaq, M.; Ibrahim, M.; Lal, R. Tillage effects on soil properties at different levels of fertilizer application in Punjab, Pakistan. *Soil Till. Res.* **2002**, *68*, 93–99. [CrossRef]
115. Mhazo, N.; Chivenge, P.; Chaplot, V. Tillage impact on soil erosion by water: Discrepancies due to climate and soil characteristics. *Agric. Ecosyst. Environ.* **2016**, *230*, 231–241. [CrossRef]
116. Nearing, M.A.; Pruski, F.F.; O'neal, M.R. Expected climate change impacts on soil erosion rates: A review. *J. Soil Water Conserv.* **2004**, *59*, 43–50.
117. Panagos, P.; Borrelli, P.; Meusburger, K.; Alewell, C.; Lugato, E.; Montanarella, L. Estimating the soil erosion cover-management factor at the European scale. *Land Use Policy* **2015**, *48*, 38–50. [CrossRef]
118. Cerdà, A.; Daliakopoulos, I.N.; Terol, E.; Novara, A.; Fatahi, Y.; Moradi, E.; Salvati, L.; Pulido, M. Long-term monitoring of soil bulk density and erosion rates in two *Prunus persica* (L.) plantations under flood irrigation and glyphosate herbicide treatment in La Ribera district, Spain. *J. Environ. Manag.* **2021**, *282*, 111965. [CrossRef]
119. Zuazo, V.H.D.; Pleguezuelo, C.R.R. Soil-erosion and runoff prevention by plant covers: A review. In *Sustainable Agriculture*; Springer: Berlin/Heidelberg, Germany, 2009; pp. 785–811. [CrossRef]
120. Lal, R. Soil degradation as a reason for inadequate human nutrition. *Food Secur.* **2009**, *1*, 45–57. [CrossRef]
121. Laflen, J.M.; Colvin, T.S. Effect of crop residue on soil loss from continuous row cropping. *Trans. ASAE* **1981**, *24*, 605–0609. [CrossRef]
122. Hass, A.; Fine, P. Sequential selective extraction procedures for the study of heavy metals in soils, sediments, and waste materials—A critical review. *Crit. Rev. Environ. Sci. Technol.* **2010**, *40*, 365–399. [CrossRef]
123. Ikenaka, Y.; Nakayama, S.M.; Muzandu, K.; Choongo, K.; Teraoka, H.; Mizuno, N.; Ishizuka, M. Heavy metal contamination of soil and sediment in Zambia. *Afr. J. Environ. Sci. technol.* **2010**, *4*, 729–739.
124. Soinne, H.; Hovi, J.; Tammeorg, P.; Turtola, E. Effect of biochar on phosphorus sorption and clay soil aggregate stability. *Geoderma* **2014**, *219*, 162–167. [CrossRef]
125. Boyle, M.; Frankenberger, W.T., Jr.; Stolzy, L.H. The influence of organic matter on soil aggregation and water infiltration. *J. Prod. Agric.* **1989**, *2*, 290–299. [CrossRef]
126. Six, J.; Paustian, K.; Elliott, E.T.; Combrink, C. Soil structure and organic matter I. Distribution of aggregate-size classes and aggregate-associated carbon. *Soil Sci. Soc. Am. J.* **2000**, *64*, 681–689. [CrossRef]
127. Sithole, N.J.; Magwaza, L.S.; Thibaud, G.R. Long-term impact of no-till conservation agriculture and N-fertilizer on soil aggregate stability, infiltration and distribution of C in different size fractions. *Soil Till. Res.* **2019**, *190*, 147–156. [CrossRef]

128. Telak, L.J.; Pereira, P.; Bogunovic, I. Management and seasonal impacts on vineyard soil properties and the hydrological response in continental Croatia. *Catena* **2021**, *202*, 105267. [CrossRef]

129. Mangalassery, S.; Kalaivanan, D.; Philip, P.S. Effect of inorganic fertilisers and organic amendments on soil aggregation and biochemical characteristics in a weathered tropical soil. *Soil Till. Res.* **2019**, *187*, 144–151. [CrossRef]

130. Dalal, R.C.; Bridge, B.J. Aggregation and organic matter storage in sub-humid and semi-arid soils. In *Structure and Organic Matter Storage in Agricultural Soils*; CRC Press: Boca Raton, FL, USA, 2020; pp. 263–307.

131. Le Bissonnais, Y.L. Aggregate stability and assessment of soil crustability and erodibility: I. Theory and methodology. *Eur. J. Soil. Sci.* **1996**, *47*, 425–437. [CrossRef]

132. Chenu, C.; Le Bissonnais, Y.; Arrouays, D. Organic matter influence on clay wettability and soil aggregate stability. *Soil Sci. Soc. Am. J.* **2000**, *64*, 1479–1486. [CrossRef]

133. Boix-Fayos, C.; Calvo-Cases, A.; Imeson, A.C.; Soriano-Soto, M.D. Influence of soil properties on the aggregation of some Mediterranean soils and the use of aggregate size and stability as land degradation indicators. *Catena* **2001**, *44*, 47–67. [CrossRef]

134. Bravo, S.; Amorós, J.A.; Pérez-De-Los-Reyes, C.; García, F.J.; Moreno, M.M.; Sánchez-Ormeño, M.; Higueras, P. Influence of the soil pH in the uptake and bioaccumulation of heavy metals (Fe, Zn, Cu, Pb and Mn) and other elements (Ca, K, Al, Sr and Ba) in vine leaves, Castilla-La Mancha (Spain). *J. Geochem. Explor.* **2017**, *174*, 79–83. [CrossRef]

135. Rees, F.; Simonnot, M.O.; Morel, J.L. Short-term effects of biochar on soil heavy metal mobility are controlled by intra-particle diffusion and soil pH increase. *Eur. J. Soil Sci.* **2014**, *65*, 149–161. [CrossRef]

136. Harter, R.D. Effect of soil pH on adsorption of lead, copper, zinc, and nickel. *Soil Sci. Soc. Am. J.* **1983**, *47*, 47–51. [CrossRef]

137. Boguta, P.; Sokolowska, Z. Influence of phosphate ions on buffer capacity of soil humic acids. *Int. Agrophys.* **2012**, *26*, 7–14. [CrossRef]

138. Kashem, M.A.; Singh, B.R. Metal availability in contaminated soils: I. Effects of flooding and organic matter on changes in Eh, pH and solubility of Cd, Ni and Zn. *Nutr. Cycl. Agroecosystems* **2001**, *61*, 247–255. [CrossRef]

139. Lampurlanés, J.; Cantero-Martínez, C. Soil bulk density and penetration resistance under different tillage and crop management systems and their relationship with barley root growth. *Agron. J.* **2003**, *95*, 526–536. [CrossRef]

140. Keesstra, S.D.; Rodrigo-Comino, J.; Novara, A.; Giménez-Morera, A.; Pulido, M.; Di Prima, S.; Cerdà, A. Straw mulch as a sustainable solution to decrease runoff and erosion in glyphosate-treated clementine plantations in Eastern Spain. An assessment using rainfall simulation experiments. *Catena* **2019**, *174*, 95–103. [CrossRef]

 *land*

Article

Straw Mulch Effect on Soil and Water Loss in Different Growth Phases of Maize Sown on Stagnosols in Croatia

Igor Bogunović [1,*], Iva Hrelja [1], Ivica Kisić [1], Ivan Dugan [1], Vedran Krevh [2], Jasmina Defterdarović [2], Vilim Filipović [2,3], Lana Filipović [2] and Paulo Pereira [4]

[1] Department of General Agronomy, Faculty of Agriculture, University of Zagreb, Svetošimunska Cesta 25, 10000 Zagreb, Croatia

[2] Department of Soil Amelioration, Faculty of Agriculture, University of Zagreb, Svetošimunska Cesta 25, 10000 Zagreb, Croatia

[3] Future Regions Research Centre, Geotechnical and Hydrogeological Engineering Research Group, Federation University, Gippsland, VIC 3841, Australia

[4] Environmental Management Laboratory, Mykolos Romeris University, Ateities St. 20, LT-08303 Vilnius, Lithuania

* Correspondence: ibogunovic@agr.hr

Citation: Bogunović, I.; Hrelja, I.; Kisić, I.; Dugan, I.; Krevh, V.; Defterdarović, J.; Filipović, V.; Filipović, L.; Pereira, P. Straw Mulch Effect on Soil and Water Loss in Different Growth Phases of Maize Sown on Stagnosols in Croatia. *Land* 2023, *12*, 765. https://doi.org/10.3390/land12040765

Academic Editors: Kleomenis Kalogeropoulos, Andreas Tsatsaris, Nikolaos Stathopoulos, Demetrios E. Tsesmelis, Nilanchal Patel and Xiao Huang

Received: 1 March 2023
Revised: 24 March 2023
Accepted: 27 March 2023
Published: 28 March 2023

Abstract: Soil and water loss due to traditional intensive types of agricultural management is widespread and unsustainable in Croatian croplands. In order to mitigate the accelerated land degradation, we studied different cropland soil management strategies to obtain feasible and sustainable agro-technical practices. A rainfall simulation experiment was conducted at 58 mm h^{-1} over 30 min on 10 paired plots (0.785 m^2), bare and straw covered (2 t ha^{-1}). The experiment was carried out in maize cultivation (Blagorodovac, Croatia) established on Stagnosols on slopes. Measurements were conducted during April (bare soil, after seeding), May (five-leaves stage), and June (intensive vegetative growth) making 60 rainfall simulations in total. Straw reduced soil and water losses significantly. The highest water, sediment loss, and sediment concentrations were identified in tillage plots during May. Straw addition resulted in delayed ponding (for 7%, 63%, and 50% during April, May and June, respectively) and runoff generation (for 37%, 32%, and 18% during April, May and June, respectively). Compared with the straw-mulched plot, tillage and bare soil increased water loss by 349%. Maize development reduced the difference between bare and straw-mulched plots. During May and June, bare plots increase water loss by 92% and 95%, respectively. The straw mulch reduced raindrop kinetic energy and sediment detachment from 9, 6, and 5 magnitude orders in April, May, and June, respectively. Overall, the straw mulch was revealed to be a highly efficient nature-based solution for soil conservation and maize cultivation protection.

Keywords: agriculture systems; clay–loam soil; artificial rainfall; nature-based solutions; soil conservation

1. Introduction

Soil is an irreplaceable resource, as it continues to provide habitat for 95% of the world's produced food. Besides enabling food security, soils are multi-functional. Soils sequester carbon; recycle nutrients; regulate and filtrate water; provide habitat support, cultural services, raw materials and food; remediate contaminants; control flooding, erosion, pests, and disease; and increase biodiversity [1]. If the soils are properly managed, many soil ecosystem outcomes are accomplished. In this context, Viana et al. [2] highlighted that UN Sustainable Development Goals (SDGs) should consider adopting innovative forms of sustainable practice to increase efficiency, build resilience, and mitigate the impact of agriculture. Besides tillage, nutrient management is vital to increase soil productivity and reduce yield variability in vulnerable environments [3]. This approach supports several SDGs, such as zero hunger (SDG 2), climate action (SDG 13), and ensuring sustainable consumption and production patterns (SDG 12); regardless, water and food security are threatened under climate change [4].

Soils are endangered by several degradation processes. Loss of organic matter, crusting, soil compaction, soil acidification, nutrient imbalance, pollution, salinization, and erosion appear as signs of agricultural intensification. Conventional agriculture practices have devastating impacts on the environment and are a major cause of land degradation, biodiversity loss, and climate change [5]. Between 60 and 70% of soils in the European Union are unhealthy [6], while at least one third of land globally is considered moderately to highly degraded [7]. Despite these risks, the importance of food security is still critical, and food production will increase in the future. Possible food crises accelerated by warfare [8] or population growth [9] are the major factors in future soil disturbance. It is expected that an increase of approximately 25–70% above current production levels may be necessary to meet 2050 crop demand [10,11], and this will increase the area subjected to agricultural soil degradation.

Maize production is one of the most important cereals both for human and animal consumption; it is grown for grain and forage. Among the main crops produced, maize is certainly one of the most important crops for humanity. According to the FAO [12], maize production covers 205 million ha, making this crop, alongside wheat, one of the most important for farmer income [13]. Maize cultivation is organized in a wide diversity of soils and environments. In addition, different crop and soil management practices occur in maize cultivation; from dominantly no-tillage and other conservation-tillage systems of production in developed countries, to conventional tillage systems which occur in low-development countries with a prevalence of smallholders [13]. This diversity of management also has powerful implications for soil condition, quality, and soil degradation. Conventional agricultural practices, such as ploughing or excessive agrochemical use, are a major threat to soil system sustainability. Bare soils, for instance, are a generator of sediment, nutrient loss, and diffuse pollution [14]. In addition, previous works revealed that soils under maize are highly vulnerable to soil erosion and sediment transport on silty [15], loam [16], clay–loam [17], or sandy soils [18] in different environments. Since the cultivation of maize is in wide rows, the plant cover is not significant, making it a ineffective crop for raindrop interception.

Several erosion control measures are used in different agricultural landscapes: reduced tillage and/or no-tillage management, suitable crop rotations, mulching, cover crops, strip and/or contour cropping, and terracing [19]. Recently, nature-based solutions (NBS) have become popular measures to control and reverse land degradation. Their introduction to practical crop production is essential for sustainable development [20]. Application of organic mulch materials is one example of a NBS for restoring degraded ecosystems and delivering vital ecosystem services, since it has a positive/neutral effect on several land degradation processes in agricultural soils [21]. Therefore, this strategy should be considered a primary practice to halt soil erosion; furthermore, it supports major EU policy priorities, in particular the European Green Deal, Biodiversity strategy and Climate adaptation strategy. Globally, mulching has been tested as a solution for controlling soil erosion [14,22–24]. However, the precise contribution of mulching at particular maize growing stages for controlling soil erosion has not been singled out as a separate variable. This topic is especially important due to reports indicating that a single high-magnitude rainstorm can be responsible for 93% of the total annual soil losses [25]. Besides the favourable impact on hydrological response, organic mulch application has other benefits for the soil system as well. Mulch was proven to improve soil structure; conserve soil moisture; reduce soil compaction; and increase soil organic matter, infiltration, nutrient concentration, and cycling [26–29]. Despite all mentioned positives, in countries with poor agricultural sectors, a clear tradition about conservation practices is missing, so the use of mulch in practical crop production is rare and its impact in most agricultural systems is not well understood. Maize cropping systems in Central Europe have mostly focused on testing the effects of different tillage directions or no-tillage implementation in recent decades, thanks to pioneers who developed new strategies in soil conservation. However, most of the mentioned research focused on the effect of tillage systems and crops on yearly soil erosion proportions [30–33], without considering the effect of single rainstorms on different crop cultures in especially vulnerable growth stages.

New challenges are confronting Croatian croplands, resulting from the potential for organic residues in agricultural production to improve soil systems and ensure more sustainable management. Therefore, the aim of this work is to (1) investigate the use of straw mulch as an NBS to reduce soil and water loss in maize cropping systems, (2) to determine the critical stages of maize development when the soil degradation is greatest, and (3) to assess the short-term impact of straw mulch on soil health. We hypothesize that organic mulching will preserve soil quality and promote the recovery of soil services in poor-quality stagnosols on slopes.

2. Materials and Methods

2.1. Location, Climate and Soil

The experiment was set up in the municipality of Dežanovac, in the southern part of Bjelovar-Bilogora County, continental Croatia (Figure 1). Bjelovar-Bilogora County is a large Croatian rainfed cropland production zone, with maize, wheat, barley, soybean, and rapeseed being the dominant crop cultures. The study area is characterized by gentle hills area with dominantly silty soils developed on loess [15]. The studied plot has been arable land for more than two centuries, with conventional agricultural practices located at 129 m a.s.l. on a 9° slope. The climate is temperate continental, with an average rainfall of 889 mm, but with erratic distribution, particularly in the spring and autumn, when most of the high-intensity rains occur. The mean annual temperature is 10.7 °C, ranging from −0.4 °C in January to 20.6 °C in July [15]. Soil is silty stagnosol with low organic matter concentration and poor structure, which makes this soil type prone to compaction and erosion [34]. General soil properties are presented in Table 1.

Figure 1. (**a**) Study site Location, (**b**) view of the study site, (**c**) soil profile, (**d**) sampling strategy, (**e**) rainfall simulator during installation, and (**f**) view of the plot size and shape.

Table 1. Soil profile characteristics of the study site. Values following $\pm$ indicate standard deviation.

Horizons	Ap + Eg	Eg + Btg	Btg
Depth range (cm)	0–24	24–35	35–95
pH in KCl (*w/w* 1:2.5)	4.21 ± 0.15	4.20 ± 0.18	4.81 ± 0.23
Organic matter (g kg^{-1})	16 ± 3.3	14 ± 4.2	6 ± 3.8
Available P$_2$O$_5$ (g kg^{-1})	172 ± 18	65 ± 4	244 ± 24
Available K$_2$O (g kg^{-1})	308 ± 6	123 ± 8	502 ± 12
Clay (<0.002 mm) (g kg^{-1})	235.8 ± 9.0	241.3 ± 6.9	230.3 ± 11.7
Silt (0.02–0.002 mm) (g kg^{-1})	291.2 ± 42.1	273.4 ± 7.2	289.0 ± 39.9
Fine sand (0.2–0.02 mm) (g kg^{-1})	465.1 ± 47.9	479.8 ± 8.4	476.1 ± 40.5
Coarse sand (2–0.2 mm) (g kg^{-1})	7.9 ± 1.4	5.6 ± 1.0	4.6 ± 1.1
Texture classification	Clay Loam	Clay Loam	Clay Loam

The annual management consists of ploughing during October or November, followed by disking or cultivation in spring (April or March) prior to sowing. Maize is usually fertilized with urea 46% (150 kg ha^{-1}), NPK 7:20:30 (400 kg ha^{-1}), and calcium ammonium nitrate (KAN) 27% (300 kg ha^{-1}). Farmyard manure or slurry was not part of the regular management, while organic residues are regularly incorporated in soil by tillage interventions. The preceding culture was winter wheat, and straw was mixed with soil during primary tillage. Maize was sowed by a JD 750A planter (John Deere, Moline, IL, USA) on 15 April 2020 (73,000 grains ha^{-1}; inter row spacing 70 cm, sowing depth 4 cm), and herbicide application ("Adengo", dosage 0.44 L ha^{-1}) was performed on 9 May. Inter-row cultivation intervention was not performed during 2020.

2.2. Experimental Design and Field Observations

The experimental setup consisted of 10 paired plots (Figure 1), named straw and bare (control). Each paired plot consisted of two plots of 0.785 m^2 (metal ring of 1 m diameter), with a 3 m distance between them. The bare plot is a control treatment. The surface of this area is without vegetation due to conventional tillage management (tilled), while the straw plot was manually covered with 2 t ha^{-1} of barley straw; the mulch covered approximately 80% of the topsoil (straw).

Rainfall simulation experiments were carried out during the three growth stages of maize during April (after seeding—VE stage), May (five leaves—V3 stage), and June (intensive vegetative growth—V5 stage) by using a pressurized type of rainfall simulator (UGT Rainmaker, Müncheberg, Germany). In total, 60 rainfall simulation experiments (10 per treatment, 20 per growing stage) were performed using a rainfall intensity of 58 mm h^{-1} for a duration of half an hour; storms such as that simulated have a return ratio in the area of research every 7 years [35]. Plots were established in non-traffic areas, and a plastic vessel was used above the plots to ensure calibration before the simulation commenced. Before each rainfall simulation experiment, soil samples (0–10 cm depth) were collected in the close vicinity of a circular metal ring used for overland flow collection. Samples were collected using soil cores (for bulk density—BD, water holding capacity—WHC, and soil water content—SWC) and by shovel for soil structural characteristics (mean weight diameter—MWD, and water stable aggregates—WSA). A photo of the plot surface and measurement of inclination were noted to obtain vegetation/mulch cover and slope, respectively. Finally, a chronometer was used to determine the time to ponding (TP) and time to runoff generation (TR). To collect the overland flow during rainfall simulation experiments, a plastic canister was connected to the metal ring (plot) for the collection of overland flow. The collected surface flow was weighed and filtered to obtain runoff and soil loss (SL) after drying on a filter paper. Sediment concentration (SC) was calculated by dividing the mass of SL by the mass of the runoff [36].

2.3. Laboratory Analysis

Soil cores were weighted before and after wetting for determination of WHC, and dried in an oven at 105 °C for 48 h and weighed obtain the BD and SWC, according to Black's method in [37]. Undisturbed soil collected by shovel and stored in rectangular boxes was used for gentle hand preparation of soil aggregates, following instructions of Diaz-Zorita et al. [38], before soil aggregates were subjected to dry sieving for the duration of 30 s to obtain aggregate size fractions [39] and calculate MWD. A 4 g amount of the size fraction 0.4–0.5 mm diameter was used for soaking to obtain WSA, following the method of Kemper & Rosenau [39].

2.4. Statistical Analyses

The normality of data distribution was assessed with Shapiro–Wilk test ($p > 0.05$). Several variables did not respect the Gaussian distribution, so they were normalized with a Box–Cox transformation. A two-way ANOVA was used to identify significant differences among plots and growth stages. If significant differences were found (at $p < 0.05$), the Tukey HSD post-hoc test was applied. A principal component analysis (PCA) (based on the correlation matrix) was performed on Box–Cox data to identify the intrinsic relationships between the variables. Data analyses were carried out with Statistica 12.0 (StatSoft, Tulsa, OK, USA) [40]. Figures were elaborated with Plotly [41] to present the original data.

3. Results

3.1. Management and Growth-Stage Impact on Soil and Hydrological Response

Results of straw and growth-stage impact on soil are presented in Table 2. BD ranged from 1.15 g cm^{-3} to 1.59 g cm^{-3} in tilled plots, and from 1.29 g cm^{-3} to 1.57 g cm^{-3} in straw plots. BD was significantly lower in WE and W3 stage compared with the cropland W7 stage in tillage plots. WHC and SWC of the soils ranged from 28.70% to 44.00%, and from 20.20% to 34.90%, respectively. SWC was significantly higher in the tilled plots than in the straw plots at W3 stage. Temporal patterns reveal significantly higher SWC at W3 stage than in the other stages in tillage plots, while in straw plots, the W7 stage showed significantly lower SWC than at other stages.

Table 2. Results of two-way ANOVA analysis considering soil properties. Different letters after mean values in the columns represent significant difference at $p < 0.05$. Capital letters show statistical difference between stage growth. Lower case letters show statistical differences between treatments. Abbreviations: WE—after seeding; V3—five leaves stage; V5—intensive vegetative growth; BD, bulk density; SWC, soil water content; WHC, water holding capacity; MWD, mean weight diameter; WSA, water-stable aggregates.

Growth Stage	Treatment	BD (g cm^{-3})		SWC (%)		WHC (%)		MWD (mm)		WSA (%)	
		Mean	SD	Mean	SD	Mean	SD	Mean	SD	Mean	SD
VE	Tilled	1.35 Ba	0.11	27.61 Aa	1.50	40.84 Ba	1.50	3.51 Aa	0.20	24.51 Aa	4.57
	Straw	1.38 Aa	0.06	28.47 Aa	2.53	41.44 Aa	1.46	3.52 Aa	0.15	26.04 Aa	3.68
V3	Tilled	1.43 Ba	0.05	33.05 Aa	1.23	38.69 Aa	3.60	3.32 Aa	0.23	21.50 Aa	4.52
	Straw	1.45 Aa	0.07	29.07 Aa	1.43	40.05 Ab	0.80	3.31 Aa	0.22	24.61 Aa	5.24
V5	Tilled	1.46 Aa	0.08	25.80 Aa	1.57	39.23 Ba	1.56	2.88 Ba	0.12	24.92 Aa	4.52
	Straw	1.45 Aa	0.07	25.36 Aa	3.54	39.44 Ba	2.63	2.85 Ba	0.27	28.03 Aa	5.24

MWD and WSA varied between 2.23–3.91 mm and 13.16–35.26%, respectively. At W7 stage, MWD was significantly lower in both treatments than at WE and W3 stage. The highest WSA was marked at W7 stage in both treatments. In all cases, no significant difference was identified.

The mean TP in the tilled plots was found to be 126.4 s, with a maximum value of 360 s and a minimum value of only 5 s (Table 3). In straw plots, the mean was found to be 189.7 s, with a maximum value of 420 s and a minimum value of only 20 s. TP was significantly higher in the WE stage, compared to the W3 stage in both treatments, while the absolute values were higher in straw plots than in tilled. For TR, the values of 360 s were registered in tilled plots, reaching 780 and 60 s as maximum and minimum values, respectively. In straw plots, TR mean was 510 s, reaching 1800 and 120 s as maximum and minimum values, respectively. TR was significantly higher in the WE and W7 stages, compared to the W3 stage in both treatments, while the absolute values were higher in straw plots than in tilled. The runoff values ranged from 7.34 to 32.98 m^3 ha^{-1} (mean 20.16 m^3 ha^{-1}) in the WE stage, from 67.50 to 129.35 m^3 ha^{-1} (mean 98.42 m^3 ha^{-1}) in the W3 stage, and from 55.48 to 108.73 m^3 ha^{-1} (mean 82.10 m^3 ha^{-1}) in the W7 stage. In all growth stages, a significantly higher runoff was identified in tilled plots. In both treatments, significantly lower runoff was identified during WE stage than during other stages.

Table 3. Results of two-way ANOVA analysis considering overland flow properties. Different letters after mean values in the columns represent significant difference at $p < 0.05$. Capital letters show statistical difference between stage growth. Lower case letters show statistical differences between treatments. Abbreviations: WE—after seeding; V3—five-leaves stage; V5—intensive vegetative growth; PT, time to ponding; RT, time to runoff; SC, sediment concentration; SL, sediment loss.Growth stage.

	Treatment	TP (s)		TR (s)		Runoff (m^3 ha^{-1})		SC (g kg^{-1})		SL (kg ha^{-1})	
		Mean	SD	Mean	SD	Mean	SD	Mean	SD	Mean	SD
VE	Tilled	234 Aa	71.8	456 Aa	158.0	33.0 Ba	14.0	18.5 Aa	11.8	639.5 Ba	662.1
	Straw	252 Aa	73.8	720 Aa	434.5	7.3 Bb	7.8	11.2 Aa	7.5	74.5 Bb	64.6
V3	Tilled	35 Ba	37.2	192 Ba	97.2	129.3 Aa	28.4	25.4 Aa	8.3	3283.8 Aa	1292.6
	Straw	95 Ba	83.2	282 Ba	113.3	67.5 Ab	37.7	8.8 Ab	2.9	586.5 Ab	314.3
V5	Tilled	110 Aba	35.5	432 Aa	132.1	108.7 Aa	40.7	18.4 Aa	5.6	2030.2 Aa	1108.2
	Straw	222 ABa	85.1	528 Aa	205.5	55.5 Ab	43.9	16.5 Aa	28.1	418.5 Ab	229.4

SC values ranged from 5.13 g L^{-1} to 41.46 g L^{-1} in tilled plots and from 4.45 g L^{-1} to 95.65 g L^{-1} in straw plots. SC showed different results, exhibiting significantly higher SC values in tilled plots in comparison with straw plots during the W3 stage. Temporal trends indicate non-significant behavior among growth stages at both treatments. The SL values ranged from 201.85 to 5067.29 kg ha^{-1} (mean 1984.48 kg ha^{-1}) in the tilled plots, and from 0.00 to 1047.77 kg ha^{-1} (mean 359.83 kg ha^{-1}) in the straw plots. Significant differences were observed at all growth stages among treatments. Tilled plots show significantly higher SL in every growth stage. Among growth stages, soil losses were significantly lower at WE than at W3 and W7.

3.2. Interrelation of the Variables

The first four factors explained 71.4% of the total variance. Factor 1 explained 38.89%, Factor 2 explained 14.84%, and Factors 3 and 4 explained 9.22% and 8.46%, respectively. Factor 1 had high positive loadings in WHC, TP, and TR, and high negative for runoff, SL, and SC (Table 4). Factor 2 had high positive loadings for SWC and MWD, and high negative loadings for WSA. Finally, Factor 3 and Factor 4 had high negative loadings in BD and slope, respectively. The intersection between Factor 1 and Factor 2 shows that runoff, SL, BD, SC, slope, and SWC are inversely related to the majority of the other variables, especially to the TP, TR, WHC, WSA and MWD (Figure 2A). The land management practices and time of measurement had different impacts on studied variables in tilled and straw treatments. The variability is lower in the W7 stage compared with the WE and W3 stages (Figure 2B).

Table 4. Loadings matrix considering the first four factors extracted from the Principal Component Analysis. Eigenvalues retained in each factor are in bold.

Variable	Factor 1	Factor 2	Factor 3	Factor 4
Slope (°)	−0.267470	0.222276	0.305639	**−0.670179**
Bulk density (g cm^{-3})	−0.401055	−0.406840	**−0.423273**	0.007005
Soil water content (%)	−0.352835	**0.601550**	−0.341593	0.063127
Water holding capacity (%)	**0.515545**	0.314809	0.154563	−0.369870
Mean weight diameter (mm)	0.156117	**0.801851**	−0.132513	0.244787
Water stable aggregates (%)	0.323820	**−0.453285**	0.072216	0.350031
Time to ponding (s)	**0.804425**	−0.212584	0.100781	−0.121820
Time to runoff generation (s)	**0.788527**	−0.285131	0.189527	−0.096806
Runoff (m^3 ha^{-1})	**−0.909082**	−0.232817	−0.032220	−0.163268
Sediment concentration (g L^{-1})	−0.426888	0.179558	0.717140	0.398189
Sediment loss (kg ha^{-1})	**−0.913082**	−0.051735	0.324428	0.073881

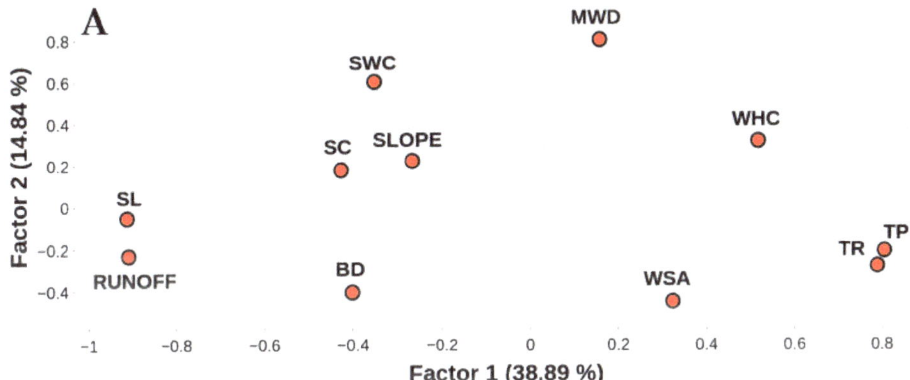

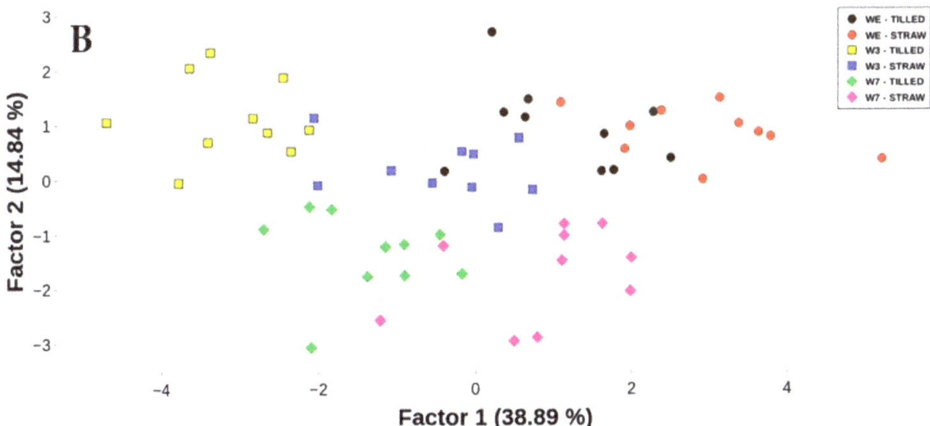

Figure 2. Relation between Factor 1 and Factor 2: (**A**) variables and (**B**) cases. Abbreviations: BD indicates bulk density; SWC, soil water content; WHC, water holding capacity; MWD, mean weight diameter; WSA, water stable aggregates; TP, time to ponding; TR, time to runoff generation; SC, sediment concentration; SL, sediment loss; SWC, soil water content. WE—after seeding; V3—five leaves stage; V5—intensive vegetative growth.

4. Discussion

4.1. Soil Properties

The results revealed that different land management partially changed some soil properties. The soil compaction between treatments was not significantly different at any growth stage. This is in agreement with the observations of previous studies. Dugan et al. [14] did not find difference between mulched and tilled treatments in the first few months after mulch application in hazelnut orchards. Similar results were noted in studies of Mulumba & Lal [42] and Głąb & Kulig [43] in croplands, indicating that a longer period of mulch application is needed, or the dosage of mulch application is too low to create a positive effect on soil compaction [21,44]. However, such criteria are not always solid, since there is proof that under reduced or no-tillage systems, mulching can decrease BD in the intermediate period [43,45]. The present study was performed under conventional tillage management; a long period with a high mulch application dose is needed. Tilled plots showed an increase in compaction over time, while in straw plots, the highest compaction was determined at the W3 stage of maize. The temporal increase of the compaction can be attributed to soil consolidation and tractor traffic. Moreover, the soil on the research site is silty with poor physical and chemical characteristics [34], which makes it very susceptible to consolidation and compaction [15]. Other findings also reported a significant increase in soil compaction under natural conditions in a few months after tillage interventions [46–48].

WHC, WSA, and MWD was not affected by treatment during all three growth stages. This can be attributed to the short duration of mulching practice and the low dosage of straw in the experiment. Usually, mulch decomposes in the soil and gradually increases the soil organic matter concentration, which improves the soil's ability to hold water [49]. The current dosage of 2.5 t ha^{-1} of mulch is too low, which is also confirmed in the study of Nzeyimana et al. [44]. However, the gradual increase of aggregate stability over time on straw plots indicates the positive impact of straw mulch on soil structural characteristics, as is proved in other works [45,50], despite the fact that MWD decreased on both treatments in the W7 stage. This reduction is very likely due to the structural deterioration by kinetic energy from rainfall, which easily breaks the artificial tillage-created clods [47]. Jordán et al. [51] in their study of 11 years of straw application also did not detect a significant increase in MWD, even in treatment with mulch dose of 16 t ha^{-1}.

SWC in the studied area showed high values. It is known that the effects of mulching on soil moisture depend on precipitation and climatic factors. Mulch effect on water conservation is usually more marked in arid and semiarid conditions, since mulching influences the soil moisture regime by controlling the surface evaporation rate [52]. Such conditions were absent during the research period, since the precipitation was normal and the high temperatures were not as high as those during summer. Therefore, positive mulch impact on soil moisture conservation by reducing the evaporation rate was absent in studied plots. Moreover, during the W5 stage, the tillage plots showed higher SWC than straw plots. The unusual finding could be explained by the recent heavy rain that fell just before the rainfall simulation experiment was performed. Precipitation on bare soil completely entered the soil, while in straw plots, some parts remained on the cover. This is not an uncommon occurrence, as other studies [53–55] have also showed similar or higher SWC in wet conditions on straw or no-tillage plots compared with bare conventional plots. However, it is important to further investigate this finding in future research to gain a better understanding of its implications.

4.2. Hydrological Behaviour

Soil erosion in cropland under wide-row spring cultures such as maize is recurrent and considerable during intense rainfall events. The main factors responsible for the high sediment yields in continental Croatia are poor structural stability of the stagnosol under conventional agricultural management and the lack of vegetation cover under frequent and intense tillage, which are common in this region [15,56]. The absence of strong crop rotations and lack of cover crop prevalence of wide-row cash crops like maize, potato, sugar beet and soybean, together with the lack of farmyard manure application, increases the rate of erosion because these management practices result in greater sediment detachment in croplands [57].

The usual strategies to fight soil erosion involve soil conditioning to improve soil structure, using mulches, cover crops, no-tillage systems, contouring, grass margins, and wide crop rotations with a prevalence of high-density crops and perennial grasses, which effectively decrease runoff and soil and nutrient loss, as is proven in other works [30–33,35] on similar pedological, geomorphological, and environmental conditions. Nevertheless, the main strategy to reduce soil erosion in Croatia is tillage direction in conventional systems, which significantly reduces the soil and nutrient losses if performed across the slope, or using no-tillage [58,59], since other agricultural conservation practices are rarely used in agricultural systems in Croatia and other Central European countries [60–63]. Moreover, the majority of these unused strategies for reducing soil erosion are impractical or expensive because of the time and labour involved, or the treatment and origin of the materials, which often need to be transported or manufactured (i.e., mulches, farmyard manure, lime, gypsum, biochar). Farmers also have an issue with structural problems, such as small land parcels, poor economic strength, limited access to special machinery for conservation tillage, or not having the knowledge to implement conservation management [57,64–66]. Coupled with high input prices, slow administration, and low market food prices, this creates a challenging business environment [67].

Our research has shown the early maize development stage has a large effect on soil erosion and hydrological processes in sloped croplands at the pedon scale. By using a high rainfall-intensity rainfall simulation with use of a rainfall simulator on almost-bare soil in the early stage of maize development, and by studying runoff generation development and measuring soil losses under bare and straw-mulched soil, the present study demonstrated that runoff coefficients depended on a management decision: a straw mulch cover. The study site was situated on stagnosols under conventional agriculture on a slope with an average of $9°$, where high runoff and sediment discharges were already confirmed in a long-term study [31]. However, the step forward achieved in this work was obtaining the conservation potential of straw mulch as a nature-based solution to mitigate soil degradation in conventionally managed soils. Straw cover delays TP and TR, and significantly decrease runoff and SL in WE, W3 and W7 maize stages. This has also been reported by other authors in other European dry farming areas [65,68,69]. Our results indicate that farmers should use straw in the early development stages of maize cultivation because it contributes to reducing runoff and soil loss. However, in the later development stages of maize, it did not reduce the overall erosion under high-magnitude rainstorms. This is because the crop canopy is still relatively small for protecting the soil during V3 and V7 maize stages. Other works confirm the need for greater crop cover to obtain a significant reduction of soil loss [70,71]. Similar loss under the same pedological conditions were reported by Kisić [35] under wide-row cultures such as maize or soybean.

The measurements carried out in Central Croatia simulated high-magnitude storms and showed that croplands under maize can lose as much as 5.1 t ha^{-1} of soil in 30 min when measured at the plot scale, which makes conventional soil management practices on stagnosols unsustainable. Soil loss under straw-covered plots are six magnitude orders lower. Such results indicate that conventional tillage negatively affects soil sustainability. Tillage has been recognized as a major cause of soil erosion since agriculture was developed and acts as a driving factor for an acceleration of soil loss in agricultural landscapes [72]. Our rainfall simulation experiments prove that such hydrological behavior on stagnosols occurs mostly because the straw cover is embedded in the soil and acts as a barrier. Soil cover protects the soil from raindrop impacts and is a key factor in controlling erosion, and the present study results confirm this idea. Our study agrees with other research, e.g., in Austria, where mulch tillage in silt–loam croplands seems to be responsible for 20–55% lower runoff and 73–91% lower soil loss in comparison with conventional tillage [73]. In clay soil, Nishigaki [74] reported a 47% lower erosion rate on mulched cropland in comparison with bare plots. These findings confirm the fact that mulch on the soil surface plays a key role in soil conservation in the early maize development stage in the studied area. In each growth stage, we can notice a clear trend—a delay in the time to ponding and time to runoff generation with a straw cover. As a result, more rain infiltrates in the soil profile, which confirms the runoff behavior in the current study (Table 3).

4.3. Interrelations between Properties

The effectiveness of soil management, maize growing stage, and related properties are shown in the PCA results. The grouping of the variables in PCA indicates that BD, SWC, SL, slope, SC, and runoff are positively associated, and they are inversely associated with the opposite group, consisting of WSA, MWD, WHC, TP, and TR. Such results prove the fact that soil aggregation and compaction dominate the soil erosion response [38]. Soil pores are crucial for controlling infiltration, and bigger and more stable aggregates are responsible for lower compaction levels [75]. Positive interrelation between BD, SWC, SL, and runoff in our study reveals that compaction levels modify soil pore characteristics, which increases the overland flow and sediment loss. Compaction usually increases soil loss and runoff generation [75] and decreases aggregate size [38], as shown in the present study. Topsoil BD and SWC had a negative effect, while MWD and WSA had positive effects on ponding time and runoff time. On bare soil, lower BD through higher porosity contributed to accelerating the duration required for ponding. This was expected because soil compaction reduces water infiltration, while larger and more stable aggregates reduce runoff generation and soil loss [76].

Finally, a negative relationship between WSA, MWD, and SWC may be explained by cohesion forces. When aggregates are dry, their stability is higher [77]. Cohesion forces hold aggregates until SWC increases. However, when soil has a high SWC, it indicates a high proportion of small and medium pores in the total porosity, since the water cannot fill the pores of large dimensions. Such soil behavior clearly indicates that SWC is an important factor in unsustainable soil erosion in the later stages of maize development in the present experiment (Tables 2 and 3), despite the fact that maize canopy cover is higher at W3 and W7 stages in comparison with WE stage. Soil management had a significant impact on runoff rates and soil erosion risk after a simulated high-intensity storm. The soil and water loss in maize croplands in Central Croatia are not sustainable when they are conventionally tilled unless mulching is also performed.

5. Conclusions

Soil management had a significant impact on runoff rates and soil erosion risk after a simulated high-intensity storm. The soil and water loss in maize croplands in Central Croatia are not sustainable when traditionally tilled. Fast soil re-compaction after tillage intervention modifies soil structural and hydraulic properties, which in turn decreases the time to runoff generation and increases water and sediment loss. Although later maize growing stages had higher canopy cover, the physical status of the soil and the soil water content increases the erosion rate to an unsustainable level in comparison with mulched plots. Straw mulch in all studied maize growing stages is a significant measure for controlling soil and water loss. These findings show that from a soil erosion perspective, the conservation management strategies in maize croplands in Croatia need to be developed. An efficient reduction of runoff and soil erosion in early maize growth stages should be achieved through straw mulching, or by developing other soil conservation measures. It is crucial to increase land-use sustainability. The present study contributes to better soil use management in Croatian croplands.

Author Contributions: Conceptualization, I.B. and P.P.; methodology, I.B., I.D. and P.P.; software, I.B. and I.D.; validation, I.B., I.H., I.K., I.D., V.K., J.D., V.F., L.F. and P.P.; formal analysis, I.B. and I.D.; investigation, I.B., I.H. and I.D.; resources, I.B.; data curation, I.B.; writing—original draft preparation, I.B.; writing—review and editing, I.B., I.H., I.K., I.D., V.K., J.D., V.F., L.F. and P.P.; visualization, I.D.; supervision, I.B., I.K. and P.P.; project administration, I.B.; funding acquisition, I.B. All authors have read and agreed to the published version of the manuscript.

Funding: This work was supported by the Croatian Science Foundation through the "Soil erosion and degradation in Croatia" project (UIP-2017-05-7834) (SEDCRO).

Data Availability Statement: The data presented in this study are available on reasonable request from the corresponding author.

Acknowledgments: The authors are grateful for the help from Leon Josip Telak during field work.

Conflicts of Interest: The authors declare no conflict of interest. The funders had no role in the design of the study; in the collection, analyses, or interpretation of data; in the writing of the manuscript, or in the decision to publish the result.

References

1. Panagos, P.; Montanarella, L.; Barbero, M.; Schneegans, A.; Aguglia, L.; Jones, A. Soil priorities in the European Union. *Geoderma Reg.* **2022**, *29*, e00510. [CrossRef]
2. Viana, C.M.; Freire, D.; Abrantes, P.; Rocha, J.; Pereira, P. Agricultural land systems importance for supporting food security and sustainable development goals: A systematic review. *Sci. Total Environ.* **2022**, *806*, 150718. [CrossRef] [PubMed]
3. Bedeke, S.; Vanhove, W.; Gezahegn, M.; Natarajan, K.; Van Damme, P. Adoption of climate change adaptation strategies by maize-dependent smallholders in Ethiopia. *NJAS-Wagen J. Life Sci.* **2019**, *88*, 96–104. [CrossRef]
4. Corwin, D.L. Climate change impacts on soil salinity in agricultural areas. *Eur. J. Soil Sci.* **2021**, *72*, 842–862. [CrossRef]
5. European Commission. EU Soil Strategy for 2030. Reaping the Benefits of Healthy Soils for People, Food, Nature and Climate. Brussels. 2021. Available online: https://eur-lex.europa.eu/legal-content/EN/TXT/PDF/?uri=CELEX:52021DC0699&from=EN (accessed on 23 November 2022).
6. European Commission. Caring for Soil Is Caring for Life. Brussels. 2020. Available online: https://op.europa.eu/en/publication-detail/-/publication/4ebd2586-fc85-11ea-b44f-01aa75ed71a1 (accessed on 22 November 2022).
7. Löbmann, M.T.; Maring, L.; Prokop, G.; Brils, J.; Bender, J.; Bispo, A.; Helming, K. Systems knowledge for sustainable soil and land management. *Sci. Total Environ.* **2022**, *822*, 153389. [CrossRef]
8. Pereira, P.; Bašić, F.; Bogunovic, I.; Barcelo, D. Russian–Ukrainian war impacts the total environment. *Sci. Total Environ.* **2022**, *837*, 155865. [CrossRef]
9. United Nations. World Population Prospects 2022. Summary of Results. United Nations, Department of Economic and Social Affairs, Population Division, New York, USA. 2022. Available online: https://www.un.org/development/desa/pd/sites/www.un.org.development.desa.pd/files/wpp2022_summary_of_results.pdf (accessed on 22 November 2022).
10. UNCCD. Land Degradation Neutrality: Resilience at Local, National and Regional Levels. United Nations Convention to Combat Desertification. Bonn, Germany. 2014. Available online: https://catalogue.unccd.int/858_V2_UNCCD_BRO_.pdf (accessed on 23 November 2022).
11. Hunter, M.C.; Smith, R.G.; Schipanski, M.E.; Atwood, L.W.; Mortensen, D.A. Agriculture in 2050: Recalibrating targets for sustainable intensification. *Bioscience* **2017**, *67*, 386–391. [CrossRef]
12. FAOSTAT. Food and Agriculture Data. 2023. Available online: https://www.fao.org/faostat/en/#home (accessed on 23 November 2022).
13. Tanumihardjo, S.A.; McCulley, L.; Roh, R.; Lopez-Ridaura, S.; Palacios-Rojas, N.; Gunaratna, N.S. Maize agro-food systems to ensure food and nutrition security in reference to the Sustainable Development Goals. *Glob. Food Sec.* **2020**, *25*, 100327. [CrossRef]
14. Dugan, I.; Pereira, P.; Barcelo, D.; Telak, L.J.; Filipovic, V.; Filipovic, L.; Kisic, I.; Bogunovic, I. Agriculture management and seasonal impact on soil properties, water, sediment and chemicals transport in a hazelnut orchard (Croatia). *Sci. Total Environ.* **2022**, *839*, 156346. [CrossRef]
15. Bogunovic, I.; Pereira, P.; Kisic, I.; Sajko, K.; Sraka, M. Tillage management impacts on soil compaction, erosion and crop yield in Stagnosols (Croatia). *Catena* **2018**, *160*, 376–384. [CrossRef]
16. Doan, T.T.; Henry-des-Tureaux, T.; Rumpel, C.; Janeau, J.L.; Jouquet, P. Impact of compost, vermicompost and biochar on soil fertility, maize yield and soil erosion in Northern Vietnam: A three year mesocosm experiment. *Sci. Total Environ.* **2015**, *514*, 147–154. [CrossRef]
17. Lenka, N.K.; Satapathy, K.K.; Lal, R.; Singh, R.K.; Singh, N.A.K.; Agrawal, P.K.; Choudhury, P.; Rathore, A. Weed strip management for minimizing soil erosion and enhancing productivity in the sloping lands of north-eastern India. *Soil Till. Res.* **2017**, *170*, 104–113. [CrossRef]

18. Agbede, T.M.; Adekiya, A.O. Influence of biochar on soil physicochemical properties, erosion potential, and maize (*Zea mays* L.) grain yield under sandy soil condition. Commun. *Soil Sci. Plant Anal.* **2020**, *51*, 2559–2568. [CrossRef]

19. Vogel, E.; Deumlich, D.; Kaupenjohann, M. Bioenergy maize and soil erosion—Risk assessment and erosion control concepts. *Geoderma* **2016**, *261*, 80–92. [CrossRef]

20. Albert, C.; Brillinger, M.; Guerrero, P.; Gottwald, S.; Henze, J.; Schmidt, S.; Ott, E.; Schröter, B. Planning nature-based solutions: Principles, steps, and insights. *Ambio* **2021**, *50*, 1446–1461. [CrossRef]

21. Bogunović, I.; Filipović, V. Mulch as a nature-based solution to halt and reverse land degradation in agricultural areas. *Curr. Opin. Environ. Sci. Health*, 2023; *under review*.

22. Kavian, A.; Kalehhouei, M.; Gholami, L.; Jafarian, Z.; Mohammadi, M.; Rodrigo-Comino, J. The use of straw mulches to mitigate soil erosion under different antecedent soil moistures. *Water* **2020**, *12*, 2518. [CrossRef]

23. Li, R.; Li, Q.; Pan, L. Review of organic mulching effects on soil and water loss. *Arch. Agron. Soil Sci.* **2021**, *67*, 136–151. [CrossRef]

24. Telak, L.J.; Dugan, I.; Bogunovic, I. Soil management and slope impacts on soil properties, hydrological response, and erosion in hazelnut orchard. *Soil Syst.* **2021**, *5*, 5. [CrossRef]

25. Biddoccu, M.; Ferraris, S.; Opsi, F.; Cavallo, E. Long-term monitoring of soil management effects on runoff and soil erosion in sloping vineyards in Alto Monferrato (North–West Italy). *Soil Till. Res.* **2016**, *155*, 176–189. [CrossRef]

26. Truong, T.H.H.; Kristiansen, P.; Marschner, P. Influence of mulch C/N ratio and decomposition stage on plant N uptake and N availability in soil with or without wheat straw. *J. Plant Nutr. Soil Sci.* **2019**, *182*, 879–887. [CrossRef]

27. Tu, A.; Xie, S.; Zheng, H.; Li, H.; Li, Y.; Mo, M. Long-term effects of living grass mulching on soil and water conservation and fruit yield of citrus orchard in south China. *Agric. Water Manag.* **2021**, *252*, 106897. [CrossRef]

28. Duan, C.; Chen, J.; Li, J.; Feng, H.; Wu, S.; Meng, Q.; Siddique, K.H. Effects of organic amendments and ridge–furrow mulching system on soil properties and economic benefits of wolfberry orchards on the Tibetan Plateau. *Sci. Total Environ.* **2022**, *827*, 154317. [CrossRef]

29. Zhou, W.; Sun, X.; Li, S.; Du, T.; Zheng, Y.; Fan, Z. Effects of organic mulching on soil aggregate stability and aggregate binding agents in an urban forest in Beijing, China. *J. For. Res.* **2022**, *33*, 1083–1094. [CrossRef]

30. Kisic, I.; Basic, F.; Nestroy, O.; Mesic, M.; Butorac, A. Chemical properties of eroded soil material. *J. Agron. Crop Sci.* **2002**, *188*, 323–334. [CrossRef]

31. Kisic, I.; Bogunovic, I.; Birkás, M.; Jurisic, A.; Spalevic, V. The role of tillage and crops on a soil loss of an arable Stagnic Luvisol. *Arch. Agron. Soil Sci.* **2017**, *63*, 403–413. [CrossRef]

32. Basic, F.; Kisic, I.; Butorac, A.; Nestroy, O.; Mesic, M. Runoff and soil loss under different tillage methods on Stagnic Luvisols in central Croatia. *Soil Till. Res.* **2001**, *62*, 145–151. [CrossRef]

33. Basic, F.; Kisic, I.; Mesic, M.; Nestroy, O.; Butorac, A. Tillage and crop management effects on soil erosion in central Croatia. *Soil Till. Res.* **2004**, *78*, 197–206. [CrossRef]

34. IUSS—WRB. *World Reference Base for Soil Resources 2014: International Soil Classification System for Naming Soils and Creating Legends for Soil Maps*; Word Soil Resources Reports No. 106; FAO: Rome, Italy, 2014.

35. Kisić, I. *Anthropogenic Soil Erosion*; University of Zagreb, Faculty of Agriculture: Zagreb, Croatia, 2016.

36. Bogunovic, I.; Telak, L.J.; Pereira, P. Agriculture management impacts on soil properties and hydrological response in Istria (Croatia). *Agronomy* **2020**, *10*, 282. [CrossRef]

37. Black, C.A. *Method of Soil Analysis: Part 2. Chemical and Microbiological Properties*; American Society of Agronomy, Inc.: Madison, WI, USA, 1965; p. 1159.

38. Diaz-Zorita, M.; Perfect, E.; Grove, J.H. Disruptive methods for assessing soil structure. *Soil Till. Res.* **2002**, *64*, 3–22. [CrossRef]

39. Kemper, W.D.; Rosenau, R.C. Aggregate stability and size distribution. In *Methods of Soil Analysis*; Klute, A., Ed.; American Society of Agronomy, Inc.: Madison, WI, USA, 1986; pp. 425–442.

40. Statsoft. *Statistica 12.0 Software*, StatSoft Inc.: Hamburg, Germany, 2015.

41. Plotly Chart Studio. Available online: https://chart-studio.plotly.com/ (accessed on 1 November 2022).

42. Mulumba, L.N.; Lal, R. Mulching effects on selected soil physical properties. *Soil Till. Res.* **2008**, *98*, 106–111. [CrossRef]

43. Głąb, T.; Kulig, B. Effect of mulch and tillage system on soil porosity under wheat (*Triticum aestivum*). *Soil Till. Res.* **2008**, *99*, 169–178. [CrossRef]

44. Nzeyimana, I.; Hartemink, A.E.; Ritsema, C.; Stroosnijder, L.; Lwanga, E.H.; Geissen, V. Mulching as a strategy to improve soil properties and reduce soil erodibility in coffee farming systems of Rwanda. *Catena* **2017**, *149*, 43–51. [CrossRef]

45. Blanco-Canqui, H.; Lal, R. Soil structure and organic carbon relationships following 10 years of wheat straw management in no-till. *Soil Till. Res.* **2007**, *95*, 240–254. [CrossRef]

46. Busscher, W.J.; Bauer, P.J.; Frederick, J.R. Recompaction of a coastal loamy sand after deep tillage as a function of subsequent cumulative rainfall. *Soil Till. Res.* **2002**, *68*, 49–57. [CrossRef]

47. Birkas, M.; Szemők, A.; Antos, G.; Neményi, M. *Environmentally-Sound Adaptable Tillage*; Akadémiai Kiadó: Budapest, Hungary, 2008.

48. Saroa, G.S.; Lal, R. Soil restorative effects of mulching on aggregation and carbon sequestration in a Miamian soil in central Ohio. *Land Degrad. Dev.* **2003**, *14*, 481–493. [CrossRef]

49. Telak, L.J.; Pereira, P.; Ferreira, C.S.; Filipovic, V.; Filipovic, L.; Bogunovic, I. Short-term impact of tillage on soil and the hydrological response within a fig (Ficus Carica) orchard in Croatia. *Water* **2020**, *12*, 3295. [CrossRef]

50. El-Beltagi, H.S.; Basit, A.; Mohamed, H.I.; Ali, I.; Ullah, S.; Kamel, E.A.R.; Shalaby, T.A.; Ramadan, K.M.A.; Alkhateeb, A.A.; Ghazzawy, H.S. Mulching as a Sustainable Water and Soil Saving Practice in Agriculture: A Review. *Agronomy* **2022**, *12*, 1881. [CrossRef]

51. Jordán, A.; Zavala, L.M.; Gil, J. Effects of mulching on soil physical properties and runoff under semi-arid conditions in southern Spain. *Catena* **2010**, *81*, 77–85. [CrossRef]

52. Kader, M.A.; Senge, M.; Mojid, M.A.; Ito, K. Recent advances in mulching materials and methods for modifying soil environment. *Soil Till. Res.* **2017**, *168*, 155–166. [CrossRef]

53. Chen, Y.; Liu, S.; Li, H.; Li, X.F.; Song, C.Y.; Cruse, R.M.; Zhang, X.Y. Effects of conservation tillage on corn and soybean yield in the humid continental climate region of Northeast China. *Soil Till. Res.* **2011**, *115*, 56–61. [CrossRef]

54. TerAvest, D.; Carpenter-Boggs, L.; Thierfelder, C.; Reganold, J.P. Crop production and soil water management in conservation agriculture, no-till, and conventional tillage systems in Malawi. *Agric. Ecosyst. Environ.* **2015**, *212*, 285–296. [CrossRef]

55. Brezinščak, L.; Bogunović, I. Tillage and straw management impact on soil structure, compaction and soybean yield on fluvisol. *J. Cent. Eur. Agric.* **2021**, *22*, 133–145. [CrossRef]

56. Dekemati, I.; Simon, B.; Bogunovic, I.; Vinogradov, S.; Modiba, M.M.; Gyuricza, C.; Birkás, M. Three-year investigation of tillage management on the soil physical environment, earthworm populations and crop yields in Croatia. *Agronomy* **2021**, *11*, 825. [CrossRef]

57. Bašić, F. *The Soils of Croatia*; Springer: Dordrecht, The Netherlands, 2013; p. 179.

58. Kisić, I.; Bogunović, I.; Bilandžija, D. The influence of tillage and crops on particle size distribution of water-eroded soil sediment on Stagnosol. *Soil Water Res.* **2017**, *12*, 170–176. [CrossRef]

59. Kisic, I.; Bogunovic, I.; Zgorelec, Z.; Bilandzija, D. Effects of soil erosion by water under different tillage treatments on distribution of soil chemical parameters. *Soil Water Res.* **2018**, *13*, 36–43. [CrossRef]

60. Jones, R.J.A.; Le Bissonnais, Y.L.; Diaz, J.S.; Du Wel, O.; Øygarden, L.; Bazzoffi, P.; Prasuhn, V.; Yordanov, Y.; Strauss, P.; Rydell, B.; et al. *Work Package 2: Nature and Extent of Soil Erosion in Europe. Interim Report Version 3.31. EU Soil Thematic Strategy: Technical Working Group on Erosion*; European Commission—GD Environment: Brussels, Belgium, 2003.

61. Kertész, Á.; Madarász, B. Conservation agriculture in Europe. *Int. Soil Water Conserv. Res.* **2014**, *2*, 91–96. [CrossRef]

62. Panagos, P.; Imeson, A.; Meusburger, K.; Borrelli, P.; Poesen, J.; Alewell, C. Soil conservation in Europe: Wish or reality? *Land Degrad. Dev.* **2016**, *27*, 1547–1551. [CrossRef]

63. Kassam, A.; Friedrich, T.; Derpsch, R. Global spread of conservation agriculture. *Int. J. Environ. Stud.* **2019**, *76*, 29–51. [CrossRef]

64. Žutinić, Đ.; Grgić, I. Family farm inheritance in Slavonia region, Croatia. *Agric. Econ.* **2010**, *56*, 522–531. [CrossRef]

65. Hösl, R.; Strauss, P. Conservation tillage practices in the alpine forelands of Austria—Are they effective? *Catena* **2016**, *137*, 44–51. [CrossRef]

66. Blaće, A.; Čuka, A.; Šiljković, Ž. How dynamic is organic? Spatial analysis of adopting new trends in Croatian agriculture. *Land Use Policy* **2020**, *99*, 105036. [CrossRef]

67. Lončarić, R.; Lončarić, Z.; Tolušić, Z. What Croatian farmers think about situation in agriculture. *Eur. Sci. J.* **2016**, *12*, 335–347.

68. Döring, T.F.; Brandt, M.; Heß, J.; Finckh, M.R.; Saucke, H. Effects of straw mulch on soil nitrate dynamics, weeds, yield and soil erosion in organically grown potatoes. *Field Crops Res.* **2005**, *94*, 238–249. [CrossRef]

69. Maetens, W.; Poesen, J.; Vanmaercke, M. How effective are soil conservation techniques in reducing plot runoff and soil loss in Europe and the Mediterranean? *Earth-Sci. Rev.* **2012**, *115*, 21–36. [CrossRef]

70. Loch, R.J. Effects of vegetation cover on runoff and erosion under simulated rain and overland flow on a rehabilitated site on the Meandu Mine, Tarong, Queensland. *Soil Res.* **2000**, *38*, 299–312. [CrossRef]

71. Hou, G.; Bi, H.; Huo, Y.; Wei, X.; Zhu, Y.; Wang, X.; Liao, W. Determining the optimal vegetation coverage for controlling soil erosion in *Cynodon dactylon* grassland in North China. *J. Clean. Prod.* **2020**, *244*, 118771. [CrossRef]

72. Bogunovic, I.; Telak, L.J.; Pereira, P.; Filipovic, V.; Filipovic, L.; Durdevic, B.; Percin, A.; Birkas, M.; Dekemati, I.; Comino, J.R. Land management impacts on soil properties and initial soil erosion processes in olives and vegetable crops. *J. Hydrol. Hydromech.* **2020**, *68*, 328–337. [CrossRef]

73. Klik, A.; Rosner, J. Long-term experience with conservation tillage practices in Austria: Impacts on soil erosion processes. *Soil Till. Res.* **2020**, *203*, 104669. [CrossRef]

74. Nishigaki, T.; Shibata, M.; Sugihara, S.; Mvondo-Ze, A.D.; Araki, S.; Funakawa, S. Effect of mulching with vegetative residues on soil water erosion and water balance in an oxisol cropped by cassava in east cameroon. *Land Degrad. Dev.* **2017**, *28*, 682–690. [CrossRef]

75. Hamza, M.A.; Anderson, W.K. Soil compaction in cropping systems: A review of the nature, causes and possible solutions. *Soil Till. Res.* **2005**, *82*, 121–145. [CrossRef]

76. Cerdà, A.; Keesstra, S.D.; Rodrigo-Comino, J.; Novara, A.; Pereira, P.; Brevik, E.; Giménez-Morera, A.; Fernández-Raga, M.; Pulido, M.; Di Prima, S.; et al. Runoff initiation, soil detachment and connectivity are enhanced as a consequence of vineyards plantations. *J. Environ. Manag.* **2017**, *202*, 268–275. [CrossRef] [PubMed]
77. Haynes, R.J.; Swift, R.S. Stability of soil aggregates in relation to organic constituents and soil water content. *J. Soil Sci.* **1990**, *41*, 73–83. [CrossRef]

Article

Restoring Degraded Landscapes through an Integrated Approach Using Geospatial Technologies in the Context of the Humanitarian Crisis in Cox's Bazar, Bangladesh

Rashed Jalal [1,*], Rajib Mahamud [1], Md. Tanjimul Alam Arif [1], Saimunnahar Ritu [1], Mondal Falgoonee Kumar [1], Bayes Ahmed [2], Md. Humayun Kabir [3], Mohammad Sohal Rana [3], Howlader Nazmul Huda [4], Marco DeGaetano [1], Peter John Agnew [1], Amit Ghosh [1], Fatima Mushtaq [1], Pablo Martín-Ortega [1], Andreas Vollrath [1], Yelena Finegold [1], Gianluca Franceschini [1], Rémi d'Annunzio [1], Inge Jonckheere [1] and Matieu Henry [1]

[1]   Food and Agriculture Organization of the United Nations, 00153 Rome, Italy
[2]   Institute for Risk and Disaster Reduction (IRDR), University College London (UCL), Gower Street, London WC1E 6BT, UK
[3]   Bangladesh Forest Department, Ministry of Environment Forest and Climate Change, Sherebangla Nagar, Agargaon, Dhaka 1207, Bangladesh
[4]   International Organization for Migration, House 13A, Road 136, Gulshan 1, Dhaka 1212, Bangladesh
*   Correspondence: rashed.jalal@fao.org

Citation: Jalal, R.; Mahamud, R.; Arif, M.T.A.; Ritu, S.; Kumar, M.F.; Ahmed, B.; Kabir, M.H.; Rana, M.S.; Huda, H.N.; DeGaetano, M.; et al. Restoring Degraded Landscapes through an Integrated Approach Using Geospatial Technologies in the Context of the Humanitarian Crisis in Cox's Bazar, Bangladesh. *Land* **2023**, *12*, 352. https://doi.org/10.3390/land12020352

Academic Editors: Kleomenis Kalogeropoulos, Andreas Tsatsaris, Nikolaos Stathopoulos, Demetrios E. Tsesmelis, Nilanchal Patel and Xiao Huang

Received: 22 December 2022
Revised: 22 January 2023
Accepted: 23 January 2023
Published: 28 January 2023

Abstract: The influx of nearly a million refugees from Myanmar's Rakhine state to Cox's Bazar, Bangladesh, in August 2017 put significant pressure on the regional landscape leading to land degradation due to biomass removal to provide shelter and fuel energy and posed critical challenges for both host and displaced population. This article emphasizes geospatial applications at different stages of addressing land degradation in Cox's Bazar. A wide range of data and methods were used to delineate land tenure, estimate wood fuel demand and supply, assess land degradation, evaluate land restoration suitability, and monitor restoration activities. The quantitative and spatially explicit information from these geospatial assessments integrated with the technical guidelines for sustainable land management and an adaptive management strategy was critical in enabling a collaborative, multi-disciplinary and evidence-based approach to successfully restoring degraded landscapes in a displacement setting.

Keywords: land degradation; sustainable land management; emergency; earth observation; sustainable development goals; Rohingya

1. Introduction

Productive land is the foundation of global food security and environmental health, zero hunger, poverty eradication and energy for all. However, this finite resource is under continuous threat. Anthropic activities affects more than 70% of the global ice-free land surface [1]. Globally, the biophysical status of 5670 million hectares (ha) of land is declining, of which 1660 million ha (29%) is attributed to human-induced land degradation [2]. With up to 40% of the planet's land degraded, which would reach the size of South America by 2050, land degradation is recognized as one of the significant environmental threats to society, directly affecting half of humanity and threatening roughly half of the global gross domestic product (GDP) [3]. Addressing land degradation through sustainable management of natural resources and socioeconomic development, as well as "strengthen cooperation on desertification, dust storms, land degradation and drought and promote resilience and disaster risk reduction", is recognized in the 2030 Agenda for Sustainable Development [4].

Land degradation is intrinsically related to other environmental challenges, including climate change, loss of biodiversity, and humanitarian crises. In 2020, more than 70 million forcibly displaced persons were scattered globally [5], and the number continues to grow. Since 2008, climate refugees have been growing by more than 20 million people annually [6] and could reach 140–200 million people by 2050 [7]. This expanding humanitarian crisis potentially affects every people and ecosystem, putting refugees as both a cause and victim of environmental and land degradation [8]. Restoring degraded land in and around refugee camps can bring positive externalities to displaced populations and host communities, and is increasingly being taken into consideration given the associated environmental challenges affecting humanitarian settings.

In Bangladesh, one of the most vulnerable countries to climate change [9], the sudden influx of nearly a million Rohingya refugees/Forcibly Displaced Myanmar Nationals (FDMNs) from the Rakhine state, Myanmar in August 2017, made Cox's Bazar (the southernmost coastal hill district of Bangladesh) home to one of the largest refugee settlements in the world. The August 2017 influx, the largest and fastest refugee influx into Bangladesh, has put substantial additional pressure on natural resources and increased already existing challenges to human health, food security, nutrition, water supply and sanitation, shelter, education, access to energy and environmental services, not only for the people displaced but also for their host communities. To address this situation, the major stakeholders have made a joint effort to rehabilitate the degraded landscapes inside and outside the camp area since 2018. As a result, a total of 450 ha (approximately) of degraded areas were brought under different restoration activities by different agencies, and an additional 2000 ha of degraded forestland was maintained jointly with the Bangladesh Forest Department (BFD) of the Government of Bangladesh (GoB).

Assessment and monitoring of restoration projects are often complex due to the challenges related to accessibility, lack of affordable and appropriate methodologies, difficulty in obtaining long-term data and lack of funds, together with capacity constraints in general [10]. Advancements in geospatial and earth observation technology and the availability of higher resolution satellite data have considerable potential in effectively delivering timely, cost-effective, reliable, and homogeneous information. However, only some examples of the use of geospatial technologies to assess restoration interventions are available [11]. In general, there is a lack of evaluation and dissemination of the restoration results, representing a constraint on applying the best technologies and approaches available [12]. There is also a broad consensus on the need for innovative approaches to systematically evaluate the effectiveness of restoration efforts [10,12,13].

In this context, this article aims to present the ongoing geospatial approach that integrated and facilitated different aspects of addressing land degradation in a displacement setting in Cox's Bazar, Bangladesh. The overall approach to land degradation assessment and implementation of restoration activities evolved over time, taking into consideration the availability of data and methods, need, capacity, expertise, and lessons learned. The approach included delineating forest land boundaries, land cover mapping, wood fuel supply and demand assessment, land degradation assessment, preparation of technical specifications and suitability analysis for restoration activities, and implementation and monitoring of restoration activities.

2. Materials and Methods

A wide range of data was collected or prepared using different methods to support the process of addressing land degradation. Table 1 summarizes the list of data used for various applications. After a brief overview of the study area, descriptions of the methods are provided in subsequent sub-sections.

Table 1. Data used in different stages of addressing land degradation.

Data	Accessed from [1]	Date of Data	Application
Areas of interest			
Rohingya refugee camps	HDX	2018	Area of interest
Cox's Bazar south forest division	BFD	1920s (CS sheets), updated and digitized in 2018	Area of interest
Forest land boundaries	BFD	1920s (CS sheets), updated and digitized in 2018	Area of interest and land suitability assessment
Satellite image			
Sentinel 2 images	GEE	2017 to 2019	Land degradation assessment
Landsat 4, 5 and 8 images	GEE	2003 to 2021	Restoration monitoring
Other			
Buildings, roads, water body footprints	HDX	2019	Land suitability assessment (inside the camps)
Protected areas	BFD	2018	Land suitability assessment
Land cover 2015	BFD	2015	Land suitability assessment
Digital elevation model (0.5 m resolution)	IOM-NPM	2019	Land suitability assessment (inside the camps)
SRTM Digital elevation model (30 m resolution)	GEE	2000	Land suitability assessment (outside the camps)
Elephant path	BFD	2016	Land suitability assessment
Restoration activity areas	FAO	2018 and 2019	Restoration monitoring
Wood fuel supply and demand	FAO-IOM	2016 and 2017	Restoration planning

[1] BFD—Bangladesh Forest Department; CS—Cadastral survey; FAO—Food and Agriculture Organization of the United Nations; GEE—Google Earth Engine; HDX—Humanitarian Data Exchange; IOM—International Organization for Migration; NPM—Needs and Population Monitoring.

2.1. Study Area and Context

Cox's Bazar district, with an area of about 2492 km^2, is located between 20°43′ and 21°56′ north latitudes and 91°50′ and 92°23′ east longitudes. The district is located at the fringe of the Bay of Bengal with an unbroken sea beach, the longest one in the world. It is bounded by Chattogram district to the north, Bandarban district and Myanmar to the east, and the Bay of Bengal to the south and west (Figure 1).

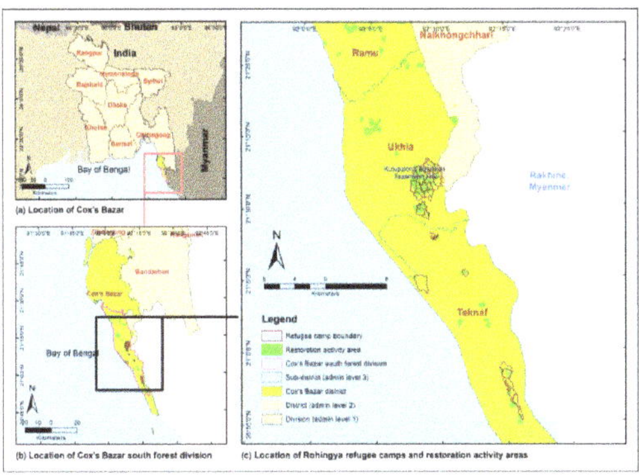

Figure 1. Location of Cox's Bazar district (**a**), Cox's Bazar south forest division (**b**), refugee camps and areas of restoration activities (**c**).

As of October 2022, over 943,000 Rohingya refugees/FDMNs reside in the Ukhiya and Teknaf sub-districts [14] in Cox's Bazar south forest division, an administrative area for the management of forest land by the BFD. The Cox's Bazar south forest division covers a significant part of the hill forests of the country, representing features of tropical evergreen and semi-evergreen forests, and has one of the most species-rich and productive reserve forests. However, these natural resources are becoming degraded through illegal logging, encroachment, hill-cutting, forest fires, shifting cultivation, human settlement, agriculture and horticulture expansion, and clear-felling followed by commercial plantation with short rotation of exotic species [15,16].

Such prevailing land degradation dynamics in the area are further exacerbated by fluctuating but persistent arrivals of Rohingya refugees/FDMNs, with a massive influx of around 742,000 Rohingya refugees/FDMNs since 25 August 2017 [5]. The vast majority live in 34 extremely congested refugee camps, including the largest single site, the Kutupalong-Balukhali Expansion Site, which accommodates more than 635,000 Rohingya refugees/FDMNs [14]. All the camps are in Cox's Bazar south forest division. This put significant pressure on the regional landscape resulting from removal of trees roots and cover grass to provide shelter and fuel for this forcefully displaced population. Figure 2 depicts the loss of vegetation from February 2017 to February 2018 due to the expansion of the Kutupalong-Balukhali refugee camp.

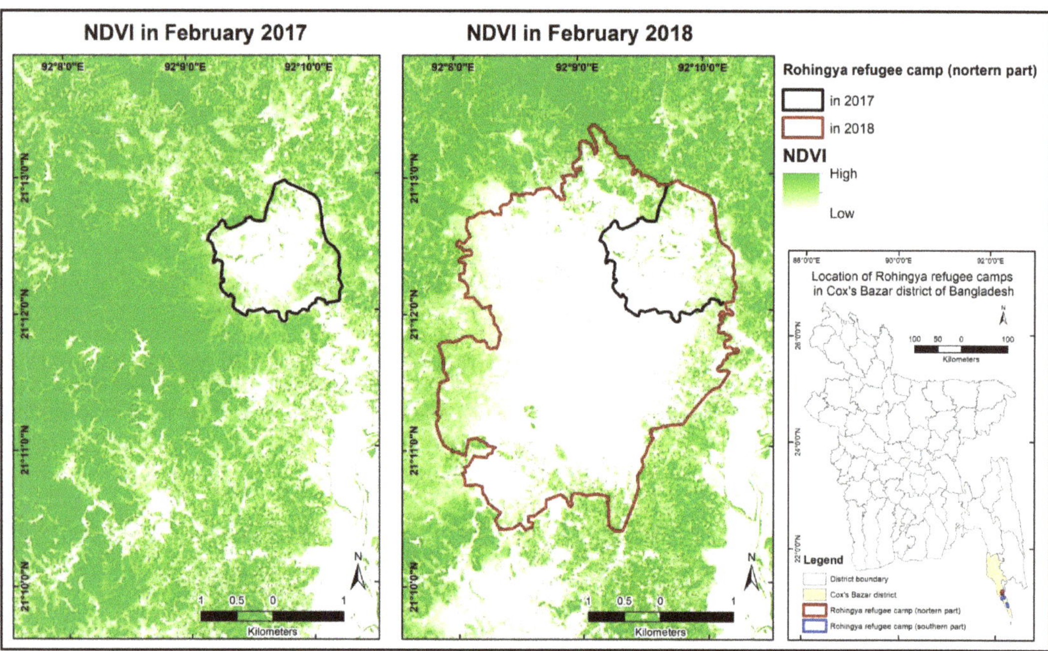

Figure 2. Change of vegetation between 2017 and 2018 as depicted by a decreased normalized difference vegetation index (NDVI). A lower NDVI means less vegetation cover.

In response, various national and international agencies coordinated by the Energy and Environment Technical Working Group (EETWG), in close collaboration with the BFD and the local host communities and Rohingya refugees/FDMNs, have been working together to implement an integrated land restoration approach to rehabilitate the degraded lands inside and outside the camp area since 2018. This article considered restoration activities in about 531 ha of land for which geographic boundaries were available. Table 2 presents the distribution of different restoration activities' areas, times, and locations.

Table 2. Area distribution and location of different restoration activities.

Type of Activity	Activity Started	Location	Area (ha)
Forest restoration	2018	Outside refugee camp	0.36
		Inside refugee camp	27.51
	2019	Outside refugee camp	298.23
		Inside refugee camp	66.53
Land stabilization	2018	Inside refugee camp	9.33
	2019	Outside refugee camp	0.78
		Inside refugee camp	4.96
Reforestation	2018	Outside refugee camp	11.36
		Inside refugee camp	33.97
	2019	Outside refugee camp	38.08
		Inside refugee camp	40.12
		Total	531.23

2.2. Delineation of Forest Land Boundaries and Land Cover Mapping

The forest land boundaries of Cox's Bazar south forest division were delineated from available cadastral survey (CS) sheets. The CS sheets were scanned and geo-referenced using differential global positioning systems (DGPS). The geo-referenced images were digitized to prepare the GIS layer and were further corrected using IKONOS (acquired in 2012), RapidEye (acquired in 2012) and IRS PAN (acquired in 2004) satellite images of the Cox's Bazar south forest division available in the BFD archive. The edges of each sheet map were matched with adjacent ones, and positional accuracy was compared with reference points collected from the field using real-time kinematic (RTK) positioning [17].

The 2015 national land cover map [18] was used as the baseline land information. It was developed using multi-spectral ortho (Level 3) SPOT6/7 four-band images of 6-m spatial resolution with a maximum of 10% cloud coverage. An object-based image analysis (OBIA) approach was adopted to create image objects, followed by a visual image interpretation technique to classify land cover. The overall accuracy of the 2015 national land cover map was estimated at 89% [19].

2.3. Wood Fuel Supply and Demand Assessment

An assessment of wood fuel supply and demand was conducted in 2016 and was later updated in 2017 [20] after the August 2017 influx. The assessment combined field and remote sensing data following the recommended approach by d'Annunzio, R. et al. [21] for assessing wood fuel supply and demand in displacement settings. The process included an assessment of standing woody biomass available for use as fuel (fuel wood supply), the changes they had undergone over a given period, consumption over the same time (assuming wood fuel consumption is equal to wood fuel demand) and the gap between demand and supply.

The assessment of supply was performed by combining field measurements for above-ground biomass stock with land cover changes based on historical satellite image time series analysis. For assessing the biomass stock in close proximity of refugee camps, samples were taken randomly from different land covers (based on the 2015 land cover map), having the potential for supplying wood fuel. A total of 15 plots were measured. The plot design and wood fuel assessment followed the same procedures as in the Bangladesh Forest Inventory [22], where each sample plot consists of 5 sub-plots (1 in the center and the remaining four in four cardinal directions), each with a radius of 19 meters. For demand assessment, household interviews, participatory rural appraisals (PRA) and focused group discussions (FGDs) were conducted to assess the fuel wood energy consumption of varying social units inhabiting the area.

2.4. Land Degradation Assessment

Land degradation mapping was performed employing a before and after land cover change analysis using Sentinel 2 multispectral 10 m images for February 2017 and 2018. Five broad land cover classes (i.e., water, settlement, bare land, sparse vegetation and dense vegetation) were delineated for 2017 and 2018 based on NDVI thresholds defined through expert judgements. Different levels of land degradation were identified throughout Cox's Bazar south forest division based on the land cover change from 2017 to 2018 as follows:

- High degradation: if the dense vegetation class in February 2017 was converted to bare land, settlement or water in February 2018.
- Medium degradation: if the sparse vegetation class in February 2017 was converted to bare land, settlement or water in February 2018.
- Low degradation: if the dense vegetation class in February 2017 was converted to sparse vegetation in February 2018.

2.5. Technical Specifications for Restoration Activities

Technical specifications for the restoration activities were prepared using the World Overview of Conservation Approaches and Technologies (WOCAT) and documented in a living report [23] in consultation with experts and stakeholders. Particular attention was paid to the sustainability of plantation activities, considering aspects such as the supply of seedlings from nurseries, vegetation growth, vegetation layers, inputs and workforce, as well as ensuring the maintenance of plant biodiversity and promoting the use of native species.

2.6. Suitability Analysis for Restoration Activities

Various spatial data (e.g., land cover, forest land boundary, slope, altitude, roads, river, elephant path, flood risk due to low elevation and protected areas) were integrated with land degradation to perform the land suitability assessment based on criteria identified in consultation with local and national experts to identify potential and priority areas for restoration. The process is continuously being updated as more data become available.

2.7. Implementation of Restoration Activities

A collaborative process was established by the Energy Environmental Technical Working Group (EETWG) and the Inter Sector Coordination Group (ISCG) to support the coordination, planning and implementation of restoration activities inside the camps. The site management and site development (SMSD) team assisted in overall camp planning and management. The BFD played a crucial role in providing technical guidance in the design and implementation process (e.g., plant selection, plantation management, and logistics) of restoration activities inside and outside the camps.

In coordination with EETWG, implementing partner organizations mapped the available areas for restoration interventions in respective camps. Field area mapping was conducted using GPS. Plantation targets were set depending on the budget and human resources available. Specific areas were allocated for each organization, and documents (including maps) were maintained to avoid overlapping and gaps. The degradation map and potential restoration areas were verified on the ground jointly by relevant stakeholders and implementing agencies and approved by the authorized government departments before the initiation of restoration works. Considering the importance of quality planting materials for the success of any restoration initiative, plant nurseries were developed with support from the BFD and the host communities around the camps.

2.8. Monitoring Restoration Activities

The productivity state, one of the three metrics for calculating the land productivity sub-indicator for the sustainable development goals (SDG) indicator 15.3.1—Proportion of land that is degraded over total land area [24], was used to assess the performance of restoration activities. The productivity state in the monitoring periods can be calculated from the 16 most recent years of annual net primary productivity (NPP) of vegetation data up to and including the most recent year in the monitoring period. The mean of the most recent three years is compared to the distribution of annual NPP values in the preceding 13 years. For calculating the land productivity sub-indicator and reporting on the SDG indicator 15.3.1, the good practice guidance for the SDG indicator 15.3.1 [24] recommended that only the areas of the lowest negative Z score (<-1.96) be considered as degraded and other areas as not degraded.

In this study, the productivity states of two monitoring periods of 2019 to 2021 and 2016 to 2018 were calculated by comparing the mean annual normalized difference vegetation index (NDVI), as a proxy of NPP, to the distribution of annual NDVI values observed in 2003 to 2015. A fixed baseline period of 2003 to 2015 was used to facilitate a direct comparison of the Z scores between the monitoring periods. Annual NDVI estimates for the restoration areas were retrieved from Landsat 4, 5 and 8. Productivity states were calculated as follows: calculate the mean (μ) and standard deviation (σ) of the annual NDVI estimates from 2003 to 2015 inclusive (Equations (1) and (2), respectively), calculate the means of the yearly NDVI estimates for the monitoring periods (Equation (3)) and calculate the Z statistics for the monitoring periods (Equation (4)).

$$\mu = \frac{\sum_{2003}^{2015} x}{13} \tag{1}$$

$$\sigma = \sqrt{\frac{\sum_{2003}^{2015} (x - \mu)^2}{13}} \tag{2}$$

$$\bar{x} = \frac{\sum_{y-2}^{y} x}{3} \tag{3}$$

$$z = \frac{\bar{x} - \mu}{\sigma/\sqrt{3}} \tag{4}$$

where x is the annual NDVI and y is the end year of the monitoring period. Pixel-wise differences of Z scores between 2019 to 2021 and 2016 to 2018 (subtracting the Z score of 2016 to 2018 from the Z score of 2019 to 2021) were calculated. Hence, positive difference indicates improvement of state in 2019 to 2021 compared with 2016 to 2018 and vice versa. t-tests were conducted to assess statistical significances and effect sizes [25] of the changes and directions in productivity states considering different aspects (i.e., type, time and location) of restoration activities. Z scores were calculated using the SDG 15.3.1 module of SEPAL (https://sepal.io/, accessed on 20 January 2023), and t-tests were performed using R.

3. Results

3.1. Forest Land Delineation

From the forest land delineation, 67,692 ha of land boundary in Cox's Bazar south forest division was demarcated, of which 42,686 ha (63%) were forest land under the management of the BFD. 35,801 ha of the forest land were identified as reserved and 6884 ha as protected forest. Of the 2639 ha of refugee camp area, about 75% was in the forest land (69% reserved forest and 6% protected forest), as shown in Figure 3.

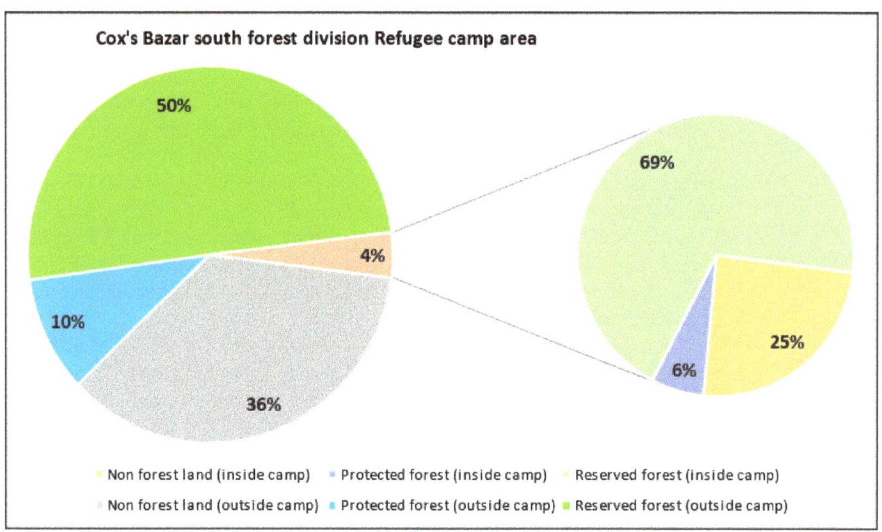

Figure 3. Distribution of forest land area with Cox's Bazar south forest division and refugee camp area.

3.2. Wood Fuel Supply and Demand

The fuelwood demand assessment revealed a six-fold increase in fuelwood demand, from 54,542 tons per year in 2016 to 312,807 tons per year in 2017 (estimated from the total number of refugee households), while the entire available stock was estimated as 331,266 tons of dry biomass. This revealed immense pressure on existing forest resources, indicating a complete loss of forestlands if the land degradation continued for a few years.

3.3. Land Degradation

In total, 7220 ha of land (about 11% of the total area) were degraded within one year, of which about 74% was BFD forest land (i.e., protected or reserved forest). BFD forest lands, especially within and near the refugee camp area, were highly impacted by different levels of degradation. As shown in Table 3, about 494 ha and 836 ha of land inside the camp were high and medium degraded, of which 99% and 90% were BFD forest land, respectively. Land degradation maps [26–28] were prepared and published for wider dissemination and sensitization among the stakeholders.

Table 3. Land degradation in Cox's Bazar south forest division (areas are in ha).

	Forest Type	Degradation			Other (Enhancement or No Change)	Total
		High	Medium	Low		
	Non forest land	4	79	22	543	648
Inside camp	Protected forest	5	56	4	101	165
	Reserved forest	484	701	25	616	1826
	Non forest land	9	107	42	1533	1691
within 1 km from camp boundary	Protected forest	2	32	8	220	262
	Reserved forest	315	411	349	2684	3760

Table 3. *Cont.*

	Forest Type	Degradation			Other (Enhancement or No Change)	Total
		High	Medium	Low		
within 1–5 km from camp boundary	Non forest land	33	348	202	7742	8325
	Protected forest	3	64	38	1133	1237
	Reserved forest	173	536	930	14,484	16,122
5 km further from the camp boundary	Non forest land	48	656	300	13,338	14,342
	Protected forest	12	200	166	4843	5220
	Reserved forest	38	279	542	13,235	14,094
Total		1127	3468	2625	60,472	67,692

3.4. Technical Specifications for Restoration Activities

The technical specifications [23] were prepared in consultation with stakeholders and experts. They were followed to avoid unplanned activities, protect plant biodiversity, allocate resources efficiently and improve enabling conditions to implement landscape restoration in Cox's Bazar south forest division. The activities and technical specifications were updated over time based on experiences, lessons learned and feedback from the different national and international agencies involved in landscape restoration activities. Figure 4 illustrates a representative example of schematic technical specifications and implementation of land stabilization activity on the ground. Specifically, the example demonstrates the use of multiple vegetation layers, including long-rooted grass species to stabilize topsoil, leguminous shrubs for increasing soil fertility, bamboo as living reinforcement on vulnerable slopes and fast-growing tree species for quick vegetation cover to reduce the risk of rainfall-induced landslides.

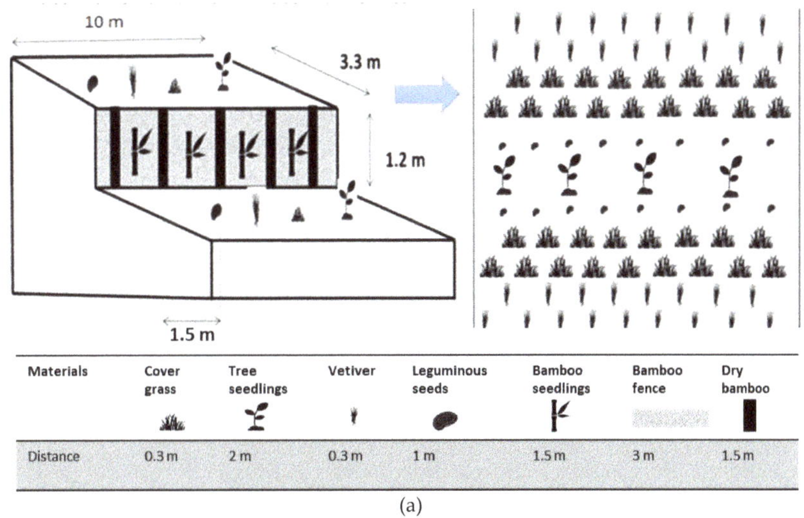

Figure 4. *Cont.*

(b)

Figure 4. Schematic technical specification (**a**) and on-the-ground implementation of land stabilization activity (**b**) using multiple vegetation layers to reduce the risk of rainfall-induced landslides and facilitate post-landslide slope rehabilitation (photo credit: Saikat Mazumder, FAO).

3.5. Suitability Analysis for Restoration Activities

Suitable areas for landscape restoration were identified and published [29–31]. Depending on the emerging needs and updated data, the approach is under continuous revision. For instance, settlement footprints inside the camps and high-resolution (0.5 m) digital elevation models (DEM) were used for mapping suitable areas in 2019 inside the camp and new activities such as riparian plantation and roadside plantation were added, which were not included in the 2018 restoration plan. Table 4 presents the criteria for land suitability analysis for restoration activities inside the camps in 2019. Figure 5 shows an example map of a land restoration plan for a camp in 2019.

Table 4. Criteria for land suitability analysis inside the camps in 2019.

Criteria	Restoration Activities
Bare land in January 2019 and high slope ($\geq$30°)	Land stabilization (biological and mechanical)
Bare land in January 2019 and low slope (<30°)	Land stabilization (biological)
Sparse vegetation in January 2019 and non-forest in 2015	Afforestation/reforestation
Sparse vegetation in January 2019 and forest in 2015	Forest restoration
Dense vegetation in January 2019	Maintenance and protection
Land within 5 meters from rivers and streams and within 1 meter from other water bodies	Land under use (waterside)
Land within 1 meter from settlements and roads	Land under use (waterside)
Land within 2 meters from the land under use (roadside)	Roadside plantation
Land within the 5 meters from the land under use (waterside)	Riparian plantation
Plantation in 2018	Plantation in 2018

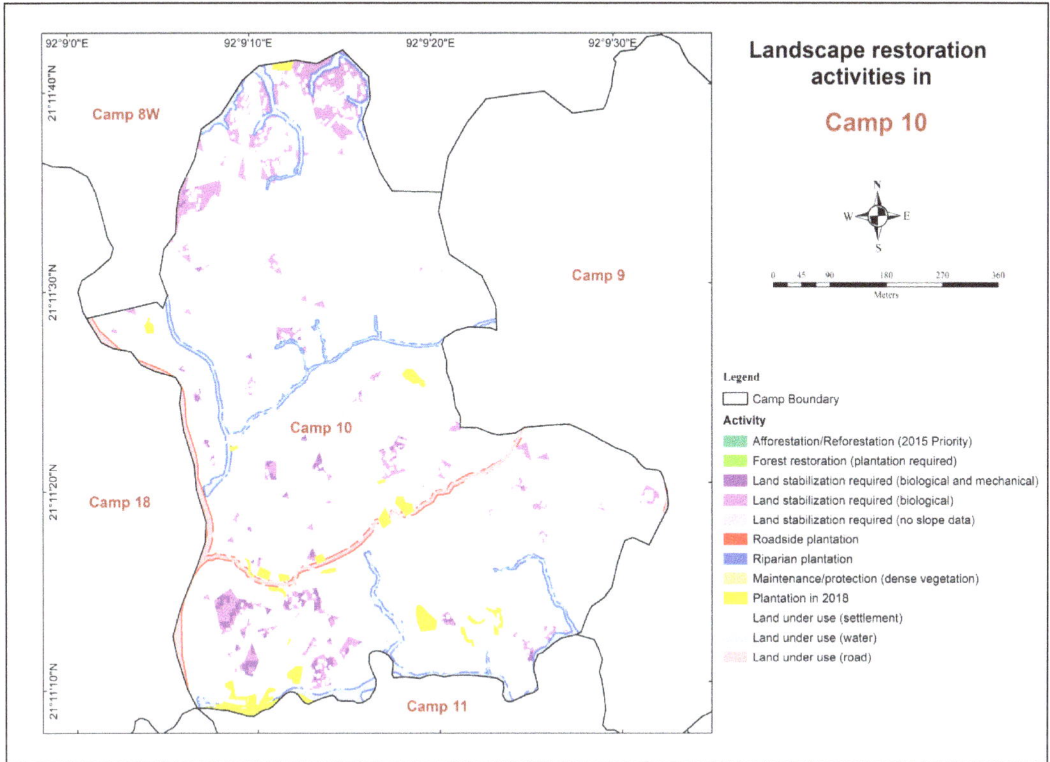

Figure 5. Land restoration plan map for Camp 10 in 2019.

3.6. Implementation of Restoration Activities

The major stakeholders have made a joint effort to engage about twenty organizations to rehabilitate the degraded forestlands inside and outside the camp area since 2018. Approximately 450 ha of degraded areas have been brought under different restoration activities by various agencies across 34 camps, involving more than 100,000 person-days. While Rohingya refugees/FDMNs were engaged inside the camps, such as in site preparation, plantation management and camp maintenance, the host communities carried out activities outside the camps, and 66 nurseries were brought into a nursery management team (NMT). An additional 2000 ha of forestland and critical watershed areas outside the camp area have been subject to restoration activities in 2020.

3.7. Monitoring Restoration Activities

Overall, there was a significant increase in the productivity state in 2019 to 2021 compared to 2016 to 2018, with a small effect size for all restoration areas ($t = 16.9$, $p < 0.001$, $d = 0.22$). However, mixed performances were observed for restoration areas when disaggregated by type, location and year of activity, as presented in Table 5. For restoration activities outside the camp area (about 66% of the total), significant positive change was observed with a moderate effect size ($t = 41.6$, $p < 0.001$, $d = 0.65$). However, in the remaining areas (inside the camp), a significant decrease in productivity state with a small effect size was found ($t = -15.5$, $p < 0.001$, $d = -0.35$). For the restoration activities in 2018, the productivity state decreased significantly; however, the effect size was trivial ($t = -4.8$, $p < 0.001$, $d = -0.16$). A significant and small increase in productivity state was noted for restoration works of 2019 (representing about 84% of the restoration area). Considering the type of restoration, forest restoration and land stabilization works (representing approximately 77% of the restoration area) were found to have significant positive impacts. In contrast, there was a significant negative impact on reforestation areas with a moderate effect size (Table 5).

Table 5. Results from *t*-tests.

Restoration Area	Area (ha)	df	*t* Statistic	*p* Value	Effect Size (Cohen's *d*)	Descriptor [1]
All	531	6106	16.9	$p < 0.001$	0.22	Significant and small increase
Location						
Inside camp	182	1993	−15.5	$p < 0.001$	−0.35	Significant and small decrease
Outside camp	349	4142	41.6	$p < 0.001$	0.65	Significant and moderate increase
Year						
2018	83	892	−4.8	$p < 0.001$	−0.16	Significant and trivial decrease
2019	449	5225	21.6	$p < 0.001$	0.30	Significant and small increase
Type						
Forest restoration	393	4564	41.3	$p < 0.001$	0.61	Significant and moderate increase
Land stabilization	15	154	3.5	$p < 0.001$	0.28	Significant and small increase
Reforestation	124	1386	−22.1	$p < 0.001$	−0.59	Significant and moderate decrease

[1] Effect size is labelled as trivial, small, moderate and large when $|d| < 0.2$, $0.2 \leq |d| < 0.5$, $0.5 \leq |d| < 0.8$, $|d| \geq 0.8$, respectively.

4. Discussion

There were several rapid assessments of environmental impacts following the influx in August 2017. However, they differ in methodology, data, timeframe and area of interest; hence the results are not directly comparable. For instance, one rapid environmental assessment study [32] identified about 1500 ha of forest land encroachment due to camp establishment up to November 2017. However, the assessment was semi-quantitative, and the results were not spatially explicit. Another study used satellite imagery and applied machine learning algorithms to quantity expansion of refugee settlements and estimated about 2283 ha of degradation in forest areas [33]. The study provided spatially explicit information on forest cover degradation without consideration of legal forest land boundaries.

Overall, restoration activities showed significant small to moderate improvement in at least 66% of restored areas. A global meta-analysis of 221 studies revealed an improvement of vegetation structure between 36% and 77% by forest restoration compared with degraded ecosystems [34]. Most of the gains were found outside the camp, where long-term care and maintenance were comparatively easier to provide under the jurisdiction of the BFD. The major challenge in sustaining restoration activities within the camp areas was associated with the relocation of refugee settlements inside the camps converting restored areas into other land use. People's high dependency on forest resources (for energy and shelter, etc.) and management initiatives (e.g., species selection, plantation method, collaboration and protection, etc.) along with limited funding and lack of priority for long term maintenance also had direct implications for the success of restoration activities. These factors were more pronounced during the initial stage of the restoration activities. They could be attributed as the driving factors for not attaining the results as expected for restoration activities in 2018. Such landscape context is also identified as one of the major driving factors of restoration success globally [34]. Accordingly, restoration activities in 2019 were more organized, considering the lessons learned from restoration in 2018. A satisfactory sapling survival and growth rate was observed for plantations in 2019 in a field-based plantation assessment [35] which supports the results of this study.

The initial humanitarian responses to land degradation in the area were spontaneous, mainly due to a need for more informed decision-making and collaboration. Over the past years, an integrated approach evolved to reverse the degradation of hundreds of hectares of land. Recognizing the emergency need and the underlying driving factor for land degradation (i.e., land cover changes due to rapid vegetation loss in this case), a simple, intuitive and easy-to-implement method for land degradation assessment, based on evaluation of NDVI dynamics and land cover change, was found effective not only in identifying the degraded land but also in informing the stakeholders about the magnitude and immediate need to respond.

Integrating land tenure information with restoration planning and implementation is critical for achieving restoration outcomes, including land degradation neutrality [36]. In the context of Cox's Bazar, the integration of spatially explicit land degradation information with land tenure (i.e., legal forest land boundary) facilitated the identification and prioritization of land for restoration and other interventions through a geospatial suitability analysis. However, in the early stage of restoration works in 2018, there were instances of scattered plantation interventions with little consideration of the local context. This aggravated the crisis by contributing to conflict. A collaborative, inclusive, evidence-based approach recognizing the direct land users, prevailing complex socioecological system, and land ownership were necessary for success.

Understanding the supply and demand of fuel wood was critical for evidence-based decision-making (e.g., what stocking rates are required for successful reforestation efforts). The assessment of fuelwood supply and demand was necessary, in this regard, to determine sustainable and optimum energy supply from plantations, considering the reduction of demand due to improved energy consumption (e.g., improved cooking arrangements) and/or the use of alternative energy sources (e.g., solar energy and LPG, etc.). Notably, the immense pressure on forest resources revealed by the updated assessment was critical in mobilizing and sensitizing key stakeholders to take immediate action to prevent the complete loss of forest resources. The assessment was also essential for raising awareness of the importance of safe access to fuel and energy.

Preparing technical specifications and guided implementation helped the implementing agencies avoid unplanned activities, protect plant biodiversity and allocate resources efficiently. Stakeholder involvement at every stage of land restoration was essential, requiring substantial coordination between local authorities, community leaders, United Nations agencies, non-governmental organizations and other partner organizations. The collaborative implementation of restoration activities, with due attention to different and sometimes conflicting stakeholders' interests, was critical to the program's success.

Land degradation in an area due to rapid loss of vegetation could be easily detected through remote sensing technologies. Usually, assessing the success of restoration interventions requires long-term records [11], considering the time needed for restoration work to take effect on the ground that can be detected by remote sensing. The time elapsed since restoration began is one of the main drivers of forest restoration success [34]. The restoration works in Cox's Bazar started in less than five years, making it more challenging to assess. In this regard, the approach adopted by comparing the Z scores for the productivity state metric, being sensitive to the recent magnitude and direction of change [24], was found relevant and can be used in a similar context. In general, remote sensing and geospatial analyses were adequate for the preliminary assessment of degradation and the identification and prioritization of suitable restoration areas. However, the results needed to be validated by ground observation before commencing restoration activities. Given the emergency nature of the problem, the approach took a practical and intuitive approach, which was later enhanced with additional data, information and capacity.

5. Conclusions

The unprecedented land degradation in Cox's Bazar required an urgent response to restore degraded landscapes. The longer the degradation persisted, the more difficult it would be to restore the landscape. In the most severely affected areas, there was a substantial risk of irreversible damage, with a total loss of large tracts of forest. There had also been scattered plantation interventions with little consideration of the local context, which aggravated the crisis by contributing to conflict. Implementing and sustaining restoration activities in such an emergency displacement setting was challenging. It required an integrated, multidisciplinary and collaborative approach that considers the entire landscape, including the people and ecosystems it contains.

Within less than five years of restoration work, remote sensing analysis revealed significant positive impacts in most areas brought under restoration activities. The availability of timely information on the status of wood fuel supply and demand, spatially explicit information on land degradation and land tenure, the identification and prioritization of suitable land for restoration interventions, and a collaborative and inclusive approach to implementation were necessary preconditions for such success. Areas where restoration works did not perform as expected were identified along with possible drivers. These driving factors were carefully considered for more effective continuation of the ongoing efforts.

Over the last five years, in the transition from emergency to resiliency in a protracted displacement setting, the approach to address land degradation and manage restoration activities is continuously being updated as more data, methodologies, and capacities become available. Recognizing that every challenge is unique, the integrated approach adopted in Cox's Bazar could be applied–with proper contextualization–in similar displacement settings.

Author Contributions: Conceptualization, R.J., R.M., M.T.A.A., S.R., M.F.K. and M.H.; methodology, R.J., R.M., B.A., A.G., P.M.-O., A.V., Y.F., R.d. and M.H.; software, R.J. and A.G.; validation, R.J. and A.G.; formal analysis, R.J. and A.G.; investigation, R.M., M.T.A.A., S.R., M.F.K., M.H.K., M.S.R. and H.N.H.; data curation, R.M., M.T.A.A., S.R., M.F.K. and H.N.H.; writing—original draft preparation, R.J., R.M., M.T.A.A., S.R., M.F.K., A.G., F.M., P.M.-O., A.V., Y.F., G.F., I.J., M.H., B.A. and H.N.H.; writing—review and editing, R.J., R.M., M.T.A.A., S.R., M.F.K., M.D., P.J.A., A.G., F.M., P.M.-O., A.V., Y.F., G.F., R.d., I.J., M.H., B.A., M.H.K., M.S.R. and H.N.H.; visualization, R.J., S.R., A.G. and F.M.; supervision, I.J. and M.H.; project administration, M.D., P.J.A. and M.H.; funding acquisition, M.D., P.J.A. and M.H. All authors have read and agreed to the published version of the manuscript.

Funding: This research received no external funding.

Data Availability Statement: Not applicable.

Acknowledgments: The authors would like to thank the anonymous reviewers for their comments that helped improve the manuscript. The authors are grateful to the Rohingya refugees and host communities in Cox's Bazar, the relevant officials of Arannyak Foundation (AF), Bangladesh Forest Department (BFD), Bangladesh Rural Advancement Committee (BRAC), Center for Natural Resource Studies (CNRS), Food and Agriculture Organization of the United Nations (FAO), Helvitas, International Organization for Migration (IOM), International Union for Conservation of Nature (IUCN), office of Refugee Relief and Repatriation Commissioner (RRRC), Shushilon, United Nations High Commissioner for Refugees (UNHCR), United Nations Children's Fund (UNICEF) and World Food Programme (WFP) for their support during different stages of the work.

Conflicts of Interest: The authors declare no conflict of interest.

References

1. IPCC. *Climate Change and Land: An IPCC Special Report on Climate Change, Desertification, Land Degradation, Sustainable Land Management, Food Security, and Greenhouse Gas Fluxes in Terrestrial Ecosystems*; Shukla, P.R., Skea, J., Calvo Buendia, E., Masson-Delmotte, V., Pörtner, H.-O., Roberts, D.C., Zhai, P., Slade, R., Connors, S., van Diemen, R., et al., Eds.; The Intergovernmental Panel on Climate Change: Geneva, Switzerland, 2019.
2. FAO. *The State of the World's Land and Water Resources for Food and Agriculture—Systems at Breaking Point*; Synthesis Report 2021; FAO: Rome, Italy, 2021; p. 82.
3. UNCCD. *Land Restoration for Recovery and Resilience, in Global Land Outlook 2*; UNCCD: Bonn, Germany, 2022; p. 176.
4. Assembly, U.G. *Transforming Our World: The 2030 Agenda for Sustainable Development, 21 October 2015, A/RES/70/1*; United Nation: New York, NY, USA, 2015; p. 35.
5. UNHCR. *Global Trends: Forced Displacement in 2020*; UNHCR: Geneva, Switzerland, 2021; p. 71.

6. IDMC. *Global Report on Internal Displacement*; IDMC: Geneva, Switzerland, 2016; p. 108.

7. Rigaud, K.K.; de Sherbinin, A.; Jones, B.; Bergmann, J.; Clement, V.; Ober, K.; Schewe, J.; Adamo, S.; McCusker, B.; Heuser, S.; et al. *Groundswell: Preparing for Internal Climate Migration*; The World Bank: Washington, DC, USA, 2018; p. 256.

8. Suhrke, A. *Pressure Points: Environmental Degradation, Migration and Conflict*; University of Toronto: Toronto, ON, Canada; American Academy of Arts and Sciences: Cambridge, MA, USA, 1993.

9. MoEF. *Bangladesh Climate Change Strategy and Action Plan 2009*; Ministry of Environment and Forests (MoEF), Government of the People's Republic of Bangladesh: Dhaka, Bangladesh, 2009; p. 76.

10. Ntshotsho, P.; Esler, K.J.; Reyers, B. Identifying Challenges to Building an Evidence Base for Restoration Practice. *Sustainability* **2015**, *7*, 15871–15881. [CrossRef]

11. Meroni, M.; Schucknecht, A.; Fasbender, D.; Rembold, F.; Fava, F.; Mauclaire, M.; Goffner, D.; Di Lucchio, L.M.; Leonardi, U. Remote sensing monitoring of land restoration interventions in semi-arid environments with a before–after control-impact statistical design. *Int. J. Appl. Earth Obs. Geoinf.* **2017**, *59*, 42–52. [PubMed]

12. Bautista, S.; Aronson, J.; Vallego, R.V. *Land Restoration to Combat Desertification: Innovative Approaches, Quality Control and Project Evaluation*; Fundación Centro de Estudios Ambientales del Mediterráneo—CEAM: Valencia, Spain, 2010.

13. Benayas, J.M.R.; Newton, A.C.; Diaz, A.; Bullock, J.M. Enhancement of Biodiversity and Ecosystem Services by Ecological Restoration: A Meta-Analysis. *Science* **2009**, *325*, 1121–1124. [CrossRef] [PubMed]

14. UNOCHA. Rohingya Refugee Crisis. 2022. Available online: https://www.unocha.org/rohingya-refugee-crisis (accessed on 13 December 2022).

15. Chowdhury, J.A. *Towards Better Forest Management*; Oitijjhya: Dhaka, Bangladesh, 2006; pp. 175–188. (In Bengali)

16. Hossain, M.K. *Silviculture of Plantation Trees of Bangladesh*; Arannayk Foundation: Dhaka, Bangladesh, 2015.

17. BFD and FAO. *Forest Land Boundary Delineation of Cox's Bazar North and South Forest Division from CS Mouza Sheets*; Bangladesh Forest Department and Food and Agriculture Organization of the United Nations: Dhaka, Bangladesh, 2018.

18. GoB. *Land Cover Atlas of Bangladesh 2015 (in Support of REDD+)*; Forest Department, Ministry of Environment, Forest and Climate Change, Government of the People's Republic of Bangladesh: Dhaka, Bangladesh, 2019; p. 161.

19. Jalal, R.; Iqbal, Z.; Henry, M.; Franceschini, G.; Islam, M.S.; Akhter, M.; Khan, Z.T.; Hadi, M.A.; Hossain, M.A.; Mahboob, M.G.; et al. Toward Efficient Land Cover Mapping: An Overview of the National Land Representation System and Land Cover Map 2015 of Bangladesh. *IEEE J. Sel. Top. Appl. Earth Obs. Remote Sens.* **2019**, *12*, 3852–3861. [CrossRef]

20. IOM and FAO. *Assessment of Fuel Wood Supply and Demand in Displacement Settings and Surrounding Areas in Cox's Bazaar District*; BFD and FAO: Dhaka, Bangladesh, 2017; p. 96.

21. Thulstrup, A.W.; Henry, M.; D'Annunzio, R.; Gianvenuti, A. *Assessing Woodfuel Supply and Demand in Displacement Settings: A Technical Handbook*; Food and Agriculture Organization of the United Nations and United Nations High Commissioner for Refugees FAO: Rome, Italic, 2016.

22. Henry, M.; Iqbal, Z.; Johnson, K.; Akhter, M.; Costello, L.; Scott, C.; Jalal, R.; Hossain, A.; Chakma, N.; Kuegler, O.; et al. A multi-purpose National Forest Inventory in Bangladesh: Design, operationalisation and key results. *For. Ecosyst.* **2021**, *8*, 12. [CrossRef]

23. Mahamud, R.; Arif, T.; Jalal, R.; Billah, M.; Rahman, L.M.; Rahman, T.; Iqbal, Z.; Henry, M. *Technical Specification for Land Stabilization and Greening the Camp in Cox's Bazar South Forest Division*; Bangladesh Forest Department: Dhaka, Bangladesh, 2018; p. 52.

24. Sims, N.C.; Newnham, G.J.; England, J.R.; Guerschman, J.; Cox, S.J.D.; Roxburgh, S.H.; Viscarra Rossel, R.A.; Fritz, S.W.I.; Wheeler, I. *Good Practice Guidance. SDG Indicator 15.3.1, Proportion of Land That Is Degraded Over Total Land Area. Version 2.0*; United Nations Convention to Combat Desertification: Bonn, Germany, 2021.

25. Cohen, J. *CHAPTER 2—The t Test for Means, in Statistical Power Analysis for the Behavioral Sciences*; Cohen, J., Ed.; Academic Press: Cambridge, MA, USA, 1988; pp. 19–74.

26. Ritu, S.; Jalal, R.; Henry, M. *Land Degradation in the Refugee Camps of Northern Part of Cox's Bazar South Forest Division*; FAO Publication, Food and Agriculture Organization of the United Nations and Bangladesh Forest Department: Dhaka, Bangladesh, 2018.

27. Ritu, S.; Jalal, R.; Henry, M. *Land Degradation in the Refugee Camps of Middle Part of Cox's Bazar South Forest Division*; FAO Publication, Food and Agriculture Organization of the United Nations and Bangladesh Forest Department: Dhaka, Bangladesh, 2018.

28. Ritu, S.; Jalal, R.; Henry, M. *Land Degradation in the Refugee Camps of Southern Part of Cox's Bazar South Forest Division*; FAO Publication, Food and Agriculture Organization of the United Nations and Bangladesh Forest Department: Dhaka, Bangladesh, 2018.

29. Ritu, S.; Jalal, R.; Henry, M. *Potential Areas of Landscape Restoration Interventions within and around Refugee Camps of Southern Part of Cox's Bazar South Forest Division*; Food and Agriculture Organization of the United Nations and Bangladesh Forest Department: Dhaka, Bangladesh, 2018.

30. Ritu, S.; Jalal, R.; Henry, M. *Potential Areas of Landscape Restoration Interventions within and around Refugee Camps of Northern part of Cox's Bazar South Forest Division*; Food and Agriculture Organization of the United Nations and Bangladesh Forest Department: Dhaka, Bangladesh, 2018.

31. Ritu, S.; Jalal, R.; Henry, M. *Potential Areas of Landscape Restoration Interventions within and around Refugee Camps of Middle Part of Cox's Bazar South Forest Division*; Food and Agriculture Organization of the United Nations and Bangladesh Forest Department: Dhaka, Bangladesh, 2018.

32. UNDP and UN WOMEN. *Report on Environmental Impact of Rohingya Influx*; UNDP and UN WOMEN: Dhaka, Bangladesh, 2018; p. 106.
33. Hassan, M.M.; Smith, A.C.; Walker, K.; Rahman, M.K.; Southworth, J. Rohingya Refugee Crisis and Forest Cover Change in Teknaf, Bangladesh. *Remote Sens.* **2018**, *10*, 689. [CrossRef]
34. Crouzeilles, R.; Curran, M.; Ferreira, M.S.; Lindenmayer, D.B.; Grelle, C.E.V.; Benayas, J.M.R. A global meta-analysis on the ecological drivers of forest restoration success. *Nat. Commun.* **2016**, *7*, 11666. [CrossRef] [PubMed]
35. WFP. *Plantation Assessment Report: Reforestation in the Rohingya Camps, Cox's Bazar*; The World Food Programme: Rome, Italy, 2020.
36. Chigbu, U.E.; Chilombo, A.; Lee, C.; Mabakeng, M.R.; Alexander, L.; Simataa, N.V.; Siukuta, M.; Ricardo, P. Tenure-restoration nexus: A pertinent area of concern for land degradation neutrality. *Curr. Opin. Environ. Sustain.* **2022**, *57*, 101200. [CrossRef]

 *land*

Article

Agricultural Land Concentration in Estonia and Its Containment Possibilities

Marii Rasva and Evelin Jürgenson *

Forest and Land Management and Wood Processing Technologies, Institute of Forestry and Engineering, Estonian University of Life Sciences, 51006 Tartu, Estonia
* Correspondence: evelin.jyrgenson@emu.ee

Abstract: Land is essential to livelihoods, so it is hard to overstate its strategic significance for well-being and prosperity. It has been detected that farm size greatly influences agricultural sustainability from the viewpoints of the economy, environment, and society. Land concentration is negatively affecting the development of rural communities. Similar to other European countries, Estonia is undergoing agricultural land concentration. One way to stop the further concentration of agricultural land is to set an upper limit to land acquisition (similar to that in Latvia and Lithuania). This paper aimed to determine what kind of regulations concerning agricultural land use and ownership Estonia needs to restrain land concentration. Four sources of data were used for this research: statistical data from Statistics Estonia, the data for the land holdings of agricultural producers from the Estonian Agricultural Registers and Information Board, data from the Land Registry and available literature. The outcome of the study confirmed that Estonia requires policy direction and regulations for the agricultural land market, that would help to lighten the impact of land concentration in rural areas in the long run, similar to several other European countries.

Keywords: land concentration; sustainable land management; policy directions; acquisition of agricultural land

Citation: Rasva, M.; Jürgenson, E. Agricultural Land Concentration in Estonia and Its Containment Possibilities. *Land* **2022**, *11*, 2270. https://doi.org/10.3390/land11122270

Academic Editors: Kleomenis Kalogeropoulos, Andreas Tsatsaris, Nikolaos Stathopoulos, Demetrios E. Tsesmelis, Nilanchal Patel and Xiao Huang

Received: 18 November 2022
Accepted: 9 December 2022
Published: 12 December 2022

Publisher's Note: MDPI stays neutral with regard to jurisdictional claims in published maps and institutional affiliations.

1. Introduction

Land is fundamental to prosperity and well-being and, due to this, it is hard to overstate its strategic importance for existence. Large-scale land acquisitions transform land use and food systems in targeted districts worldwide [1–6]. The outcome of these large-scale land acquisitions is that agricultural land becomes concentrated. It has been found that large-scale land purchases are causing socio-economic destruction. [7–11].

Agricultural land concentration is a topic of discussion in different countries, but particularly current in post-Soviet countries. Land concentration is an activity by which large agricultural concerns increasingly buy, or lease, land from other agricultural producers [12–15]. Supporting small farmers remains essential for food security and to combat rural poverty. This phenomenon has affected countries like Slovakia [16], Hungary [5], Romania [17], Poland [18] and many other countries. These countries all experienced major land reforms after the Soviet Union collapsed. In the process of land reform, a lot of land came onto the market and it was possible for buyers to purchase as much land as they wanted. The prices of land are still low in these countries, compared with other European Union countries.

The process of agricultural land concentration started decades ago but has recently accelerated. It has been detected that farm size greatly influences agricultural sustainability in economic, environmental, and social aspects [19]. Small agricultural producers are vanishing rapidly, and places of employment in rural areas are decreasing [20–24]. The rural living situation is worsened by job losses, poor social infrastructure, and the fact the younger generation is moving away from rural areas. The process of land concentration is generally not reversible [25,26]. Land concentration is negatively affecting the development of rural communities. Small agricultural producers are vital for rural communities as they conserve rural cultural heritage and rural life. They enliven rural social life, produce valuable products, use natural resources sustainably and assure a range of landowners in rural regions [25,27–29]. Sustainable land use that ensures a fair and balanced distribution of land, water, biodiversity and other environmental resources between various competing claims, is necessary to secure human needs now and in the future [30].

Division of land ownership to cover a wide range is the foundation of the social market economy and social cohesion [31]. It also ensures job creation in rural areas, adds significant value to agricultural production, and is essential for ensuring peace in society. The future of the agricultural sector depends on a new generation of farmers. The will to innovate and invest in young people is vital for rural areas. The ageing of the agricultural sector can be stopped, and the continuity of rural life can be secured through this.

Estonia has undergone, over its history, considerable structural changes, affecting its agriculture. Through various streaks of occupation and simultaneous reforms, Estonia became independent in 1991 and launched the most recent, still incomplete, land reform. The laws of land reform, and agricultural reform, are inclined towards agriculture based on small farms. In the early years of the reforms, between 1993 and 2001, the number of farms in Estonia grew, and many small farms were involved [32]. However, over the period of 2001–2020, the number of small farms decreased and is continuing its downward trend.

Agricultural development is not favouring small-scale farming. The only choice for small- and medium-sized farms is to grow or go. If farms are not able to grow in size and acquire more land (move to larger sized farming groups), they are not able to survive. Larger and more competitive agricultural producers push small farmers out of business, and agricultural land becomes even further concentrated. Small farms struggle to survive in the existing market situation, where large producers have a clear advantage. Thus, the State should step in and regulate the agricultural land market so that small, medium and large producers can coexist and operate under similar conditions.

The agricultural land market cannot be regulated only by means of market principles because land genesis does not respond to prices in the same way as regular goods [33]. Several EU countries have laws with various objectives, from preserving agricultural land for agricultural use to curbing land concentration. Since 2013, Hungary, Slovakia, Latvia, Lithuania, Bulgaria, Romania and Poland have sanctioned land laws targeting unwanted developments in their land markets [34].

Agricultural land concentration can be a threat to soil use as well. Previous studies have shown that environmental damage from large-scale agricultural production includes the destruction of soil fertility, contamination of water sources, loss of biodiversity, and draining of wetlands [35,36]. Large-scale agricultural producers, whose primary purpose may be to earn as much profit as possible, might be the outcome of further agricultural land concentration. The cost of this kind of behaviour may result in severe and irreversible environmental damage and harm to the soil [17,35]. Industrialised agricultural producers are mainly interested in greater yields, which means soils are often harmed through more intensive agriculture. Healthy soils are vital to reverse biodiversity destruction, assure healthy food and guarantee everyday well-being. The European Union (EU) soil strategy for 2030 has a vision and objectives to achieve healthy soils by 2050 [37]. The EU soil strategy for 2030 supports the goals of the European Green Deal.

Besides the intensive use of agricultural land, there are several other environmental issues. One is the soil sealing that can happen through land use changes. Agricultural areas are replaced with development areas in the ongoing urban sprawl [38,39]. The consequences of this kind of land use change may be that agricultural land use becomes more complicated, as agricultural activities can disturb nearby land owners and users [40,41]. This situation emerges when there are no buffers between expanded urban areas and rural areas, or when people who are not farming move to rural areas [42]. Eventually, only a small number of agricultural producers survive near cities [43].

Agricultural land fragmentation is also not environmentally friendly, as farming is more expensive than it would be with compact land use. Extensive driving to get from one field to another field results in increased pollution [44–48]. Agricultural activity is complex and involves different aspects. Although land use changes and fragmentation are important topics it is not feasible to handle all the involved issues in one paper. Therefore, this study focused on agricultural land concentration and opportunities for restraining land concentration in Estonia. Similar to many other European countries, Estonia requires policy direction and regulations for the agricultural land market so as to relieve the influence of land concentration in rural areas for extended periods. This paper aimed to determine the kind of regulations, concerning agricultural land use and ownership, that Estonia needs to curb land concentration.

2. Materials and Methods

Four types of data sources were required for this research. Statistics Estonia was the resource for statistical data. The data on agricultural producers' land holdings were obtained from the Estonian Agricultural Registers and Information Board (ARIB). Land Registry data was used to analyse changes in land ownership of the 49 largest agricultural producers, according to 2020 ARIB data. Books, scientific papers, reports, acts of law, regulations and documents were researched.

Data from Statistics Estonia (PMS416, PMS422) was used to analyse changes in Estonian agricultural land use, including data on the number of agricultural households and agricultural land use area. This data was for the years 2001, 2003, 2005, 2007, 2010, 2013, 2016 and 2020.

To obtain an overview of changes in Estonian agricultural land users' land holdings, ARIB data for agricultural land area and number of farms in 2011 and 2020 were dissected. GIS software ArcGIS (version 10.4) was applied to summarise land users and land area per farm. Farms were divided into six groups according to the size of their land holdings: 0–<2 ha, 2–<40 ha, 40–<100 ha, 100–<400 ha, 400–<1000 ha and >1000 ha. This division (into six groups) was based on the method of the Farm Accountancy Data Network (FADN), in which agricultural land area is separated into four size groups (0–<40 ha, 40–<100 ha, 100–<400 ha, >400 ha). To obtain a better understanding of the smallest farmers, we divided the FADN size group 0–<40 ha into size groups of 0–<2 ha and 2–<40 ha. We divided FADN size group >400 ha into size groups of 400–<1000 and >1000 ha to define the largest agricultural producers. This means that two FADN size groups were changed for this study. Figure 1 presents the study area and its position in Europe.

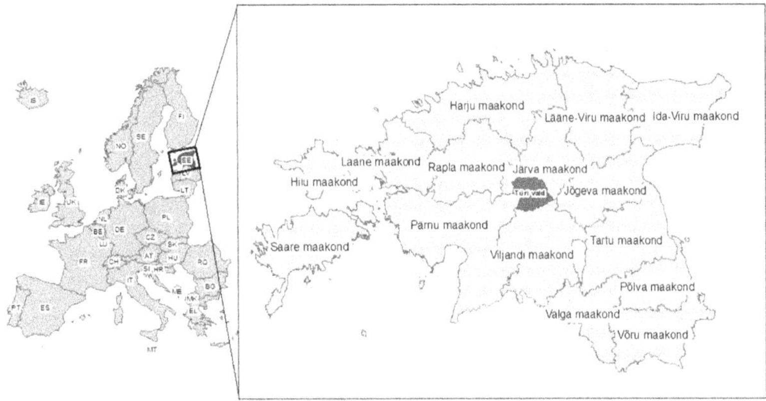

Figure 1. Position of Estonia (study area) in Europe.

Land Registry data were used to get an overview of land ownership changes among 49 land users. Land Registry data covered the years 2001, 2016 and 2021. In order to acquire data from the Land Registry, ARIB 2020 data was used to ascertain the 49 largest producers. After searching the ARIB 2020 data, an inquiry was sent to the Land Registry concerning the 49 largest agricultural producers.

The farmed land areas of the 49 largest agricultural land users were studied to compare land ownership and changes in land use area. Unfortunately, the earliest records from the ARIB concerning land use were only available from the year 2003. ARIB data from 2003, 2016 and 2021 were used to compare land ownership with land use in this study. Farms were grouped into six clusters according to the size of their land holdings: 0 ha, less than 100 ha, 101–200 ha, 201–400 ha, 401–1000 ha and more than 1000 ha. Data was applied based on these group sizes.

Available books, scientific papers, reports, acts of law, regulations and documents were studied to determine the restrictions EU countries have implemented to protect their agricultural land against concentration. Firstly, information from reports and scientific articles was used to find countries where such restrictions are implemented. Secondly, some legal acts (that were available online and in English) from these countries were studied to determine the exact regulations. Figure 2 illustrates the countries' division in the study regarding restrictions on agricultural land acquisitions.

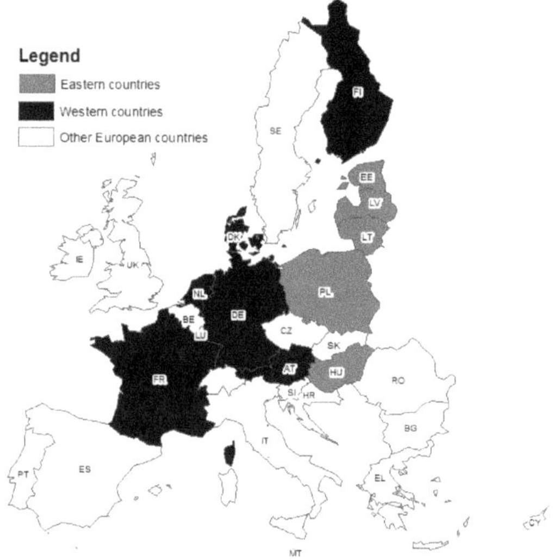

Figure 2. Division of the countries included in the study regarding restrictions on agricultural land acquisitions.

EU countries that were included in the study were divided into two groups. The first group included countries from the western part of the EU (Germany, The Netherlands, Denmark, France, Austria and Finland). The second group included post-Soviet EU countries (Estonia, Hungary, Poland, Latvia and Lithuania).

3. Results

3.1. Agricultural Land Use Changes in Estonia

The number of agricultural households has diminished year-to-year (Figure 3). In 2020, there were 11,369 farms in Estonia, a considerable decrease from that in 2001 when there were 55,748 farms in Estonia. Meanwhile, the agricultural land area has stayed nearly the same. The utilised agricultural land area was 871,213 ha and 975,323 ha in Estonia in 2001 and 2020, respectively (Figure 3).

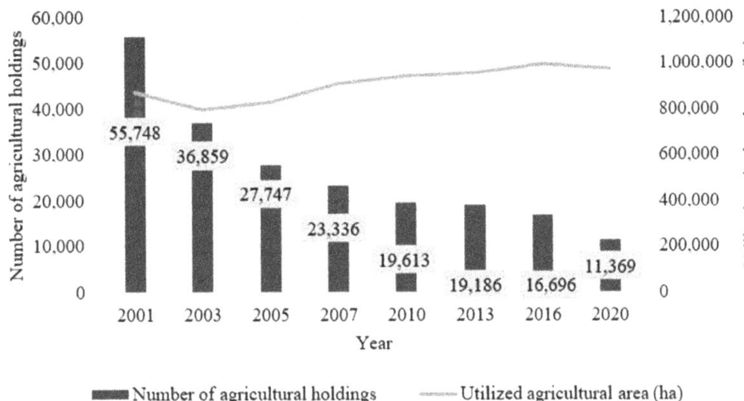

Figure 3. The number of farms and area of agricultural land in Estonia between 2001 and 2020 (Statistics Estonia).

The average land use per farm in Estonia has grown due to the decrease in the number of farms and the almost consistent farming area (Figure 4). In 2001, the area of agricultural land use per farm was 16 ha. It had grown to 86 ha by 2020. The average agricultural land use area per farm grew from 2 to 26 ha per year.

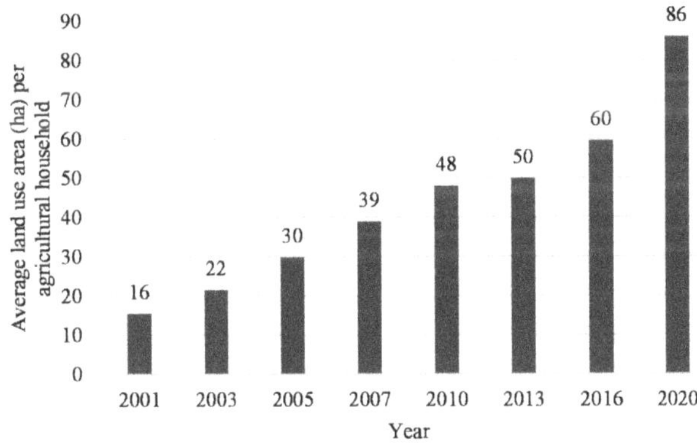

Figure 4. Average utilised land per farm in Estonia between 2001 and 2020 (Statistics Estonia).

The number of farms in the three smallest land user groups (0–2 ha, 2–<40 ha and 40–<100 ha) diminished over the years 2011 to 2020. The number of farms in the three largest (100–<400 ha, 400–<1000 ha and >1000 ha) groups grew (Table 1).

An analysis of farmers, according to the six farm size groups showed that between 2011 and 2020, area farmed by land users in the size groups 100–<400 ha and >1000 ha increased the most. Meanwhile, the area farmed by land users in the size groups 0–<2 ha and 2–<40 ha diminished. The agricultural land area used by size group 40–<100 ha stayed almost similar for the period considered.

Table 1. Data for land user groups, according to the area of farms for the years 2011 and 2020 (ARIB).

Groups (ha)	2011			2020		
	Number	Area (ha)	Area (%)	Number	Area (ha)	Area (%)
<2	1475	2140	0	1211	1778	0
2–<40	11,654	132,888	15	9785	107,119	11
40–<100	1460	91,563	10	1615	91,578	9
100–<400	1174	225,708	26	1660	275,696	28
400–<1000	337	207,844	24	556	244,574	25
>1000	126	216,893	25	212	257,964	26
Total	16,226	877,036	100	15,039	978,711	100

There were 768 farmers in Estonia, with land holdings above 400 ha in 2020, who utilised 502,539 ha, or 51%, of the agricultural land. In 2011, 463 farmers with land holdings above 400 ha utilised 424,736 ha, or 48%, of the farmed area. The agricultural land area used by larger farms increased, while that used by smaller ones diminished (Figure 5a). The number of farms in size groups 0–<2 ha and 2–<40 ha diminished (Figure 5b).

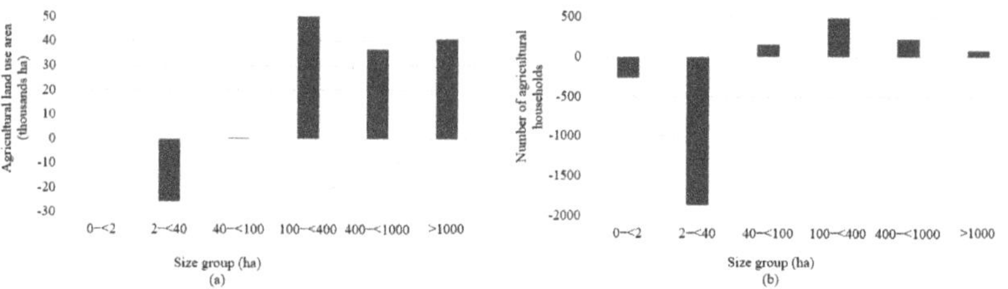

Figure 5. (**a**) The difference in the area of agricultural land use, and (**b**) the difference in the number of farms in size groups between 2011 and 2020 (ARIB).

In 2011, 1475 farms in size group 0–<2 ha used 2139.72 ha of agricultural land. In size group 2–<40 ha, 11,654 farms used 132,888.41 ha. In 2020, there were 264 fewer farms in size group 0–<2 ha using 361.57 ha less land. In size group 2–<40 ha, there were 1869 fewer farms, using 25,769.19 ha less land than in 2011.

Farms in size groups 40–<100 ha, 100–<400 ha, 400–<1000 ha and >1000 ha increased in number. In 2011 in size group 400–<1000 ha, 337 farms used 207,843.80 ha of farmed land. In size group >1000 ha, 126 farms used 216,892.61 ha. By 2020, there were 219 more farms in size group 400–<1000 ha and 86 more in size group >1000 ha. The farmed area increased by 36,730.39 ha in size group 400–<1000 ha and by 41,071.78 ha in size group >1000 ha.

In 2020, there were 275 legal persons and 936 self-employed in size group 0–<2 ha (Figure 6). In size group 2–<40 ha there were 4203 legal persons and 5582 self-employed. The self-employed formed the majority in these two size groups. There were no self-employed in size groups 400–<1000 ha and >1000 ha. In size group 400–<1000 ha, there were 556 legal persons, and 212 in size group >1000 ha.

The number of farms in size group 0–<2 ha formed 8.1% of the total number of farms in Estonia (Figure 7a), utilising 0.2% of the total land area (Figure 7b) in 2020. The number of farms in size group 2–<40 ha accounted for 65.1% of all Estonian land users, using 10.9% of all agricultural land areas.

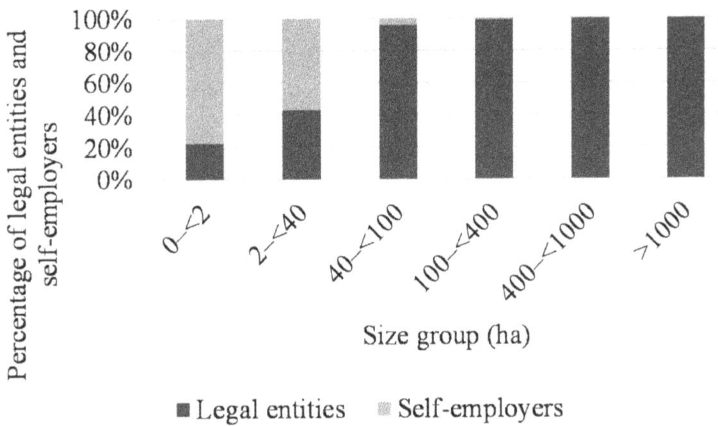

Figure 6. Percentage of legal entities and self-employed in size groups in 2020 (ARIB).

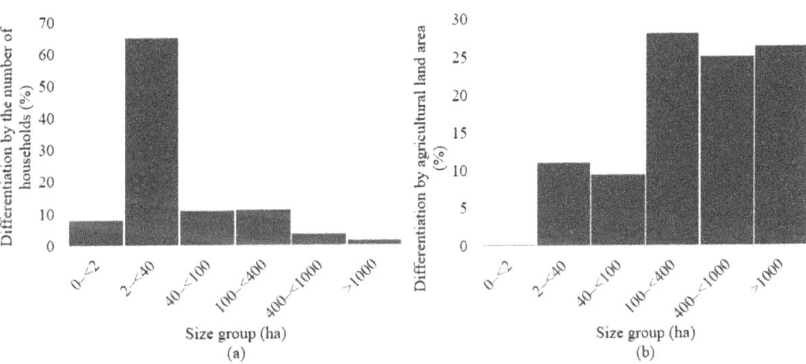

Figure 7. (**a**) Distribution of size groups by the number of farms, and (**b**) differentiation of size groups by agricultural land area in 2020 (ARIB).

Simultaneously, the number of farms in size group 400–<1000 ha accounted for 3.7% of the whole number, utilising 25% of all agricultural land in Estonia. The number of farms in size group >1000 ha accounted for 1.4% of the whole, utilising 26.4% of all used agricultural land in Estonia.

3.2. Agricultural Landownership Changes in Estonia in 2001–2021

The area of properties owned by the 47 largest farms increased between 2001 and 2021. Two producers' land ownership area decreased in the same period but increased between 2001 and 2016. One producer owned 96.04 ha of land in 2001, and 2164.94 ha in 2016. The second producer owned 76.01 ha of land in 2001, and 1116.18 ha in 2016.

In 2001, 41 producers had no land ownership or owned fewer than 100 ha of land (Figure 8 and Table 2). In 2021, all producers were landowners and only four owned less than 100 ha of land. In 2001, 20 producers owned fewer than 100 ha of land, and their average landownership area was 38 ha. In 2021, it was 53 ha.

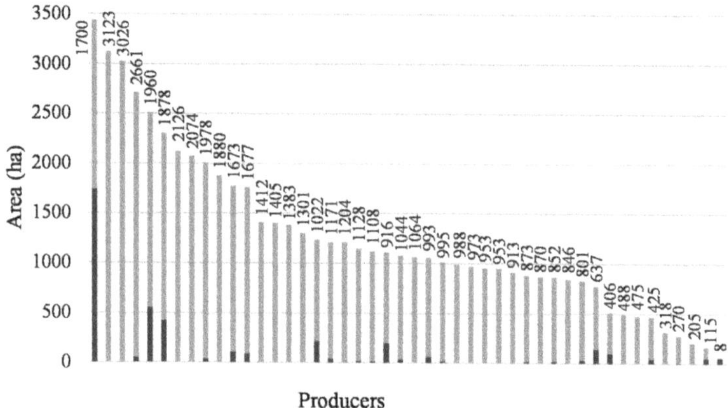

Figure 8. Area of landownership (Land Registry) of 47 agricultural producers in 2001 and 2021 (Land Registry).

Table 2. Changes in the 49 largest producers' landownership area (Land Registry) between 2001 and 2021 (Land Registry).

Groups (ha)	2001			2021		
	Number	Average Area		Number	Average Area	
		ha	%		ha	%
0	21	0	0	0	0	0
<100	20	38	1	4	53	2
101–200	4	136	5	1	164	5
201–400	1	215	8	3	267	9
401–1000	2	488	19	15	787	26
>1000	1	1741	67	26	1750	58
Total	49	2618	100	49	3021	100

The larger sized groups grew over the years (Table 2). In 2001, there were two farms in size group 401–1000 ha, and their average landownership area was 488 ha. In 2021, there were 15 producers in this size group, and their average landownership area was 787 ha. Massive changes occurred in size group >1000 ha. In 2001, one producer owned 1747 ha. In 2021, there were 26 producers with landownership larger than 1000 ha, and their average landownership area was 1750 ha.

The average landownership area of these 49 producers was 86.48 ha in 2001. In 2016, this area was 1135.80 ha, and in 2021, it was 1193.62 ha. The average landownership area of the 49 largest producers grew by an average of 1107.17 ha between 2001 and 2021. The most enormous land ownership area was 1700.14 ha, and the smallest was five hectares (Figure 8). The average growth area was 1280.96 ha. Sixteen producers' landownership area grew by more than the average. Ten producers' landownership area grew by more than 100,000%, and the most significant growth was 312,347%.

Analysing the changes in the 49 largest producers' land use area by dividing them into size groups, it was found that the number of farms in the largest size group grew between 2003 and 2021 (Table 3). Most of the producers grew in size and moved into size group >1000 ha. In 2001, there were 40 farms in size group >1000 ha, five farms in size group 401–1000 ha, two producers in size group 201–400 ha, and in size groups 101–200 ha and

<100 ha, there was one producer. In 2021, there was one producer in size group 401–1000 ha and 48 in size group >1000 ha.

Table 3. Changes in the 49 largest producers' land use area between 2003–2021 (ARIB) and proportion of their landownership (2001–2021) area (Land Registry) to land use area (ARIB).

Groups (ha)	2003		Proportion of 2001 Land Ownership Area to 2003 Land Use Area (%)	2021		Proportion of 2021 Land Ownership Area to 2021 Land Use Area (%)
	Number	Average Area (ha)		Number	Average Area (ha)	
0	0	0	0	0	0	0
<100	1	58.51	0	0	0	0
101–200	1	145.54	0	0	0	0
201–400	2	283.48	25	0	0	0
401–1000	5	764.08	0	1	847.44	103
>1000	40	3040.48	3	48	2538.01	47

Producers in size group 201–400 ha utilised the largest share (25%) of owned land in 2001. In 2021, there were no producers in size groups smaller than 401–1000 ha. All these producers had moved into larger size groups. In size group 401–1000 ha, producers had almost no owned land in 2001. In 2021, the share of owned land in this size group had grown to 103% of the total land use. Farms in size group >1000 ha owned 3% of the utilised land area in 2001. In 2021, the share of owned land in this size group was 47% of the utilised land area.

Comparing the increase in land ownership area between 2001–2021 (Land Registry) to the changes taking place in utilised land area (owned land and rented land, from ARIB) of the 49 largest land users, it was found that many producers' land use area had decreased, while the area of land ownership had increased (Figure 9).

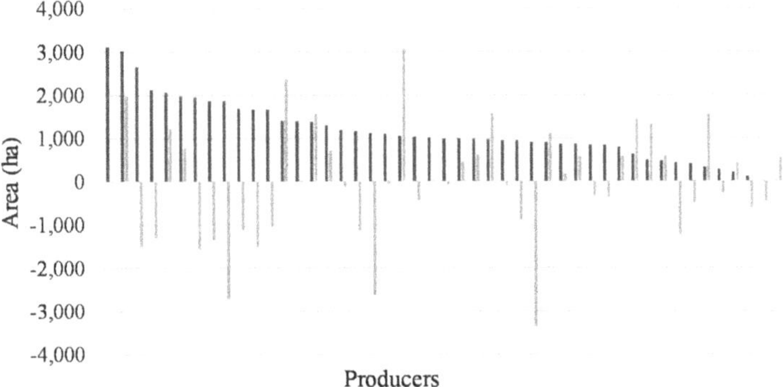

Figure 9. Increase in landownership area between 2001–2021 (Land Registry) and change in land use area between 2003–2021 (ARIB).

In 2001, the average share of landownership area from the utilised land area (2003) was 1.02%. In 2016, the average share of landownership area from utilised land area was 46%, and in 2021 it was 47%. Consequently, the share of owned land increased in the case of the largest land users.

3.3. Restrictions to Agricultural Land Purchase in European Union

The agricultural land market is subject to different regulations in the countries of the world. The importance of a well-functioning agricultural land market is difficult to over-emphasise.

Restrictions on the acquisition of agricultural land vary in different EU countries. The member states decide on adopting and implementing agricultural land market regulations, as certain land market regulations are missing at the EU level. However, EU treaties disallow restrictions on the movement of capital [49].

In Germany, the legislation concerning the ownership of agricultural land favours people engaged in farming. The approach aims to protect agricultural land from being turned into development areas, to protect nature and the environment, and to assure food security [50]. There is a permit obligation before any agricultural land transaction. Local municipalities also possess a pre-emptive right to purchase agricultural land, and the magistrate can appoint inheritable agricultural land to one particular heir in the case of inheritance. In Germany, there is also a minimum area of agricultural land that is subject to permit obligation.

In the Netherlands and Finland, there are no restrictions on acquiring agricultural land based on the buyer's legal form or citizenship [49]. However, there is a permit obligation in Finland for persons from certain third countries. In Denmark, there are no longer any specific restrictions on acquiring agricultural land.

There is a need to apply for a specific permit if persons from third countries wish to acquire land in France. An obligation has to be approved by *Sociétés d'Aménagement Foncier et d'Etablissement Rural* to purchase agricultural land [49]. In Austria, there is also an obligation for approval from the *Grundverkehrskommisson*. However, in Austria, there are exceptions to this rule.

In Hungary, there is an obligation to qualify as a farmer to purchase more than one hectare of agricultural land [49]. To qualify as a farmer, a person has to be a citizen of Hungary or another EU country [51]. A person who does not have the qualification mentioned earlier must first be able to prove that they have been engaged in agriculture for at least the previous three years. Secondly, this person must prove that they received an income from agriculture over the previous three years.

In Hungary, there is a restriction on third persons using acquired agricultural land. The owner must use this land only for agricultural purposes for at least five years from the purchase [49]. The agricultural land area that one person can purchase in Hungary is limited to 300 ha [20], and a maximum of 1200 ha of agricultural land can be in the ownership of one farmer [49,52]. Corporations have no right to own land in Hungary, but there are exceptions to this rule. It is not easy for a third-world person to obtain a farmer's qualification in Hungary.

In Poland, there is an obligation for a person from Poland or the EU to qualify as a private farmer when purchasing agricultural land. A private farmer is a subject who owns or uses a maximum of 300 ha of agricultural land and is registered to live in the local municipality [49]. Purchasable agricultural land and already-owned land cannot exceed 300 hectares in Poland; although exceptions exist to this rule [49,53]. Persons not qualified as private farmers must acquire approval from the National Support Centre for Agriculture to purchase agricultural land in Poland.

A person from Latvia or another EU country must be registered to conduct business there to acquire agricultural land in Latvia. A self-employed person must confirm in writing that they will start agricultural activity there within one year of purchasing the land [49]. From 2017, a person cannot acquire more than 2000 hectares of land, and related persons cannot acquire more than a further 4000 hectares of land [49,54]. A corporate body must also prove that agricultural activities will commence on the purchased land and indicate the actual profit recipients. Persons from third countries are not permitted to purchase land in Latvia; although exceptions exist.

In Lithuania, there are also restrictions on how much agricultural land can be acquired and by whom. Similar to restrictions in Latvia, these are important to prevent further agricultural land concentration [20]. In Lithuania, in the case of agricultural land purchase, it is mandatory to prove that the person will use the land only for agricultural purposes for at least the next five years [49]. A person cannot own more than 500 hectares of total agricultural land in Lithuania [55]. Persons from third countries cannot acquire land in Lithuania.

Estonia has no specific restrictions on acquiring agricultural land for citizens of Estonia or the EU [56]. Corporate bodies from EU countries must be involved in agriculture in the EU for at least three years prior to purchasing land in Estonia that exceeds 10 ha. Corporate bodies must also be involved in agriculture to purchase agricultural land, and its affiliate has to be registered in Estonia. Persons from third countries have the right to purchase agricultural land in Estonia only with permission from the local government and provided the person has lived in Estonia for at least six months [56].

4. Discussion

A growing population and an aim to decarbonise the economy mean that agricultural land is in demand for a broader range of uses than ever before [57–59]. Agricultural land is a unique asset exposed to pressure from non-agricultural uses, increasing demand for food, energy and biomass. Agriculture is a significant source of greenhouse gases, and as the world's greenhouse gas levels continue to heighten, climate change is appearing much quicker than foreseen [60,61]. There is a need for productive, yet sustainable, agriculture to ensure future food security for the world's increasing population [32,62]. The European Green Deal and Sustainable Development Goals set some of the goals needed to move towards sustainability [12,39,63,64]. Land is a finite resource, and more cannot be produced. Growth in farm size is connected with a statistically significant decrease in fertilizer and pesticide use per hectare, showing clear gains for environmental conservation [19]. Small agricultural producers are the core of European agriculture, and increasing concentration makes it harder for family farmers to access land.

The phenomenon of land concentration in the EU and many parts of the world is one of the most severe land matters. This phenomenon started to emerge decades ago and has recently accelerated. The ongoing agricultural land concentration affects Europe's small farms and hinterlands. Some EU countries have taken steps to prevent and reverse agricultural land concentration. For example, Hungary, Poland, Latvia, and Lithuania have adopted regulations against excessive land concentration and other undesirable patterns in their land markets.

Utilised agricultural land area in Estonia has remained almost the same over 20 years (871,213 to 975,323 ha) or grown a little. The number of farms diminished by almost five times (from 55,748 to 11,369) within 20 years, while the area of agricultural land use per farm grew five times (from 16 ha to 86 ha). While average land use per farm in Estonia has grown, the agricultural land in Estonia has become progressively concentrated in legal entities' hands.

The whole number of farms in Estonia has diminished, and farms that have shut down their activities are primarily in size groups 0–<2 ha (−264 producers) and 2–<40 ha (−1869 producers). The most extensive increase in the number of farms between 2011 and 2020 appeared in the size group 100–<400 ha (486 producers).

Most of the self-employed were farming in size groups 0–<2 ha (77%) and 2–<40 ha (57%) in 2020. Legal entities dominated in size groups over 40 ha. Some self-employed farmed in size group 40–<100 ha (4%) and a few in 100–<400 ha (1%). Farmers in size groups over 400 ha (100%) were legal entities. The largest portion of agricultural land is concentrated in the usage of large corporate users in Estonia in Lääne-Viru and Järva counties [32,65], in which the most fertile soils in Estonia are located. Therefore, the largest concentration of agricultural land occurs in regions where soils are most fertile. This phenomenon has also been seen elsewhere in the world.

Agricultural development in Estonia is not favouring small-scale farming. Yet small agricultural producers are preferable for environmental sustainability, protection of traditional values and economic flexibility [23]. One reason for the decline of small-scale producers is that the CAP does not cater to the needs of small-scale farmers. Land users with enormous domains receive more significant subsidies, which means they can obtain more land. Secondly, large agricultural businesses are deluging markets with cheap food and agricultural products. A situation has been created wherein small-scale agricultural producers cannot compete in the marketplace.

It is indicated [66–70] that agricultural producers' holdings will grow in the future, with small farms disappearing. If small farms are not able to grow in size and acquire more land (move to larger sized groups), they will not be able to survive. Larger and more compatible agricultural producers will push them out of business, and agricultural land will become even further concentrated. Small farms struggle to survive in the existing market situation where large producers have a clear advantage. Thus, the State should step in and regulate the agricultural land market so that small, medium and large producers can coexist and operate under similar conditions.

The results of this paper indicate that the area of land ownership of large land users is growing alongside the increase in land use area. This conclusion was made by analysing the land ownership of the 49 largest land users. Further research on changes in landownership is needed to make firm conclusions. Nonetheless, this paper indicates that land ownership is concentrated beside land use, and this is a dangerous sign. Control over land is concentrated increasingly in the hands of a small number of large corporations, and there is a need to take action against this development in Estonia.

Like many other European countries, Estonia requires policy direction, strategy and regulations for handling the agricultural land market to relieve the impact of land concentration in remote areas in the long run [20]. Restrictions on acquiring agricultural land in Estonia are necessary to stop further concentration and reverse the current situation, where small and medium farms cannot compete with large corporate bodies. Small and medium farms require more support from the State. The State should also create conditions for newcomers entering the sector. Farmers have pointed out that they need support from the State to acquire agricultural land [71].

An upper limit should be set on how much agricultural land one person, or related persons, can own in Estonia to prevent further agricultural land concentration. To restrain agricultural land from ending up in the possession of large business with no relation to agriculture, a portfolio obligation to have a particular qualification for purchasing agricultural land is also necessary in Estonia. A prior right of purchasing agricultural land should be enacted to guarantee that newcomers and small farms can acquire the necessary land. The possibility of fair market competition for all farming types should be assured.

Before setting restrictions on obtaining agricultural land, there is also a need to create a clear structure of enterprises in Estonia to determine how much land one enterprise owns or rents. Without this, grounds for the circumvention of restrictions are possible.

5. Conclusions

Circumstances regarding agricultural land concentration are similar in Estonia to other post-socialist EU countries. The number of farms has dropped, and the agricultural land area per farm has increased. Surviving farms are growing in size. The size of agricultural holding plays an essential role in the environment, including in agricultural sustainability. Small farms are disappearing, although these producers are believed to contribute more to environmental sustainability, preservation of traditional values, and economic resilience than large ones.

The Estonian case study showed that agricultural land use concentration is happening along with land ownership concentration. A large area of land is already concentrated in the ownership of a small number of large farms. Restrictions on acquiring agricultural land in Estonia are needed to restrain further concentration and reverse the current situation, where small and medium farms cannot compete with large corporate bodies.

There is a need to regulate how much agricultural land one person or related persons can own in Estonia. Agricultural land should be owned only by those who have a particular qualification. A prior right of purchasing agricultural land should be enacted to guarantee that newcomers and small farms can acquire the necessary land.

The direction of agricultural land use and ownership in Estonia is a topic for studies and disputes over relevant regulations, potential limitations for possession and the usage of pre-emptive rights. Measures concerning agricultural land concentration in Estonia should be implemented in an interplay between agricultural producers and the government to encourage green development. The balance between large agricultural producers and small farms in Estonia must ensure that both farming types remain in fair market competition. There is undoubtedly a need for transparency in the structure of enterprises in Estonia.

Author Contributions: Conceptualization, M.R. and E.J.; methodology, M.R. and E.J.; software, M.R.; validation, M.R. and E.J.; writing—original draft preparation, M.R.; writing—review and editing, E.J.; visualization, M.R.; supervision, E.J. All authors have read and agreed to the published version of the manuscript.

Funding: This work was supported by the European Union, European Regional Development Fund (Estonian University of Life Sciences ASTRA project "Value-chain based bio-economy". Funding number: 2014-2020.4.01.16-0036) and "Public sector (Estonia)" project P210157MIMP (Governing land take and agricultural land use in Estonia and providing implementation of the European Green Deal).

Data Availability Statement: Not applicable.

Conflicts of Interest: The authors declare no conflict of interest.

References

1. Oberlack, C.; Giger, M.; Anseeuw, W.; Adelle, C. Why do large-scale agricultural investments induce different socio economic, food security, and environmental impacts? Evidence from Kenya, Madagascar, and Mozambique. *Ecol. Soc.* **2021**, *4*, 18. [CrossRef]
2. Cristina Rulli, M.; D'Odorico, P. Food appropriation through large scale land acquisitions. *Environ. Res. Lett.* **2014**, *9*, 064030. [CrossRef]
3. Gironde, C.; Golay, C. Large-Scale Land Acquisitions, Livelihoods and Human Rights in South-East Asia. *Int. Dev. Policy* **2015**, *6*, 275–292. [CrossRef]
4. Porsani, J.; Caretta, M.A.; Lehtilä, K. Large-scale land acquisitions aggravate the feminization of poverty: Findings from a case study in Mozambique. *GeoJournal* **2019**, *84*, 215–236. [CrossRef]
5. Juhasz, J. Large-scale and small-scale farming in Hungarian agriculture: Present situation and future prospects. *Eur. Rev. Agric. Econ.* **1991**, *18*, 399–415. [CrossRef]
6. Claeys, P.; Vanloqueren, G. The Minimum Human Rights Principles Applicable to Large-Scale Land Acquisitions or Leases. *Globalizations* **2013**, *10*, 193–198. [CrossRef]
7. Chahongnao, S. Customary Tenure and Large-Scale Land Acquisitions in the Global South: Issues and Redressal Approaches in Governance Policy. *Acad. Lett.* **2021**, *12*, 850. [CrossRef]
8. Moreda, T. Large-scale land acquisitions, state authority and indigenous local communities: Insights from Ethiopia. *Third World Q.* **2017**, *38*, 698–716. [CrossRef]
9. Roudart, L.; Mazoyer, M. Large-Scale Land Acquisitions: A Historical Perspective. *Int. Dev. Policy* **2015**, *6*. [CrossRef]
10. Roudart, L.; Dave, B. Land policy, family farms, food production and livelihoods in the Office du Niger area, Mali. *Land Use Policy* **2017**, *60*, 313–323. [CrossRef]
11. Yengoh, G.T.; Steen, K.; Armah, F.A.; Ness, B. Factors of vulnerability: How large-scale land acquisitions take advantage of local and national weaknesses in Sierra Leone. *Land Use Policy* **2016**, *50*, 328–340. [CrossRef]
12. Rasva, M.; Jürgenson, E. Europe's Large-Scale Land Acquisitions and Bibliometric Analysis. *Agriculture* **2022**, *12*, 850. [CrossRef]
13. De Schutter, O. *Large-Scale Land Acquisitions and Leases: A Set of Core Principles and Measures to Address the Human Rights Challenge*; United Nations: New York, NY, USA, 2009.
14. Foreign land purchases for agriculture: What impact on sustainable development? *Sustain. Dev. Innov. Briefs* **2010**, *8*, 1–8.
15. Liu, P. *Impacts of Foreign Agricultural Investment on Developing Countries: Evidence from Case Studies*; Food and Agriculture Organization: Rome, Italy, 2014.
16. Palsova, L.; Bandlerova, A.; Machnicova, Z. Land Concentration and Land Grabbing Processes-Evidence from Slovakia. *Land* **2021**, *10*, 873. [CrossRef]

17. Constantin, C.; Luminița, C.; Vasile, A.J. Land grabbing: A review of extent and possible consequences in Romania. *Land Use Policy* **2017**, *62*, 143–150. [CrossRef]

18. Stacherzak, A.; Hełdak, M.; Hájek, L.; Przybyła, K. State interventionism in agricultural land turnover in Poland. *Sustainability* **2019**, *11*, 1534. [CrossRef]

19. Ren, C.; Liu, S.; van Grinsven, H.; Reis, S.; Jin, S.; Liu, H.; Gu, B. The impact of farm size on agricultural sustainability. *J. Clean. Prod.* **2019**, *220*, 357–367. [CrossRef]

20. Jürgenson, E.; Rasva, M. The changing structure and concentration of agricultural land holdings in Estonia and possible threat for rural areas. *Land* **2020**, *9*, 41. [CrossRef]

21. van Vliet, J.A.; Schut, A.G.T.; Reidsma, P.; Descheemaeker, K.; Slingerland, M.; van de Ven, G.W.J.; Giller, K.E. De-mystifying family farming: Features, diversity and trends across the globe. *Glob. Food Sec.* **2015**, *5*, 11–18. [CrossRef]

22. Bojnec, Š.; Latruffe, L. Farm size, agricultural subsidies and farm performance in Slovenia. *Land Use Policy* **2013**, *32*, 207–217. [CrossRef]

23. Wuepper, D.; Wimmer, S.; Sauer, J. Is small family farming more environmentally sustainable? Evidence from a spatial regression discontinuity design in Germany. *Land Use Policy* **2020**, *90*, 104360. [CrossRef]

24. von Braun, J.; Mirzabaev, A. *Small Farms: Changing Structures and Roles in Economic Development*; Zentrum für Entwicklungsforschung (ZEF), Center for Development Research: Bonn, Germany, 2015.

25. European Economic and Social Committee Land grabbing—A Warning for Europe and a Threat To Family Farming (Own-Initiative Opinion) 2015. Available online: https://eur-lex.europa.eu/legal-content/EN/TXT/PDF/?uri=CELEX:52014IE0926&from=PL (accessed on 8 December 2022).

26. Van Der Ploeg, J.D.; Franco, J.C.; Borras, S.M. Land concentration and land grabbing in Europe: A preliminary analysis. *Can. J. Dev. Stud.* **2015**, *36*, 147–162. [CrossRef]

27. FAO. *Empowering Smallholders and Strengthening Family Farms for Improved Rural Livelihoods and Poverty Reduction*; FAO: Rome, Italy, 2016.

28. Llambí, L. The Future of Small Farms. *World Dev.* **2010**, *38*, 1341–1348. [CrossRef]

29. Graeub, B.E.; Chappell, M.J.; Wittman, H.; Ledermann, S.; Kerr, R.B.; Gemmill-Herren, B. The State of Family Farms in the World. *World Dev.* **2016**, *87*, 1–15. [CrossRef]

30. Schneider, F.; Feurer, M.; Lundsgaard-Hansen, L.M.; Myint, W.; Nuam, C.D.; Nydegger, K.; Oberlack, C.; Tun, N.N.; Zähringer, J.G.; Tun, A.M.; et al. Sustainable Development Under Competing Claims on Land: Three Pathways Between Land-Use Changes, Ecosystem Services and Human Well-Being. *Eur. J. Dev. Res.* **2020**, *32*, 316–337. [CrossRef]

31. European Parliament. *The State of Play of Farmland Concentration in the EU: How to Facilitate the Access to Land for Farmers*; European Parliament: Strasbourg, France, 2017.

32. Rasva, M.; Jürgenson, E. Changes of agricultural producers in Estonia according to the size of land use. *Agron. Res.* **2020**, *18*, 516–528.

33. Courleux, F. *Regulating Agricultural Land Markets: The Main Economic Arguments*; Agriculture Strategies: online, 2019.

34. EC Commission Interpretative Communication on the Acquisition of Farmland and European Union Law (2017/C 350/05). *Off. J. Eur. Union* **2017**. Available online: https://eur-lex.europa.eu/legal-content/EN/TXT/PDF/?uri=CELEX:52017XC1018(01)&from=EN (accessed on 8 December 2022).

35. Grant, E.; Das, O. Land grabbing, sustainable development and human rights. *Transnatl. Environ. Law* **2015**, *4*, 289–317. [CrossRef]

36. Falkinger, J.; Grossmann, V. Oligarchic land ownership, entrepreneurship, and economic development. *J. Dev. Econ.* **2013**, *101*, 206–215. [CrossRef]

37. *Soil Strategy for 2030*; European commission: Brussels, Belgium . Available online: https://environment.ec.europa.eu/topics/soil-and-land/soil-strategy_en (accessed on 8 December 2022).

38. Maasikamäe, S.; Jürgenson, E.; Mandel, M.; Veeroja, P. Determination of Valuable Agricultural Land in the Frame of Preparation of Countywide Spatial Plans: Estonian Experiences and Challenges. *Econ. Sci. Rural Dev. Int. Sci. Conf. Econ. Sci. Rural Dev. Locat.* **2014**, *36*, 77–85.

39. Marquard, E.; Bartke, S.; Gifreu i Font, J.; Humer, A.; Jonkman, A.; Jürgenson, E.; Marot, N.; Poelmans, L.; Repe, B.; Rybski, R.; et al. Land Consumption and Land Take: Enhancing Conceptual Clarity for Evaluating Spatial Governance in the EU Context. *Sustainability* **2020**, *12*, 8269. [CrossRef]

40. Jürgenson, E.; Maasikamäe, S.; Sikk, K.; Hass, H. Methodology for the Determination of Peri Urban Areas on the Basis of Data of Land Type and Use by Example of the Town of Tartu. *Econ. Sci. Rural Dev.* **2017**, *45*, 110–117.

41. Jürgenson, E.; Maasikamäe, S.; Hass, H. *Land Reform Implementation in Estonia: Comparison of Land Reform Activity and Land Stock Characteristics Between Municipalities*; Estonian University of Life Sciences: Tartu, Estonia, 2011; p. 5.

42. Tomalty, R. *Farmland at Risk: How Better Land Use Planning Could Help Ensure a Healthy Future for Agriculture in the Greater Golden Horseshoe*; FAO: Rome, Italy, 2015.

43. Hunter, M.; Sorensen, A.; Nogeire-McRae, T.; Beck, S.; Shutts, S.; Murphy, R. *Farms Under Threat 2040: Choosing an Abundant Future*; American Farmland Trust: Washington, DC, USA, 2022.

44. Arslan, F.; Değirmenci, H.; Rasva, M.; Jürgenson, E. Finding least fragmented holdings with factor analysis and a new methodology: A case study of kargılı land consolidation project from Turkey. *Agron. Res.* **2019**, *17*, 1556–1572. [CrossRef]

45. Jürgenson, E. Land reform, land fragmentation and perspectives for future land consolidation in Estonia. *Land Use Policy* **2016**, *57*, 34–43. [CrossRef]

46. Looga, J.; Jürgenson, E.; Sikk, K.; Matveev, E.; Maasikamäe, S. Land Use Policy Land fragmentation and other determinants of agricultural farm productivity: The case of Estonia. *Land Use Policy* **2018**, *79*, 285–292. [CrossRef]

47. Hartvigsen, M. *Land Reform in Central and Eastern Europe after 1989 and its Outcome in the Form of Farm Structures and Land Fragmentation*; FAO Land Tenure Working Paper; FAO: Rome, Italy, 2013.

48. Viira, A.-H.; Ariva, J.; Kall, K.; Jürgenson, E.; Maasikamäe, S.; Põldaru, R. Restricting the eligible maintenance practices of permanent grassland—A realistic way towards more active farming? *Agron. Res.* **2020**, *18*. [CrossRef]

49. Vranken, L.; Tabeau, E.; Roebeling, P. *Agricultural Land Market Regulations in the EU Members*; Publications Office of the European Union: Luxembourg, 2021; ISBN 9789276419907.

50. Analüüs Euroopa Liidu Lepinguriikides Kehtestatud Põllumajandusmaa Kaitsemeetmetest Ja Põllumajandus-Ja Metsamaa Omandamise Kitsendustest. 2019. Available online: https://docplayer.ee/178289012-Anal%C3%BC%C3%BCs-euroopa-liidu-lepinguriikides-kehtestatud-p%C3%B5llumajandusmaa-kaitsemeetmetest-ja-p%C3%B5llumajandus-ja-metsamaa-omandamise-kitsendustest.html (accessed on 8 December 2022).

51. Balogh, T. Expanded Opportunities for EU Citizens to Acquire Agricultural Land. Available online: http://roadmap2015.schoenherr.eu/expanded-opportunities-eu-citizens-acquire-agricultural-land/ (accessed on 8 December 2022).

52. Csák, C. The Regulation of Agricultural Land Ownership in Hungary After Land Moratorium. *Zb. Rad. Pravnog Fak. Novi Sad* **2017**, *51*, 1125–1135. [CrossRef]

53. Źróbek-Różńska, A.; Zielińska-Szczepkowska, J. National Land Use Policy against the Misuse of the Agricultural Land—Causes and Effects. Evidence from Poland. *Sustainability* **2019**, *11*, 6403. [CrossRef]

54. Par Zemes Privatizāciju Lauku Apvidos. Available online: https://likumi.lv/ta/id/74241-par-zemes-privatizaciju-lauku-apvidos (accessed on 8 December 2022).

55. Republic of Lithuania Law on the Acquisition of Agricultural Land. Available online: https://e-seimas.lrs.lt/portal/legalAct/lt/TAD/d39d8f32a5b911e68987e8320e9a5185?jfwid= (accessed on 8 December 2022).

56. Riigikogu Kinnisasja Omandamise Kitsendamise Seadus 2021. Available online: https://www.riigiteataja.ee/akt/KAOKS (accessed on 8 December 2022).

57. EC. *EU Agricultural Outlook for Markets, Income and Enviroment 2021–2031*; European Commission: Luxembourg, 2021.

58. Savills. *The Farmland Market*; Savills: London, UK, 2022.

59. Tscharntke, T.; Clough, Y.; Wanger, T.C.; Jackson, L.; Motzke, I.; Perfecto, I.; Vandermeer, J.; Whitbread, A. Global food security, biodiversity conservation and the future of agricultural intensification. *Biol. Conserv.* **2012**, *151*, 53–59. [CrossRef]

60. United Nations. *The Sustainable Development Goals Report*; United Nations: New York, NY, USA, 2019.

61. GIZ. *What Is Sustainable Agriculture*; GIZ: Bonn, Germany, 2012.

62. Ricciardi, V.; Ramankutty, N.; Mehrabi, Z.; Jarvis, L.; Chookolingo, B. How much of the world's food do smallholders produce? *Glob. Food Sec.* **2018**, *17*, 64–72. [CrossRef]

63. Green Deal: Key to a Climate-Neutral and Sustainable EU. Available online: https://www.europarl.europa.eu/news/en/headlines/society/20200618STO81513/green-deal-key-to-a-climate-neutral-and-sustainable-eu (accessed on 8 December 2022).

64. Zoomers, A.; van Noorloos, F.; Otsuki, K.; Steel, G.; van Westen, G. The Rush for Land in an Urbanizing World: From Land Grabbing Toward Developing Safe, Resilient, and Sustainable Cities and Landscapes. *World Dev.* **2017**, *92*, 242–252. [CrossRef]

65. Rasva, M. *Land grabbing in Brazil and Africa, Chinese and European Participation in Land Grabbing and the Concentration of Estonian Agricultural Land Use*; Estonian University of Life Sciences: Tartu, Estonia, 2015.

66. Beckers, V.; Beckers, J.; Vanmaercke, M.; Van Hecke, E.; Van Rompaey, A.; Dendoncker, N. Modelling Farm Growth and Its Impact on Agricultural Land Use: A Country Scale Application of an Agent-Based Model. *Land* **2018**, *7*, 109. [CrossRef]

67. Lowder, S.K.; Skoet, J.; Raney, T. The Number, Size, and Distribution of Farms, Smallholder Farms, and Family Farms Worldwide. *World Dev.* **2016**, *87*, 16–29. [CrossRef]

68. Akimowicz, M.; Magrini, M.-B.; Ridier, A.; Bergez, J.-E.; Requier-Desjardins, D. What Influences Farm Size Growth? An Illustration in Southwestern France. *Appl. Econ. Perspect. Policy* **2013**, *35*, 242–269. [CrossRef]

69. FAO. *Global Agriculture towards 2050*; FAO: Rome, Italy, 2009.

70. Gollin, D. *Farm Size and Productivity. Lessons from Recent Literature*; CGIAR Independent Advisory and Evaluation Services (IAES): Rome, Italy, 2018.

71. Sammler, L. Põllumajandus on Kiires Uuenemises: Tulevikus Võiks Linnainimesel Olla Terve Digifarm. Available online: https://www.delfi.ee/teema/96846556 (accessed on 8 December 2022).

 land

Article

Carbon Sequestration Potentials of Different Land Uses in Wondo Genet Sub-Catchment, Southern Ethiopia

Habitamu Taddese [1]**, Mesele Negash** [1]**, Tariku Geda** [2] **and Gebiaw T. Ayele** [3,*]

[1] Wondo Genet College of Forestry and Natural Resources, Hawassa University, Shashemene P.O. Box 128, Ethiopia

[2] Ethiopian Biodiversity Institute, Ministry of Agriculture, Addis Ababa P.O. Box 30726, Ethiopia

[3] Australian Rivers Institute and School of Engineering and Built Environment, Griffith University, Nathan, QLD 4111, Australia

[*] Correspondence: gebiaw.ayele@griffithuni.edu.au or gebeyaw21@gmail.com

Citation: Taddese, H.; Negash, M.; Geda, T.; Ayele, G.T. Carbon Sequestration Potentials of Different Land Uses in Wondo Genet Sub-Catchment, Southern Ethiopia. *Land* **2022**, *11*, 2252. https://doi.org/10.3390/land11122252

Academic Editors: Kleomenis Kalogeropoulos, Andreas Tsatsaris, Nikolaos Stathopoulos, Demetrios E. Tsesmelis, Nilanchal Patel and Xiao Huang

Received: 6 October 2022
Accepted: 7 December 2022
Published: 9 December 2022

Publisher's Note: MDPI stays neutral with regard to jurisdictional claims in published maps and institutional affiliations.

Abstract: Forests play an important role in combating the challenges posed by changing climate through sequestering carbon in their living biomasses and the soil. Tropical forests, which harbour a large number of species, are anticipated to play a great role in this regard due to the favourable growing environments. However, there is limited knowledge of the variability in carbon stock among land use types and its relationship with biodiversity. Therefore, this study assessed the variability in storing the different carbon pools among natural forest, woodland and khat plantation land use types. It also explored the relationship between biodiversity and carbon storage in the different carbon pools. Plant inventory and sample collection were undertaken following standard methods. In addition, soil samples were taken at three depth profile classes of 0–30 cm (top layer), 30–60 cm (middle layer) and 60–100 cm (bottom layer). Results of the study revealed that there was no statistically significant relationship between biodiversity and total biomass carbon, soil organic carbon or total carbon stock at a 95% level of confidence. The results indicated that the natural forest had the highest plant biomass (456.93 Mg ha^{-1}) followed by woodland (19.78 Mg ha^{-1}) and khat plantation (2.46 Mg ha^{-1}). Consequently, the total carbon stock estimate of the natural forest (366.47 Mg ha^{-1}) was significantly larger than that of the woodland (141.85 Mg ha^{-1}) and khat plantation (125.86 Mg ha^{-1}). The variation in total carbon stock among land use types arises from the variation in the total biomass carbon stock. The study results also revealed that soil organic carbon stock decreased with soil depth in all the land-use types. The findings of this study have implication of improving topsoil management in monoculture crops such as khat plantation and conserving natural forests for enhancing carbon sequestration potentials.

Keywords: carbon stock; dry Afromontane forest; biodiversity; soil organic carbon; biomass carbon; soil depth

1. Introduction

Climate change has posed tremendous challenges to the global environment. In the past several decades, we have experienced many violent weather events that have disrupted the lifestyle of communities across the world [1]. The prevalence of recurrent floods, irregularity of rainfall, rising temperature and melting of ice have caused severe infrastructure losses, agricultural yield reductions, infectious disease outbreaks, famine, and loss of human life in different parts of the world [2–6]. Climate projections show that the change will keep on imposing its multifaceted impacts unless it is quickly unabated.

International conventions on climate change have indicated the importance of using mitigation measures to reduce such impacts. One of the most effective measures of mitigating climate change is managing forests due to their enormous potentials for regulating the global environment [7]. Forests play a substantial role as carbon sinks to mitigate the impacts of the changing climate [8,9]. Tropical forests, in particular, are more productive and cover a large part of the terrestrial ecosystem [10] that make them the focus of REDD+ (reducing emission from deforestation and forest degradation, plus the conservation and enhancement of forest carbon stocks through sustainable forest management) mechanisms. The United Nations initiative of the REDD+ mechanism has given attention to protect tropical forests for they store a large amount of carbon in their biomass [10,11] and are under continuous disturbance [12]. Tropical forests have faced problems of degradation and conversion to other land uses. Consequently, the tropical forest carbon stock in South America and Africa has been decreasing since 1990 [9]. This has been happening regardless of the increasing importance of forest biomass due to its applications in renewable energy, bio-based products development and reducing greenhouse gas emissions.

The extent of change on the carbon stock dynamics in the process of forest degradation and conversion to other land-uses required detail studies, particularly in tropical forests where the process is extensive [13]. Therefore, efforts have been made to quantify the amount of carbon stock in forests and other land-use systems. Two broad categories of carbon storage are measured in forest ecosystems; namely the biomass carbon pool and the soil organic carbon (SOC) pool [14]. Carbon pool is a component of an ecosystem that serves as a reservoir to accumulate or release carbon [14]. Thus, biomass carbon pool is the component that stores carbon in the plant biomass through the process of photosynthesis and releases it to the atmosphere when ecosystems are disturbed [15]. The biomass component is further divided into aboveground biomass (AGB) and belowground biomass (BGB) categories [14]. AGB is the weight of all living materials of plants above the soil surface including the stem, stump, branches, bark, seeds and leaves [14]. The BGB mainly refers to the dry weight of live roots [14]. Biomass of non-living plant components such as litter and deadwood have the potential to store carbon too. Biomass plays a great role in the global carbon balance [16]. Carbon stock, which is the quantity of carbon in a carbon pool, was assessed in each of the carbon pools mentioned above.

Biodiversity conservation is a simultaneously important function of forests. Tropical forests, particularly, harbour many species and support two-thirds of the world's biodiversity [17]. On the contrary, the recent increase in tropical deforestation and forest degradation has affected biodiversity [17,18]. This was particularly true since REDD+ programs have a prime concern to carbon stock enhancement rather than biodiversity conservation [19]. However, biodiversity conservation should get similar attention to carbon stock enhancement. There is interest in the possibility for carbon stock enhancement projects to incorporate biodiversity conservation objectives. This needs information about the congruency of biodiversity conservation to carbon stock enhancement. There is uncertainty in our understanding of the relationship between carbon storage and biodiversity [20]. Studies are needed to understand the relationship between carbon storage and biodiversity conservation in different land-use types. Field measurement of carbon stock provides an accurate estimate of the different carbon pools and their contribution to the total ecosystem carbon stock [21]. Therefore, the objectives of this study were to: (1) estimate carbon stock in the various carbon pools in natural forest, woodland and khat plantation in Wondo Genet sub-catchment, (2) explore the relationship between different components of carbon stock and biodiversity, and (3) investigate the effect of soil depth on the SOC stock.

This study tried to test the following research hypothesis.

- There is no variation in mean carbon stock of the different carbon pools among natural forest, woodland and khat plantation.
- Biodiversity does not have significant relationship with different components of carbon stocks.
- Soil depth does not have any effect on soil organic carbon stock.

2. Methods

This study assessed the variations in carbon stock potentials of different land use types in one of the tropical ecosystems and compared them with biodiversity variables. The methods used in this study are established methods for collecting and analysing carbon stock and biodiversity data in various parts of the world. Therefore, standard methods for measuring tree, shrub, soils and plant characteristics and data analysis were adopted from previous scientific works.

2.1. Study Area Description

The study was conducted in Wondo Genet sub-catchment, which used to be an intact forest some decades ago [22]. The mountain ranges are sources of streams that drain to the Chelekleka wetland and then to Lake Hawassa. Therefore, the catchment is an important terrestrial ecosystem that connects tourism, agriculture, urban livelihoods and ecosystem services including biodiversity conservation.

Wondo Genet sub-catchment is geographically located between 38°56′ and 38°73′ E longitudes and between 7°01′ and 7°15′ N latitudes (Figure 1). The altitudinal range of the study area is from 1770 to 2584 m above sea level. The dominant soil types in the sub-catchment are Luvisols (Alfisols), Andosols (Andisols) and Nitisols (Kandic alfisols, Utisols and Inceptisols) according to the FAO-UNESCO (and USDA) soil classification schemes, respectively [23]. The area has bimodal rainfall with mean annual precipitation at Wondo Genet station of about 1200 mm. The main rainy season is from June to August, while small rains come between March and April. The mean monthly temperature of the area ranges from 19 °C to 25 °C. The main land-use types in the sub-catchment are natural forest, plantation forest, woodland, khat plantation, sugarcane plantation, agriculture and settlements. The dominant tree species in the natural forest and woodland land-use types include *Celtis africana*, *Pouteria adolfi-friedericii*, *Acokanthera schimperi*, *Albizia schimperiana*, *Millettia ferruginea*, *Combretum molle*, *Croton macrostachyus* and *Afrocarpus falcatus*. Although the forests provide multiple services to the local community, they are under huge pressure due to illegal cutting of trees and land-use change for settlement and expansion of monoculture plantations, particularly khat. Consequently, the rich biodiversity in the Wondo Genet natural forest has dwindled in recent decades due to illegal logging and conversion of the forests to monoculture of khat and sugarcane plantations [24,25].

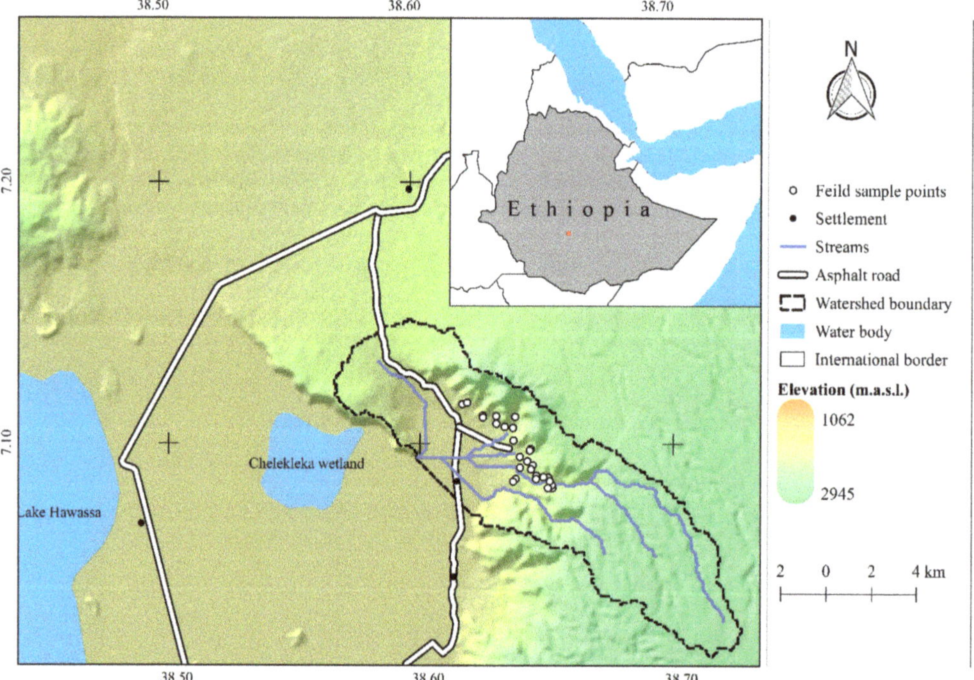

Figure 1. Location of the study area and distribution of the field plots.

2.2. Sampling Design

Stratified random sampling design was employed to collect the field data. A sketch map of the land use types was derived from high-resolution Google Earth images supported with field validation. The strata used for this study were dense remnant forest patches, woodland and khat cultivated land uses. Ten samples were surveyed in the natural forest and woodland while it was possible to collect seven samples in the khat plantations due to restrictions to access the remaining three plots.

A nested plot design was used to collect data of the different carbon pools from each sample plot location. The total tree height and diameter at breast height (i.e., at 1.3 m above the ground) of individual trees $\geq$50 cm were measured in the sampling plots of 35 m × 35m (1225 m^2) area using hypsometer and calliper or diameter tape based on the size of the trees, respectively. Shrubs and trees with smaller DBH (diameter at breast height), which is the diameter at 1.3 m above the ground, were measured in the nested plots as indicated in Table 1.

Table 1. Size and shape of sampling plots for biophysical data collection.

Stem DBH (cm)	Plot Size	Components Measured
<5	2 m × 2 m	Saplings
5–20	7 m × 7 m	Small trees and shrubs
20–50	25 m × 25 m	Medium size trees
>50	35 m × 35 m	Big trees
—	0.5 m × 0.5 m (3 sub-plots per plot location)	Grasses and litter
—	Soil samples (3 sub-plots per plot location)	Soil carbon and bulk density
—	100 m line intersect per plot with 10 m width	Deadwood

In each of the sample plots in the natural forest and woodland, three subplots were identified for grass, litter and soil sampling. These plots fall were established along the diagonal of the big sample plot at an interval of 12 m. In each of the sub-plots, the herbaceous plants were harvested, the litter separately sampled, and fresh weights measured. Then, at each of the sub-plots, a soil profile was dug to collect soil samples for bulk density and carbon content analysis. Soil samples were taken at three depth class of 0–30 cm, 30–60 cm and 60–100 cm.

Data regarding deadwood biomass was collected in the natural forest and woodland sampling plots along a North–South transect line that passed through the plot centre. The line had a length of 100 m (50 m on to the North and South directions). Deadwood found within a width of 5 m on each side of the line was measured for diameter and length. Besides, wood samples were brought to the laboratory for carbon fraction analysis.

2.3. Plant Diversity Analysis

The list and number of all woody plants was recorded during the field survey for biodiversity assessment. Different biodiversity indicators were calculated to identify the ones that better characterize carbon stock. Below are the equations for calculating the diversity indices used in this study [26].

$$D_S = 1 - \frac{\sum_{i=1}^{s} n_i(n_i - 1)}{N(N - 1)} \tag{1}$$

$$H' = -\sum_{i=1}^{s} p_i \times \ln(p_i) \tag{2}$$

$$R = \frac{s}{\sqrt{N}} \tag{3}$$

where D_S is Simpson's index (entropy); H' is Shannon–Weiner diversity index; R is the species richness index; n_i is the total number of individuals of one species (i); s is the number of species; N is the total number of individuals of all the species; p_i is the species abundance (proportion of the total number of individuals of a species from the community or the probability that a given individual belongs to the species) and is calculated as $p_i = \frac{n_i}{N}$.

2.4. Estimation of Tree and Shrub Biomass Carbon

Live biomass of trees and shrubs were estimated using non-destructive field survey sampling and allometric equations. Tree AGB was estimated from DBH and tree height measurements using the generalized biomass regression model for tropical forests suggested by Chave, Réjou-Méchain [27].

$$y_{tree} = 0.0673 \times \left(\rho \times DBH^2 \times H \right)^{0.976} \tag{4}$$

where y_{tree} is the tree biomass (Kg); ρ is wood basic density (g cm^{-3}); DBH is the diameter at breast height (cm), and H is total tree height (m).

The wood basic density data of individual tree species were obtained from different sources including the global wood density database [28] and ICRAF wood density database (http://db.worldagroforestry.org/wd; accessed on 10 October 2022). Those tree species for which we found local research findings, we used the basic wood density values from the local research findings in Ethiopia [29–32].

The general model of shrub biomass estimation developed for Australia by [33] was adopted for estimating AGB of shrubs and saplings.

$$y_{sh} = 1.128 \times e^{(2.428 \times \ln(D_{10}) - 3.007)} \tag{5}$$

where, y_{sh} is the dry weight of the plant (kg); e is Euler's number, which is a mathematical constant approximately equal to 2.71828; D_{10} is diameter at 10 cm height.

The individual plant biomass in each sample plot (expressed in kg/m^2) was converted to plot-level AGB$_j$ (Mg ha^{-1}) using Equation (6).

$$AGB_j = \frac{\sum_{i=1}^{n} AGB_{i,j}}{a_j} \times \frac{10,000}{1000} \tag{6}$$

where AGB$_{i,j}$ is the aboveground biomass (kg) of a single tree or shrub (i) in plot j; n is the number of measured trees or shrubs in sample plot j; and a$_j$ is the area of plot j in which the tree or shrub was measured; factor 1000 is the conversion of sample units of kg to Mg, and factor 10,000 is the conversion of the area in m^2 to a hectare.

In addition to the AGB, BGB of woody plants is an important carbon pool for many vegetation and land use types. Therefore, the amount of BGB was estimated from root-shoot ratio estimates, which account for about 20–24% of the AGB in tropical forests based on the amount of AGB (Cairns et al., 1997; Mokany et al., 2006). This conversion factor was adopted since BGB estimation is more difficult and time-consuming than estimating AGB. Equation (7) was used to calculate BGB of trees and shrubs based on Ponce-Hernandez (2004) and Gibbs et al. (2007).

$$BGB = \begin{cases} 0.20 \times AGB, & \text{for plots with AGB} < 125 \text{ Mg ha}^{-1} \\ 0.24 \times AGB, & \text{for plots with AGB} \geq 125 \text{ Mg ha}^{-1} \end{cases} \tag{7}$$

Carbon stock content in the total biomass (TB) of trees and shrubs was estimated by multiplying the TB (i.e., the sum of AGB and BGB) by 0.47, which is a default carbon fraction accepted by the international panel on climate change [34,35].

$$BCD = TB \times 0.47 \tag{8}$$

where BCD is the tree and/or shrub biomass carbon density (Mg ha^{-1}).

2.5. Biomass Carbon of Herbaceous Vegetation

For determining biomass carbon stock of herbaceous vegetation (which includes different grass species and seedlings of trees or shrubs), samples collected using destructive techniques were oven-dried at a temperature of 70 °C until constant weights were achieved. Then, the oven-dried samples were ignited at 550 °C for three hours in a furnace for carbon content determination. The ash was weighed and carbon fraction calculated. The carbon stock of herbaceous vegetation samples at each sample plot j (HBC$_j$, Mg ha^{-1}) was calculated using Equation (9), which was recommended by Pearson et al. (2005) [36].

$$HBC_j = \frac{\sum_{i=1}^{3} \left(\frac{W_{field(i,\,j)}}{a} \times \frac{W_{dry(i,\,j)}}{W_{fresh(i,\,j)}} \right)}{3 \times 100} \times cf \tag{9}$$

where $W_{dry(i,\,j)}$, $W_{fresh(i,j)}$ and $W_{field(i,\,j)}$ are weights of the oven-dried sub-sample (ignited), fresh sub-sample (ignited), and fresh sample (field sample) of the ith sample of the jth sample plot, respectively; i is the index of the samples collected in the sampling plot j (1, 2 or 3); cf is the carbon fraction (%); and a is the area of the sampling quadrants, which was 0.25 m^2 in this study.

2.6. Deadwood Carbon

Deadwood carbon was estimated by applying the general log volume estimation technique using Smalian formula (Equation (10)), and converting the estimated volume to biomass and then to carbon [37].

$$V = f\left(Ds^2 + Db^2\right) * L/2 \tag{10}$$

where V = wood volume (m^3), Ds = small diameter (cm), Db = large diameter (cm), L = length (m), and f (i.e., adjustment factor) = 0.00007854.

The wood volume was converted to biomass of deadwood by multiplying it with the density of the wood samples obtained from laboratory analysis [38] (Equation (11)).

$$DWC = V \times \rho \times cf \tag{11}$$

where DWC is the deadwood carbon density (Mg ha^{-1}), ρ is wood density (g m^{-3}), and cf is the carbon fraction.

2.7. Carbon Stock in Litter Biomass

Litter layer is defined as all dead organic surface material on top of the mineral soil. It includes recognizable materials such as dead leaves, twigs, dead grasses and small branches. Equation (9) was used to determine litter biomass carbon (LBC) stock. The total dry weight was determined in the laboratory after oven drying of the sample. Oven-dried samples were taken into pre-weighed crucibles. The samples were ignited at 550 °C for one hour in a furnace. Then, the crucibles with ash were weighed and the percentage of organic carbon was calculated.

2.8. Soil Organic Carbon Estimation

In this study, 81 soil samples were collected from three sub-plots in each of the 27 sample plots. At each of the soil sampling sites, a soil profile to a maximum depth of 100 cm was dug depending on the presence of the impervious bedrock at a depth less than 100 cm. Samples were taken at three profile classes of 0–30 cm (top layer), 30–60 cm (middle layer) and 60–100 cm (bottom layer). Three types of soil variables were measured; namely depth, bulk density (calculated from the oven-dried weight of a known volume of sampled soil), and the concentrations of organic carbon within the sample [36].

For bulk density assessment, undisturbed mineral soil samples were collected using a core sampler and oven-dried at 105 °C for about 24 h until constant weights were attained.

The soil bulk density was calculated as the ratio of the oven-dried weight of the soil sample to the volume of the core sampler (Equation (12)).

$$\text{Bulk density}\ \left(\frac{g}{cm^3}\right) = \frac{\text{mass of oven} - \text{dried soil (g)}}{\text{total volume(cm}^3)} \tag{12}$$

A composite sample was created for each plot for chemical analysis. The Walkley and Black [39] was used to determine organic carbon content after the samples were grounded and passed through a 2 mm sieve. The SOC (Mg ha^{-1}) at each soil profile class or the total SOC was estimated as the product of organic carbon content (C, %), soil bulk density (ρ, kg cm^{-3}) and soil depth (d, cm) as shown in Equation (13) [36,40].

$$\text{SOC} = \rho \times d \times C \tag{13}$$

Hence, the total SOC density at a sampling plot was obtained by summing up the SOC of the soil profile depth classes.

2.9. Carbon Stock in Khat Plantation

We conducted plant density assessment in seven of the planned ten randomly selected khat plantation sites. We found that the average spacing between khat plants in the plots studied was 70 cm, which yields a stocking of 20,000 khat plants per hectare. We did not get permission from the owners of the remaining three sites to collect samples. Thus, 14 khat plants were harvested in the seven sample plots and sorted into stem and root sections. The fresh weight of the stem and root were measured separately. They were dried and their carbon fractions analysed. Then, the carbon stock of the stem and root parts were calculated using a modified form of Equation (9) [36].

$$\text{KBC} = \frac{W_{\text{field}}}{a} \times \frac{W_{\text{dry}(i,\,j)}}{W_{\text{fresh}(i,\,j)}} \times \frac{1}{10,000} \times cf \tag{14}$$

where KBC is khat biomass carbon (aboveground or belowground); $W_{\text{dry}(i,\,j)}$, $W_{\text{fresh}(i,j)}$ and $W_{\text{field}(i,\,j)}$ are weights of the oven-dried sub-sample (ignited), fresh sub-sample (ignited), and fresh sample (field sample) of the ith sample of the jth sample plot, respectively; i is the index of the samples collected in the sampling plot j (1, 2 or 3); cf is the carbon fraction (%); and a is the area of the sampling quadrants, which was 0.25 m^2 in this study.

The aboveground and belowground biomass carbon stocks of the khat plants in the same plot were averaged to get a typical estimate of aboveground khat biomass carbon (AKBC) and belowground khat biomass carbon (BKBC) stocks. The AKBC and BKBC of respective plots were added to obtain TBC of the khat plots.

2.10. Total Carbon Stock Estimation

The total carbon stock (TCS) at each sample plot was calculated by summing up the carbon stock densities of the individual carbon pools in the plot. Accordingly, TCS was estimated using Equation (15) [41].

$$\text{TCS} = \begin{cases} \text{TBC} + \text{HBC} + \text{DWC} + \text{LBC} + \text{SOC;} & \text{for natural forest and woodland} \\ \text{AKBC} + \text{BKBC} + \text{SOC;} & \text{for khat plantation} \end{cases} \tag{15}$$

2.11. Statistical Analyses

Shannon-Wiener, Simpson's diversity and species richness indices were compared to identify one index for discussing biodiversity in the land used types and its relationship with carbon stock. Pearson's correlation analysis was used to explore the relationship between biodiversity and biomass carbon stock.

We used analysis of variance (ANOVA) test to study the effect of forest disturbance or conversion to monoculture khat plantation on the carbon stock distribution. The fixed effects of land use (natural forest, woodland, and khat-cultivated land) on biomass and carbon stock distribution in each carbon pool were compared. For those factors that showed significant differences in biomass or carbon distributions, the mean estimates were compared using the Tukey honestly significant difference (HSD) test ($p = 0.05$). ANOVA test was also used to explore the effect of soil depth on mean estimates of SOC. Data analysis was run in python programming language.

3. Results

3.1. Plant Diversity Indices

The three indices of plant diversity used in this study were strongly intercorrelated (Table 2) suggesting the possibility to select one of them without causing a severe consequence in the overall results. Therefore, we chose the Simpson's index (Ds) in this paper since it is strongly correlated to the commonly used Shannon–Wiener index (H') and is a standardized index that helps comparison of the results between land-use types.

Table 2. Pearson's correlation coefficient between the three diversity indices used in the current study.

	Shannon-Wiener Index (H')	Simpson's Index (Ds)
Species richness (R)	0.947707	0.942293
Shannon-Wiener index (H')	—	0.979202

Results indicated that natural forest has higher diversity than woodland with mean Ds of 0.84 and 0.76, respectively (Table 3). The khat plantation has Ds of zero since it is a monoculture. Results of ANOVA test revealed that the mean Ds index among the land use types were significant at 95% confidence level. However, multiple comparisons of the means were assessed using the Tukey HSD test and showed that the mean diversity estimates of forest and woodland were not significantly different. The uncertainty in the estimates of Ds in the natural forest was less than that in the woodland indicating the presence of a more diverse species and intact distribution around the average in the forest patches than those in the woodland where the species diversity swings wider from the average.

Table 3. Mean (±standard error) of Ds of woody plants in three land-use types. Mean values followed by the same letter did not differ significantly (Tukey HSD test, $p > 0.05$).

Land-Use Type	Mean Ds
Natural forest	0.84 (±0.02) [a]
Woodland	0.76 (±0.05) [a]
Khat	0.00 (±0.00) [b]

Findings of correlation analysis indicated that there was no statistically significant relationship between biodiversity and TBC, SOC or TCS estimates at a 95% level of confidence (Table 4).

Table 4. Correlation of Ds with total biomass carbon (TBC), soil organic carbon (SOC) and total carbon stocks (TCS); all in Mg ha^{-1}.

Sources of Carbon	Correlation Coefficient	p-Value
TBC	0.304	0.123
SOC	0.242	0.225
TCS	0.320	0.104

3.2. Biomass Distribution by Land Use Type

Table 5 shows that there was a distinct plant biomass distribution in the three land-use types. The natural forest had the highest plant biomass (456.93 Mg ha^{-1}) followed by woodland (19.78 Mg ha^{-1}). Although the planting stock of khat was very dense at an average spacing of 70 cm (i.e., 20,000 plants ha^{-1}), its total biomass density was only 2.46 Mg ha^{-1}. The mean plant biomass (sum of AGB and BGB) estimates of the plots in the natural forest had larger standard error than the plots in the woodland and khat plantation.

Table 5. Mean plant biomass and its standard error among the different land-use types.

Land-Use Type	Total Live Biomass (Mg ha^{-1})	Standard Error of the Mean (Mg ha^{-1})
Natural forest	456.93	155.88
Woodland	19.78	6.29
Khat	2.46	0.59

3.3. Comparison of Carbon Stocks by Land Use Type

Estimates of carbon stock in different carbon pools by land-use types are shown in Table 6. The mean TCS ranged from 125.86 Mg ha^{-1} in the khat plantation to 366.47 Mg ha^{-1} in the natural forest. The results revealed that the proportion of carbon stock in the carbon pool components vary among land-use types. The contribution of HBC and DWC to the TCS of each land use type was not significant with less than 1% share of the TCS. More than half (58.6%) of the carbon in the natural forest was contributed by the biomass of trees and shrubs while the share of SOC was 40.8%. On the other hand, the share of SOC was more than 92% and 99% of the TCS of woodland and khat plantation, respectively.

Table 6. Mean carbon stock (Mg ha^{-1}) and percentage contribution (values in brackets) in each carbon pool component and land-use class. Percentage contributions of the soil organic carbon (SOC), total biomass carbon (TBC), litter biomass carbon (LBC), herbaceous vegetation biomass carbon (HBC) and deadwood carbon (DWC) were calculated for each land-use type. The TCS varies between natural forest and other land-uses at a significance level of 0.05.

Land Use Type	SOC	TBC	LBC	HBC	DWC	TCS
Natural forest	149.57 (40.80)	214.76 (58.60)	1.33 (0.36)	0.40 (0.11)	0.41 (0.11)	366.47 [a]
Woodland	130.53 (92.02)	9.30 (6.60)	1.56 (1.10)	0.46 (0.30)	0.00 (0.00)	141.85 [b]
Khat	124.70 (99.10)	1.16 (0.90)	0.00 (0.00)	0.00 (0.00)	0.00 (0.00)	125.86 [b]

The share of different carbon pool components in the TCS was significantly different among the land use types at 0.05 level of significance.

3.4. SOC Distribution across Soil Depth Classes

The biomass carbon stock from woodland and khat plantation was very small (Table 5), particularly as compared to the SOC stocks. The study has shown that most of the TCS of these land-use types came from the SOC. Therefore, we explored the variability of SOC across the soil depths and land use types. Figure 2 shows the vertical distribution and statistical variation of SOC stock by land-use types. The results revealed that SOC stocks did not significantly vary by land-use type. However, SOC stocks vary significantly among soil depth classes at 0.05 level of significance. The uncertainty in SOC stock estimates in the sample plots of the natural forest was the largest while that in the woodland was the smallest.

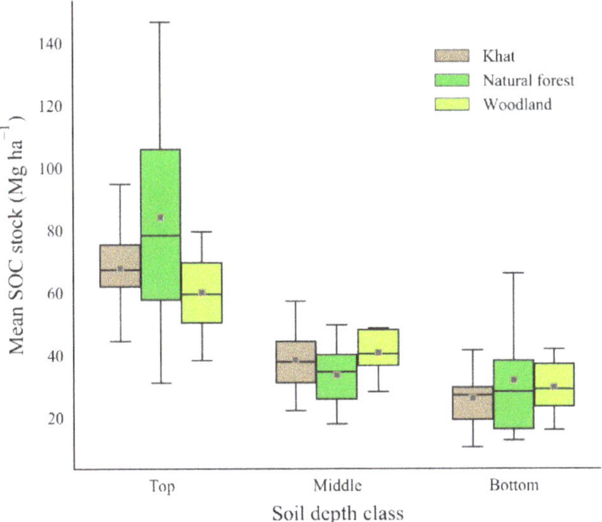

Figure 2. Mean SOC stock of each land use type across the depth classes.

Multiple pairwise comparison of means using Tukey HSD test indicated that the SOC at the top layer was significantly different from that at the middle and bottom layers (see Table 7) at 0.05 level of significance.

Table 7. Multiple comparison of means, level of significance = 0.05.

Group1	Group2	Mean Difference	Adjusted p-Value	Lower	Upper	Reject (Decision)
Bottom	Middle	8.4329	0.1499	−2.2491	19.1149	False
Bottom	Top	41.4862	0.001	30.8042	52.1682	True
Middle	Top	33.0533	0.001	22.3713	43.7353	True

4. Discussion

4.1. Relationship between Biodiversity and Carbon Stock

Assessing biodiversity, which is an indicator of the stability of an ecosystem, is useful for understanding the status of a vegetation ecosystem [42]. Three of the biodiversity indices used in this study were strongly intercorrelated and hence Simpson's diversity index was selected for further analysis since it has standardised values suited for comparison across land-use types (Table 2). As expected, the Simpson's diversity index of khat plantation was significantly lower than that of the natural forest and woodland since it was covered with a single species only (Table 3). However, the estimates of mean Simpson's diversity index for natural forest and woodland were not significantly different from each other (Table 4). This might indicate the level of disturbance in the form of illegal logging of economically important tree species and illegal settlement in the natural forest, which contribute to reducing the woody plant species diversity [24].

Correlation analysis of Simpson's diversity index with TBC, SOC and TCS revealed that plant diversity has a positive relationship with the different carbon stock values implying the potential of improved biodiversity for enhancing carbon sequestration (Table 4), which is in agreement with the findings of Strassburg, Kelly [43]. However, the relationship was not statistically significant at 95% confidence level. Hence, diversity indices did not explain the variability in biomass carbon, SOC or TCS in this study. The absence of a statistically significant difference is in line with other studies [19,44,45].

The study results indicated that the natural forest had considerably higher biomass than the woodland and khat plantation (Table 5). The high biomass distribution in the natural forest might be attributed to the fact that fuelwood collectors from the northern part of the catchment have a topographic barrier that restricted them from easily accessing to the site [22]. Similarly, government institutions such as Wondo Genet College of Forestry and Natural Resources protected the remnant forest patches in the gorges and saved big trees from illegal loggers. The higher standard error in the biomass estimates reveals this phenomenon (Figure 2). The high variability in the biomass distribution of the natural forest shows the heterogeneity in the growing environment (Figure 1) mainly due to topography [22]. The natural forests of Wondo Genet sub-catchment accumulate relatively similar TB to the findings in a dry Afromontane forest in Awi Zone of northwestern Ethiopia [46].

4.2. Carbon Pools and Their Contribution to the Total Carbon Stock

The research findings indicated that herbaceous vegetation and deadwood biomass had a very small contribution to the TCS estimates in each of the land use types under study (Table 6). The small biomass of herbaceous vegetation could be related to forest degradation due to traditional fire and illegal grazing on the undergrowth vegetation [24,47]. The intensive firewood collection because of easy access from the nearby towns and villages might be responsible for the small deadwood biomass stock in the natural forest and woodland. A similar result was observed in Pakistan [48]. In the natural forest and woodland land-use types, total biomass carbon of trees and shrubs and SOC were important in determining the TCS. The small contribution of litter biomass carbon to the TCS might be related to the extent of disturbance as observed in Taita Hills forests of Kenya in the study by Wekesa, Leley [49]. There has been frequent forest fires that could remove the litter layer [47]. Almost all of the TCS of the khat plantation came from the SOC. Examination of the variability in mean TCS estimates revealed that land use was the main factor influencing the TCS distribution. The mean estimates of TCS in the natural forests of Wondo Genet sub-catchment were higher than that of the woodland and khat plantation. The estimated TCS of the natural forest agreed with the findings by Addi et al. [50] in the moist Afromontane Gesha-Sayilem forest, which had a TCS of 362.4 Mg ha^{-1}, and that by Gebeyehu et al. [46] in the dry Afromontane forests in Awi Zone. This large TCS in the natural forest could be attributed to the presence of big trees in the gorges where moisture and nutrients are abundant (Table 6). Biomass carbon component was responsible for the differences in TCS among the land use types. Generally, the natural forest in Wondo Genet sub-catchment stores considerably large amount of carbon and plays an important role in mitigating the challenges of climate change.

4.3. Soil Organic Carbon in Different Land Uses and Its Variability across Depth Classes

The study results indicated that there was more SOC at the top layer in the natural forest than in the woodland and khat plantation (Table 6). This could be related to the level of disturbance in the land use types. The natural forest is relatively less disturbed and has more inflow of organic matter that increased the SOC at the top layer. The uncertainty in estimates of SOC at the top layer in the sample plots of the natural forest was also larger than that of the woodland and khat plantation (Figure 2), which might be associated with the topographic variation that resulted in microclimate variability and soil organic matter distribution. More SOC accumulates in the gorges while ridges have low SOC. However, the differences in SOC levels among the land use types were not statistically different (Figure 2). This shows the stability of SOC among land-use types. Among the variables explored in this study, only soil depth was the main factor that explained the variability in mean SOC estimates in the current study (Figure 2; Table 7). The topsoil had significantly larger SOC than the middle and bottom layers, the latter of which were not statistically distinct from each other (Table 7).

5. Conclusions

A weak linear relationship that exists between biodiversity metrics and carbon stock variables implies the feasibility of any forest type for carbon sequestration projects regardless of the diversity of species.

There was no statistically significant correlation between TBC, SOC and TCS and the selected biodiversity indicator in this study. Although existing literature supported the results, the study had a small sample size and limited geographic coverage to generalize about the actual relationship of biodiversity and the different carbon pools. We recommend further investigation in a wider geographic area with a larger sample size to generalize the findings in this study.

The SOC and TBC contributed to a large part of the TCS estimates for each of the land-use types used in this study. The SOC was consistent in all the land use types making it a stable carbon pool that serves as a sink of organic carbon. In addition to the SOC, a large proportion of the TCS in the natural forest was contributed by the TBC in trees and shrubs. Forest degradation and conversion to cash crop production systems such as khat plantation affect the carbon budget significantly. Therefore, REDD+ projects should focus on afforestation and sustainable forest management to sequester carbon in the woody plant biomass. On the other hand, the proportion of HBC, DWC and LBC were insignificant. Thus, carbon monitoring projects, particularly in dense forest ecosystems; need to focus on accounting for the biomass carbon in trees and shrubs as well as the SOC.

As an important carbon pool, the SOC has a significant contribution to the TCS in each land-use type. The topsoil profile has large SOC stock irrespective of the land use types. Although the variation was not significant, the results indicated that the natural forest has a higher SOC in the topsoil layer than the other land-use types. The estimated SOCs of the land use types were inseparable in the lower soil layers. This has implication of maintaining appropriate topsoil management to enhance carbon accumulation in the soil.

Author Contributions: H.T.: Conceptualization, Methodology, Software, Data gathering, Formal analysis, Investigation, Validation, Writing original draft preparation, Review and Editing; M.N. and T.G.: Review and Editing. G.T.A.: Conceptualization, Resource, Validation, Review and Editing, and acquisition of APC. All authors have read and agreed to the published version of the manuscript.

Funding: Gebiaw T. Ayele covered the APC and received funding from Griffith Graduate Research School, the Australian Rivers Institute and School of Engineering, Griffith University, Queensland, Australia. The UK Aid from the UK government has funded this research; however, the views expressed do not necessarily reflect the UK government's official policies.

Informed Consent Statement: Not applicable.

Data Availability Statement: All the data are presented in tables and figures in the manuscript.

Acknowledgments: The UK Aid from the UK government has funded this research; however, the views expressed do not necessarily reflect the UK government's official policies. The technical and logistics support from the CIRCLE programme of the African Academy of Sciences was tremendous. We would like to thank Addis Ababa and Hawassa universities for the support we got throughout the research process. We are indebted to gratitude all the members of the field crew who participated in the fieldwork for their honest contribution. Gebiaw T. Ayele received funding and acknowledges Griffith Graduate Research School, the Australian Rivers Institute and School of Engineering, Griffith University, Queensland, Australia.

Conflicts of Interest: The authors declare no conflict of interest.

References

1. Codjoe, S.N.A.; Atiglo, D.Y. The implications of extreme weather events for attaining the sustainable development goals in sub-Saharan Africa. *Front. Clim.* **2020**, *2*, 592658. [CrossRef]
2. Zhao, C.; Liu, B.; Piao, S.; Wang, X.; Lobell, D.B.; Huang, Y.; Huang, M.T.; Yao, Y.T.; Bassu, S.; Ciais, P.; et al. Temperature increase reduces global yields of major crops in four independent estimates. *Proc. Natl. Acad. Sci. USA* **2017**, *114*, 9326–9331. [CrossRef] [PubMed]

3. Wu, X.; Lu, Y.; Zhou, S.; Chen, L.; Xu, B. Impact of climate change on human infectious diseases: Empirical evidence and human adaptation. *Environ. Int.* **2016**, *86*, 14–23. [CrossRef] [PubMed]

4. Schweikert, A.; Chinowsky, P.; Espinet, X.; Tarbert, M. Climate change and infrastructure impacts: Comparing the impact on roads in ten countries through 2100. *Procedia Eng.* **2014**, *78*, 306–316.

5. Schlenker, W.; Lobell, D.B. Robust negative impacts of climate change on African agriculture. *Environ. Res. Lett.* **2010**, *5*, 14010. [CrossRef]

6. Filho, W.L.; Balogun, A.-L.; Ayal, D.Y.; Bethurem, E.M.; Murambadoro, M.; Mambo, J.; Taddese, H.; Tefera, G.W.; Nagy, G.J.; Fudjumdjum, H.; et al. Strengthening climate change adaptation capacity in Africa-case studies from six major African cities and policy implications. *Environ. Sci. Policy* **2018**, *86*, 29–37. [CrossRef]

7. Arce, J.J.C. Forests, inclusive and sustainable economic growth and employment. In *Background Study Prepared for the Fourteenth Session of the United Nations Forum on Forests*; United Nations: New York, NY, USA, 2019.

8. Canadell, J.G.; Raupach, M.R. Managing forests for climate change mitigation. *Science* **2008**, *320*, 1456–1457.

9. Köhl, M.; Lasco, R.; Cifuentes, M.; Jonsson, Ö.; Korhonen, K.T.; Mundhenk, P.; de Jesus Navar, J.; Stinson, G. Changes in forest production, biomass and carbon: Results from the 2015 UN FAO Global Forest Resource Assessment. *For. Ecol. Manag.* **2015**, *352*, 21–34.

10. Lewis, S.L.; Lopez-Gonzalez, G.; Sonké, B.; Affum-Baffoe, K.; Baker, T.R.; Ojo, L.O.; Phillips, O.L.; Reitsma, J.M.; White, L.; Comiskey, J.A.; et al. Increasing carbon storage in intact African tropical forests. *Nature* **2009**, *457*, 1003–1006. [CrossRef]

11. Baccini, A.; Goetz, S.; Walker, W.S.; Laporte, N.T.; Sun, M.; Sulla-Menashe, D.; Hackler, J.L.; Beck, P.S.A.; Dubayah, R.O.; Friedl, M.A.; et al. Estimated carbon dioxide emissions from tropical deforestation improved by carbon-density maps. *Nat. Clim. Change* **2012**, *2*, 182–185. [CrossRef]

12. Kim, D.-H.; Sexton, J.O.; Townshend, J.R. Accelerated deforestation in the humid tropics from the 1990s to the 2000s. *Geophys. Res. Lett.* **2015**, *42*, 3495–3501. [CrossRef] [PubMed]

13. Zeppetello, L.R.V.; Parsons, L.A.; Spector, J.T.; Naylor, R.L.; Battisti, D.S.; Masuda, Y.J.; Wolff, N.H. Large scale tropical deforestation drives extreme warming. *Environ. Res. Lett.* **2020**, *15*, 84012. [CrossRef]

14. Aalde, H.; Gonzalez, P.; Gytarsky, M.; Krug, T.; Kurz, W.A.; Lasco, R.D.; Martino, D.L.; McConkey, B.G.; Ogle, S.; Paustian, K.; et al. Generic methodologies applicable to multiple land-use categories. *IPCC Guidel. Natl. Greenh. Gas Invent.* **2006**, *4*, 1–59.

15. Cannell, M. Growing trees to sequester carbon in the UK: Answers to some common questions. *Forestry* **1999**, *72*, 237–247. [CrossRef]

16. Houghton, R.A.; Hall, F.; Goetz, S. Importance of biomass in the global carbon cycle. *J. Geophys. Res. Biogeosci.* **2009**, *114*, 935. [CrossRef]

17. Giam, X. Global biodiversity loss from tropical deforestation. *Proc. Natl. Acad. Sci. USA* **2017**, *114*, 5775–5777.

18. Riswan, S.; Hartanti, L. Human impacts on tropical forest dynamics. *Vegetatio* **1995**, *121*, 41–52.

19. Mandal, R.A.; Dutta, I.C.; Jha, P.K.; Karmacharya, S. Relationship between carbon stock and plant biodiversity in collaborative forests in Terai, Nepal. *Int. Sch. Res. Not.* **2013**, *2013*, 625767. [CrossRef]

20. Chisholm, R.A.; Muller-Landau, H.; Rahman, K.A.; Bebber, D.P.; Bin, Y.; Bohlman, S.A.; Bourg, N.; Brinks, J.; Bunyavejchewin, S.; Butt, N.; et al. Scale-dependent relationships between tree species richness and ecosystem function in forests. *J. Ecol.* **2013**, *101*, 1214–1224. [CrossRef]

21. Pitkänen, T.P.; Raumonen, P.; Liang, X.; Lehtomäki, M.; Kangas, A. Improving TLS-based stem volume estimates by field measurements. *Comput. Electron. Agric.* **2021**, *180*, 105882. [CrossRef]

22. Dessie, G.; Kleman, J. Pattern and magnitude of deforestation in the South Central Rift Valley Region of Ethiopia. *Mt. Res. Dev.* **2007**, *27*, 162–168.

23. Amare, T.; Hergarten, C.; Hurni, H.; Wolfgramm, B.; Yitaferu, B.; Selassie, Y.G. Prediction of soil organic carbon for Ethiopian highlands using soil spectroscopy. *Int. Sch. Res. Not.* **2013**, *2013*, 720589. [CrossRef]

24. Dessie, G.; Kinlund, P. Khat expansion and forest decline in Wondo Genet, Ethiopia. *Geogr. Ann. Ser. B Hum. Geogr.* **2008**, *90*, 187–203.

25. Kebede, M.; Yirdaw, E.; Luukkanen, O.; Lemenih, M. Plant community analysis and effect of environmental factors on the diversity of woody species in the moist Afromontane forest of Wondo Genet, South Central Ethiopia. *Biodivers. Res. Conserv.* **2013**, *29*, 63. [CrossRef]

26. Peet, R.K. The measurement of species diversity. *Annu. Rev. Ecol. Syst.* **1974**, *5*, 285–307.

27. Chave, J.; Réjou-Méchain, M.; Búrquez, A.; Chidumayo, E.; Colgan, M.S.; Delitti, W.B.; Duque, A.; Eid, T.; Fearnside, P.M.; Goodman, R.C.; et al. Improved allometric models to estimate the aboveground biomass of tropical trees. *Glob. Change Biol.* **2014**, *20*, 3177–3190.

28. Zanne, A.E.; Lopez-Gonzalez, G.; Coomes, D.A.; Ilic, J.; Jansen, S.; Lewis, S.L.; Miller, R.B.; Swenson, N.G.; Wiemann, M.C.; Chave, J. Data from: Towards a worldwide wood economics spectrum. *Dryad Dataset* **2009**. [CrossRef]

29. Belete, Y.; Abere, F.; Kebede, B.; Soromessa, T. Nondestructive Allometric Model to Estimate Aboveground Biomass: An Alternative Approach to Generic Pan-Tropical Models. 2019. Available online: https://ssrn.com/abstract=3475959 (accessed on 11 October 2022).

30. Desalegn, G.; Teketay, D.; Gezahgne, A.; Abegaz, M. *Commercial Timber Species in Ethiopia: Characteristics and Uses. A Handbook, for Forest Industries, Construction and Energy Sectors, Foresters and Other Stakeholders*; Addis Ababa University Press: Addis Ababa, Ethiopia, 2012.
31. Tadesse, G.; Zavaleta, E.; Shennan, C. Effects of land-use changes on woody species distribution and above-ground carbon storage of forest-coffee systems. *Agric. Ecosyst. Environ.* **2014**, *197*, 21–30. [CrossRef]
32. Tetemke, B.A.; Birhane, E.; Rannestad, M.M.; Eid, T. Allometric models for predicting aboveground biomass of trees in the dry afromontane forests of Northern Ethiopia. *Forests* **2019**, *10*, 1114. [CrossRef]
33. Paul, K.I.; Roxburgh, S.H.; Chave, J.; England, J.R.; Zerihun, A.; Specht, A.; Lewis, T.; Bennett, L.T.; Baker, T.G.; Adams, M.A.; et al. Testing the generality of above-ground biomass allometry across plant functional types at the continent scale. *Glob. Change Biol.* **2016**, *22*, 2106–2124. [CrossRef]
34. IPCC (International Panel on Climate Change). *2006 IPCC Guidelines for National Greenhouse Gas Inventories*; Prepared by the National Greenhouse Gas Inventories Programme; Eggleston, H.S., Buendia, L., Miwa, K., Ngara, T., Tanabe, K., Eds.; IGES: Kanagawa, Japan, 2006.
35. McGroddy, M.E.; Daufresne, T.; Hedin, L.O. Scaling of C: N: P stoichiometry in forests worldwide: Implications of terrestrial redfield-type ratios. *Ecology* **2004**, *85*, 2390–2401.
36. Pearson, T.; Walker, S.; Brown, S. *Sourcebook for Land Use, Land-Use Change and Forestry Projects*; World Bank: Washington, DC, USA, 2013.
37. Prodan, M. *Holzmesslehre*; JD Sauerlander: Frankfurt am Main, Germany, 1965.
38. Brown, S.; Lugo, A.E. Aboveground biomass estimates for tropical moist forests of the Brazilian Amazon. *Interciencia Caracas* **1992**, *17*, 8–18.
39. Walkley, A.; Black, I.A. An examination of the Degtjareff method for determining soil organic matter, and a proposed modification of the chromic acid titration method. *Soil Sci.* **1934**, *37*, 29–38. [CrossRef]
40. Kauffman, J.B.; Donato, D.C. *Protocols for the Measurement, Monitoring and Reporting of Structure, Biomass and Carbon Stocks in Mangrove Forests*; Citeseer: Bogor, Indonesia, 2012; Volume 86.
41. Pearson, T.R. *Measurement Guidelines for the Sequestration of Forest Carbon*; U.S. Department of Agriculture, Forest Service, Northern Research Station: Madison, WI, USA, 2007; Volume 18.
42. Brown, E.D.; Williams, B.K. Ecological integrity assessment as a metric of biodiversity: Are we measuring what we say we are? *Biodivers. Conserv.* **2016**, *25*, 1011–1035.
43. Strassburg, B.B.; Kelly, A.; Balmford, A.; Davies, R.G.; Gibbs, H.K.; Lovett, A.; Miles, L.; Orme, C.D.L.; Price, J.; Turner, R.K.; et al. Global congruence of carbon storage and biodiversity in terrestrial ecosystems. *Conserv. Lett.* **2010**, *3*, 98–105. [CrossRef]
44. Choliq, M.L.; Kaswanto, R. Correlation of carbon stock and biodiversity index at the small scale agroforestry landscape in Ciliwung Watershed. In *IOP Conference Series: Earth and Environmental Science*; IOP Publishing: Bristol, UK, 2017.
45. Filqisthi, T.A.; Kaswanto, R.L. Carbon stock and plants biodiversity of pekarangan in Cisadane watershed West Java. In *IOP Conference Series: Earth and Environmental Science*; IOP Publishing: Bristol, UK, 2017.
46. Gebeyehu, G.; Soromessa, T.; Bekele, T.; Teketay, D. Carbon stocks and factors affecting their storage in dry Afromontane forests of Awi Zone, northwestern Ethiopia. *J. Ecol. Environ.* **2019**, *43*, 7.
47. Kim, D.-G.; Taddese, H.; Belay, A.; Kolka, R. The impact of traditional fire management on soil carbon and nitrogen pools in a montane forest, southern Ethiopia. *Int. J. Wildland Fire* **2016**, *25*, 1110–1116. [CrossRef]
48. Ali, A.; Ashraf, M.I.; Gulzar, S.; Akmal, M. Estimation of forest carbon stocks in temperate and subtropical mountain systems of Pakistan: Implications for REDD+ and climate change mitigation. *Environ. Monit. Assess.* **2020**, *192*, 198. [CrossRef]
49. Wekesa, C.; Leley, N.; Maranga, E.; Kirui, B.; Muturi, G.; Mbuvi, M.; Chikamai, B. Effects of forest disturbance on vegetation structure and above-ground carbon in three isolated forest patches of Taita Hills. *Open J. For.* **2016**, *6*, 142–161.
50. Addi, A.; Demissew, S.; Soromessa, T.; Asfaw, Z. Carbon stock of the moist Afromontane forest in Gesha and Sayilem Districts in Kaffa Zone: An implication for climate change mitigation. *J. Ecosyst. Ecograph.* **2019**, *9*, 1–8.

 *land*

Article

Evaluating Biophysical Conservation Practices with Dynamic Land Use and Land Cover in the Highlands of Ethiopia

Meseret B. Addisie [1], Gashaw Molla [2], Menberu Teshome [3] and Gebiaw T. Ayele [4,*

[1] Guna Tana Integrated Field Research and Development Center, Debre Tabor University, Debre Tabor P.O.Box 272, Ethiopia

[2] Department of Geography and Environmental Studies, Bahir Dar University, Bahir Dar P.O.Box 79, Ethiopia

[3] Department of Geography and Environmental Studies, Debre Tabor University, Debre Tabor P.O.Box 272, Ethiopia

[4] Australian Rivers Institute, School of Engineering and Built Environment, Griffith University, Nathan, QLD 4111, Australia

[*] Correspondence: gebiaw.ayele@griffithuni.edu.au or gebeyaw21@gmail.com

Citation: Addisie, M.B.; Molla, G.; Teshome, M.; Ayele, G.T. Evaluating Biophysical Conservation Practices with Dynamic Land Use and Land Cover in the Highlands of Ethiopia. *Land* **2022**, *11*, 2187. https://doi.org/10.3390/land11122187

Academic Editors: Kleomenis Kalogeropoulos, Andreas Tsatsaris, Nikolaos Stathopoulos, Demetrios E. Tsesmelis, Nilanchal Patel and Xiao Huang

Received: 3 November 2022
Accepted: 30 November 2022
Published: 2 December 2022

Publisher's Note: MDPI stays neutral with regard to jurisdictional claims in published maps and institutional affiliations.

Abstract: Ethiopia is one of the sub-Saharan countries affected by land degradation, notably by soil erosion. The government of Ethiopia has launched an extensive biophysical soil and water conservation (SWC) effort each year to address the problem. These practices were installed on varying land use and land cover (LULC) systems. Despite the fact that the interventions covered the majority of the landmasses, there were no quantitative data on the scale of biophysical measures with the change in land use and land cover. Therefore, the objective of this study was to evaluate biophysical conservation practices with dynamic land use and land cover in the highlands of Ethiopia. The study focused on districts of the Amhara regional state's South Gondar zone. A mixed research methodology was employed to gather pertinent data for the study. The dynamics of LULC were analyzed using satellite images acquired between 1990 and 2020. Biophysical conservation measures' data and qualitative information were collected from the zonal office of agriculture. Twelve years' worth of biophysical SWC measures data were used for the study. The results indicate that cultivated land makes up the majority of land use and land cover. Bunds built on cultivated land account for 93% of conservation practices. During the study period, there was a significant decline of biophysical conservation practices implementation in each district. Although plantation was used on a wider scale, it was unable to sustain physical SWC practices or expand forest cover in the region. In addition, lack of integrated maintenance for early installed structures decreases the effectiveness of SWC measures. In conclusion, the dynamics of LULC have a significant impact on the magnitude of biophysical conservation measures. Therefore, watershed managers shall consider the spatio-temporal variation of LULC while planning conservation practices.

Keywords: land degradation; land use land cover; biophysical measures; maintenance; sustainability

1. Introduction

Land degradation is the most serious problem in sub-Saharan African countries, including Ethiopia. Land degradation in the form of soil erosion and soil fertility depletion has profound implications for agricultural gross domestic productivity for low-income countries [1–3]. People have been harming the environment for short-term gain despite the long-term benefits of sustainable management of natural resources. This goes against the idea of sustainable development goals, which guarantee that future generations will be able to utilize the same resources. For the vast majority of Ethiopians, natural resources (soil, water, and forests) are their primary means of subsistence. Natural resources are frequently degraded at the expense of being used to supply the rising demand for food. The deterioration of these resources has resulted in reduced agricultural productivity, poor environmental condition, and a subsequent decline in quality of life. The largest share of the land use is under agriculture. The cumulative actions on cultivated land have eventually resulted in land degradation [4]. Uncontrolled population growth, inappropriate farming systems, and poor land management practices exert pressure on the existing natural resource base [5,6].

Land degradation is largely associated with land use and land cover (LULC) changes. Studies have been reported to demonstrate the occurrences, causes, and effects of land degradation in various regions of Ethiopia [7–9]. Ethiopia relies on a variety of biological resources for its socioeconomic development, particularly its forests. However, there is currently an enormous strain on these resources. As the population is dependent on agricultural expansion and the clearing of forests and vegetated areas, deforestation is a key cause of land use change. Research findings indicated that factors contributing to unsustainable resource management are clearance of forests for agricultural expansion and the absence of energy-saving alternative technologies. In addition, free grazing practices are also a major source of land degradation. Both clearance of vegetated areas and poor livestock management resulted in accelerated run-off, decreased groundwater recharge, increased sediment load on rivers, siltation of reservoirs, increased flood frequency, and damage to the aquatic environment. Out of the 60 million ha of fertile agricultural land in the Amhara regional state, the assessment in [10] states that around 27 million ha are severely eroded, 14 million ha are seriously degraded, and more than 2 million ha are beyond reclamation. In order to prevent problems, a thorough degraded area restoration practice was required.

In the 1970s and 1980s, the Ethiopian government began implementing natural resource conservation measures in the highlands as a reaction to the issue of land degradation (Shiferaw and Holden, 1998). At this time, physical construction was the primary focus of the soil and water conservation (SWC) efforts. Hurni [11] set up experimental watersheds all over Ethiopia and observed soil erosion and discharge before and after implementation to determine the efficacy of these techniques. The research revealed that hillside SWC methods were successful in reducing soil losses in the first five years following implementation [12,13]. After following the construction of conservation structures for ten years, Guzman [14] conducted a new analysis of their efficacy and discovered that SWC measures had become ineffective. Physical measures being ineffectual, a strong recommendation is made for integration with biological measures. Hence, biophysical techniques are essential for land rehabilitation operations in a sustainable manner.

The South Gondar zone (SGZ) is one of the most degraded areas in the Amhara regional state. This zone has experienced rapid land use and land cover change. In contrast to other parts of Ethiopia, SGZ has a number of unproductive areas and degraded lands. The main reasons for the degradation of natural resources include soil erosion, poor grazing management, and deforestation. To address these problems, the zonal agricultural office has been practicing different land reclamation interventions using biophysical measures. The zone utilizes an integrated approach that is implemented at a watershed scale. The interventions aimed to improve social and environmental protection, decrease runoff and soil erosion, boost land production, and sustainably restore damaged ecosystems. The community-based participation that was used in the development interventions supports the empowerment of the community to manage shared resources. However, the link between such efforts and the changes in LULC has not been well-studied in the past. Therefore, the main objective of this paper is to document the spatial and temporal dynamics of development interventions in the South Gondar zone from 2011 to 2022. The study area was selected because it is an appropriate location where different biophysical conservation efforts have been made for a long time.

2. Materials and Methods

2.1. Study Area

Geographically, South Gondar zone is situated between $11°02'16''$ and $12°32'22''$ N latitude and $37°25'35''$ and $38°43'50''$ E longitude (Figure 1). The zone covers a total area of 14,064 km^2. South Gondar is bordered by zones in the Amhara regional state, including Gondar on the north, East Gojjam on the south, North Wollo on the east, Lake Tana on the west, West Gojjam on the southwest, Wag Hemra on the northeast, and South Wollo on the southeast. The major rainy season lasts between June and September in the study area, with July and August being the wettest months. The rainfall varies between 900 mm and 1599 mm, with an average annual rainfall of 1300 mm. The dominant soil types are Eutric nitisols, followed by Orthic luvisols and Haplic xerosols [9]. Mixed farming practices, including crop production and animal husbandry, are the dominant livelihood options of the people. The farming system is characterized by smallholder subsistence farming that depends on rain-fed traditional oxen-driven agriculture. The zone is characterized by high soil erosion; hence, large-scale soil and water conservation practices have been implemented since 2011 through government-sponsored campaigns.

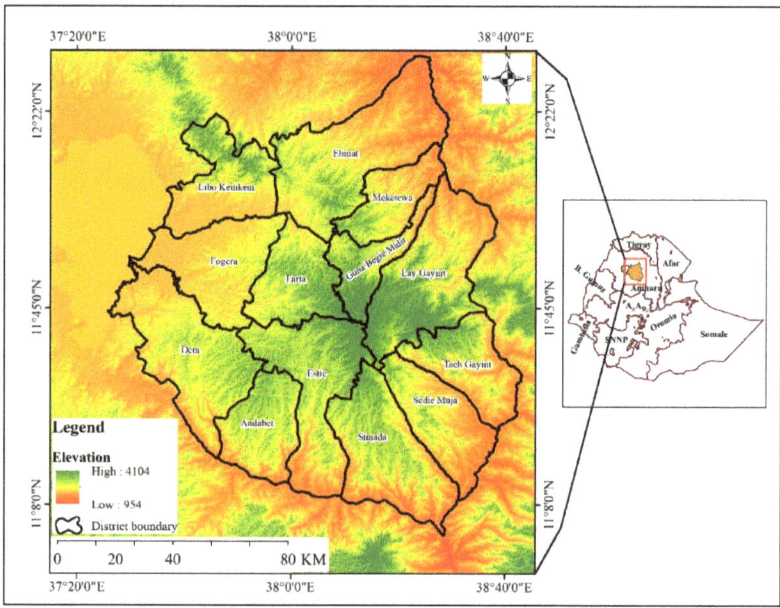

Figure 1. Location map of the study area.

2.2. Data Collection and Analysis

2.2.1. Land Use Land Cover

Five LULC types were extracted from Landsat Images of the United States Geological Survey (USGS) in the study landscape with three reference years, i.e., 1990, 2005, and 2020. The years were selected by taking into account the years of the new regime change and land redistribution, the years of conservation measures practiced, and the availability of satellite images. In order to avoid seasonal variability, the classified images were acquired after crop harvesting, which is the time the dry season started, namely February and March. In this season, the cloud cover is minimal, in order to assure similar phenology (Table S1). Three types of imagery were used for image classification, i.e., Landsat 8 Operational Land Imager/Thermal Infrared Sensor (OLI/TIRS) (30 m), Landsat-7 Enhanced Thematic Mapper Plus (ETM+) (30 m), and Landsat 5 Enhanced Thematic Mapper (ETM) derived from the USGS Landsat Archive (http://earthexplorer.usgs.gov/ accessed on 31 August 2020) for the years 2020, 2005, and 1990, respectively (Table S1). Preprocessing of the images was performed before the actual image classification. Supervised classification with a maximum likelihood algorithm was applied to obtain land use/cover information, using ERDAS IMAGINE 2015 software. This method uses selected training samples to categorize pixels of an image into predefined land use/cover classes. Accuracy assessment was used to check the compatibility of classified images with the real land cover. Hence, producer's accuracy, user's accuracy, overall accuracy, and kappa coefficient measuring techniques were used to measure the accuracies of classified images. For accuracy evaluation assessment, a total of 157, 175, and 209 ground-truth points (GCP) were used for 2020, 2005, and 1990, respectively. All images were projected to the Universal Transverse Mercator (UTM, zone 37) and the World Geodetic System 84 (WGS84) datum. Based on the field observation of the study area, five different types of land use and land cover classes were identified for this zone. The descriptions of these land use and land cover classes are shown in Table 1.

Table 1. Description of land use and land cover types in the study area.

LULC Types	General Description
Water body	It consists of both natural and artificial water features such as streams, lakes, canals, ponds, and reservoirs.
Forest	Areas covered by both mixed trees and natural and plantation forests
Cultivated land	Areas of land prepared for growing agricultural crops. This category includes areas currently under crop and land under preparation.
Grazing land	All areas of bare lands, natural grass, and small shrubs mixed with grass used for grazing purpose.
Built-up Areas	Urban areas and permanent residential areas of varied patterns, for example, cities, towns, villages, and strip developments along highways

A systematic way of analysis for land use change is first to have the total area for the different land use classes for the different durations and then compare the two maps. Using ERDAS IMAGINE 2015 software, we compared areas of different land uses cell by cell by overlaying the two maps. This can provide some basic quantitative information about the land use change that has occurred. A cross-table showing the changes from one class to another can be produced. A map of these changes can also be obtained using ERDAS IMAGINE 2015 software.

Change analysis was conducted using a post-classification image comparison technique. Images of different reference years were first independently classified, and then change detection processes were performed. The percentage of land use and land cover change detection was obtained using the formula used by Kindu [15].

2.2.2. Biophysical Conservation Measures

Experts of the agricultural development office of the South Gondar zone were interviewed as key informants because of their adequate knowledge of the study under investigation. Moreover, the study included exploring secondary data sources, such as technical reports, published works, and public statistics, to document the scale of the major interventions and their implementation.

The Zonal Agricultural Office provided the annual biophysical conservation measures' data for the South Gondar zone for the twelve-year period (2011–2022). Bunds, terraces, gully rehabilitation, drainage structures, and moisture conservation are among the physical measures gathered. Biological measures include plantation activities and area closure. Plantation is a tree planting practice used as a biophysical degraded area rehabilitation mechanism. The major types of tree species planted include Acacia Decurrence, Grevillea Robusta, Acacia Saligna, and others. The information is a part of the catalog kept by the agriculture office of the South Gondar zone. The implementation of these measures dates back to the 1970s and 1980s. This study considers only twelve years of data, with varying scales of implementation over time. Therefore, evaluating the trend of the conservation efforts made so far with large-scale community mobilization is crucial. The Mann–Kendall trend test, a non-parametric statistical analysis, was used to assess the trend. The Mann–Kendall test is used to examine the existence of monotonously increasing and decreasing trends. The test is used on the data sets to find out whether there are any trends, either positive or negative, in the time series data. Given that, the test's results are based on less of an impact from outliers.

$$S = \sum_{i=1}^{n-1} \sum_{k=i+1}^{n} sig(x_k + x_i) \tag{1}$$

where S is the Mann–Kendall test value, x_k and x_i are the sequential data values, and i, k, and n are the lengths of the data. The S value is assumed to be zero and indicates no trend; if the value of S shows a positive value, it indicates an increasing trend, whereas a negative value of S shows a decreasing trend.

Sen [16] created the Sen Slope estimator algorithm, a straightforward non-parametric test for determining the actual slope of the Mann–Kendall trend analysis. It determines the size of any significant trend that the Mann–Kendall S test reveals. In time series data, when the trend is considered to be linear, the Sen Slope estimator test is used after the Mann–Kendall test. The Sen method is not significantly impacted by outliers or single data problems. The following equation is utilized to calculate the Sen Slope estimator.

$$E = \frac{x_k}{j} - \frac{x_i}{i} \tag{2}$$

where E is the value of Sen Slope estimator; x_k and x_i are data values at time j and i.

3. Results

3.1. Land Use/Land Cover Changes

The accuracy assessment result shows the overall kappa coefficients of 0.97, 0.95, and 0.98 and overall accuracies of 98.1%, 96%, and 97.6% for the years 1990, 2005, and 2020, respectively. The accuracy result confirmed that there is high agreement between GCP and the real land use/cover (Table S2). The results reveal that there have been considerable land use and land cover changes in the last three decades (1990–2020) in the South Gondar zone. The result indicates most of the areas are covered by agricultural land (93.07%), which confirms the livelihood of the zonal community largely depends on agricultural activities. The expansion of cultivated land at the expense of other land use types is greater. Among the different LULC types, cultivated land covered areas of 83.33% (1,171,999.44 ha), 87.68% (1,233,130.32 ha), and 93.07% (1,308,910.74 ha), followed by forest covers of 10.63% (149,435 ha), 8.82% (124,015.5 ha), and 5.21% (73,293.91 ha) in the years 1990, 2005, and 2020, respectively (Table 2). This result shows that cultivated land increased by 10% in the last three decades, which is a very fast change as compared with other land use types. However, there are significant declines in forest cover by half, which are 10.63% in 1990 and 5.21% in 2020. The same is true for the grazing land use, which decreased from 5.7% in 1990 to 1.4% in 2020 (Table 2). The change in the land use and land cover in the study area was attributed to the expansion of cultivated land from 83.33% in 1999 to 93.1% in 2020.

Table 2. The extent of area covered with the five land use types (1990–2020).

LULC Types	1990		2005		2020	
	Area (ha)	(%)	Area (ha)	(%)	Area (ha)	(%)
Water Body	5347.98	0.38	2881.71	0.20	2255.24	0.16
Forest	149,435.82	10.63	124,015.5	8.82	73,293.91	5.21
Cultivated Land	1,171,999.44	83.33	1,233,130.32	87.68	1,308,910.74	93.07
Grazing Land	79,434.63	5.65	46,143.72	3.28	20,153.05	1.43
Built-Up Areas	191.79	0.01	238.41	0.02	1796.72	0.13
Total	1,406,405.13	100	1,406,405.13	1,406,405.13	1,406,405.13	100

Table 3 shows that during the past 30 years, the majority of the study area has seen a continuous change in land use and land cover. The changes were observed mainly on cultivated, forest, and grazing land cover. The rate of change in cultivated land between 2005 and 2020 is greater as compared with 1990 and 2005. However, the forest cover change was greater from 1990 to 2005 than from 2005 to 2020. The reason is that the government owned most of the forests in that period. Since 1991, with the downfall of the Derge regime, the people have aggressively cleared a significant amount of forest cover [15]). During the last 30 years, 91,423.2 ha (6.5%) of forest land was changed into agricultural land. Therefore, the forest cover of the study area decreased by 5.4% during this period (Table 2). Similarly, the grazing land cover also dramatically decreased by 4.22%, as the land use changed into cultivated land use by 3% (Tables 2 and 3).

Table 3. Change in land use and land cover from 1990 to 2005, 2005 to 2020, and 1990 to 2020.

	LULC Classes	Water Body	%	Forest	%	Cultivated	%	Grazing Land	%	Built-Up	%	Total (ha)	%
						2005							
1990	Water	2646	0.2	78.03	0	2572.7	0.2	51.1	0	0	0	5348	0.4
	Forest	119.4	0	70,281.1	5	78,408.9	5.6	624.8	0	1.6	0	149,436	11
	Cultivated	77.3	0	52,861.8	3.8	1,096,730	78	22,252	1.6	73.8	0	1,171,995	83
	Grazing	38.9	0	789.84	0.1	55,383.9	3.9	23,215	1.7	7.2	0	79,434.6	5.6
	Built-up	0	0	4.77	0	30.2	0	1	0	155.8	0	191.8	0
						2020							
2005	Water	745.5	0.1	76.5	0	1924.2	0.1	83.9	0	0	0	2830.1	0.2
	Forest	387.7	0	31,588.8	2.2	90,804.9	6.5	1617.7	0.1	155	0	124,554.1	8.9
	Cultivated	1078.4	0.1	41,008.7	2.9	1,174,200	83.5	14,955.2	1.1	1430.6	0.1	1,232,672.9	87.6
	Grazing	29.9	0	612.4	0	41,952.5	3	3488.3	0.2	26.5	0	46,109.6	3.3
	Built-up	0	0	0.4	0	47.7	0	5.8	0	184.6	0	238.4	0
						2020							
1990	Water	745.5	0.1	76.5	0	1924.2	0.1	83.9	0	0	0	2830.1	0.2
	Forest	387.7	0	31,588.8	2.2	90,804.9	6.5	1617.7	0.1	155	0	124,554.1	8.9
	Cultivated	1078	0.1	41,008.7	2.9	1,174,200	83.5	14,955.2	1.1	1430.6	0	1,232,672.9	87.6
	Grazing	29.9	0	612.4	0	41,952.5	3	3488.3	0.2	26.5	0	46,109.6	3.3
	Built-up	0	0	0.4	0	47.7	0	5.76	0	184.6	0	238.41	0

Built-up areas have been growing as the population grows at an alarming rate. The increase in built-up areas from 0.01% in 1990 to 0.13% in 2020 indicates the demand for residential areas and food. This created a concomitant pressure for cultivated land expansion at the expense of depleting untapped resources. The growing demand for agricultural land and built-up areas is decreasing the amount and size of forest cover and grazing land throughout the zone (Figure 2).

3.2. Biophysical Conservation Practices

Biophysical soil and water conservation practices have been implemented in the Amhara regional state, including the South Gondar zone (SGZ), to rehabilitate degraded areas. This study presents efforts made to treat degraded areas from 2011 to 2022. Annually, these conservation practices are implemented from January to March through large-scale government-sponsored campaigns. Conservation practices can be grouped into three major categories: (1) physical structures (bunds, terraces, gully rehabilitation, and drainage structures, including cut-off drains and waterways); (2) area closures; and (3) plantations. The implementation of these practices depends on the level of landscape degradation and available resources. For example, cut-off drains and waterways can be constructed in farmlands, hillsides, and homesteads. In addition, bench terraces may be implemented on cultivated fields or on hillslopes with deep soil.

As indicated in Table 4 and Figure 3, different biophysical measures were implemented, including soil and stone bunds; hillside and bench terraces; cutoff drains and waterways; soil moisture structures such as trenches, infiltration ditches, and others; gully rehabilitation; area closures; and plantations. The total number of these structures implemented per year is indicated in Table 4. In general, 685 thousand ha of physical structures; 13.5 thousand m^3 of drainage lines; 386 thousand ha of area closure; and 587.6 thousand ha of area are covered with plantation, using 2.5 billion seedlings (excluding 2022 plantation). Of the physical structures, more than 80% are bunds. Bunds could be soil, stone, or stone-faced soil depending on the availability of the resource. The majority of these are soil bunds.

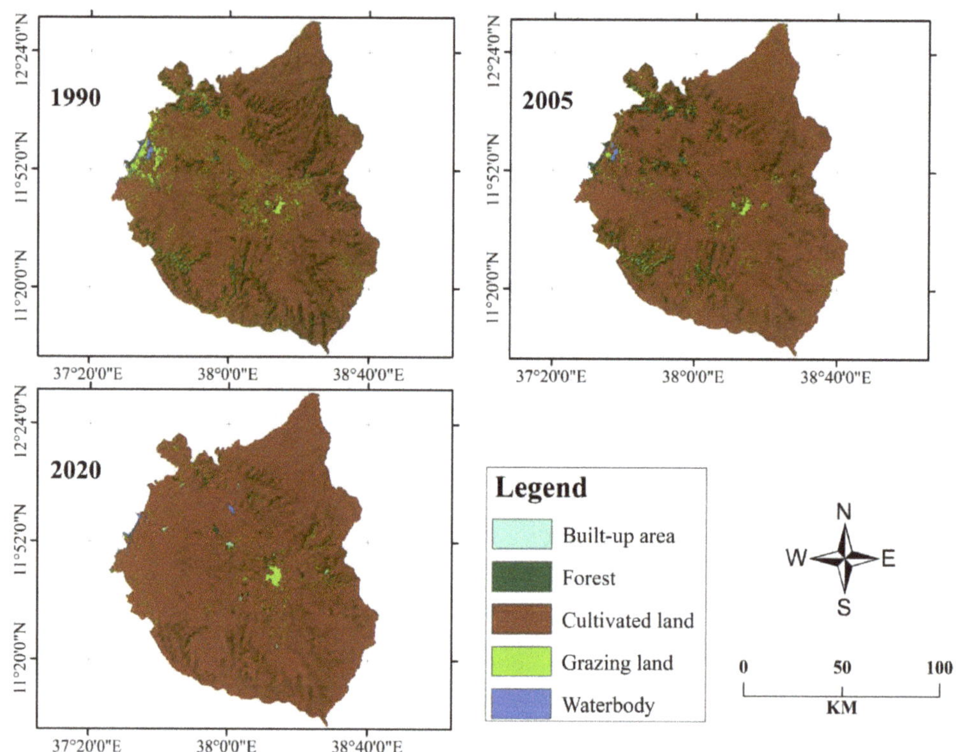

Figure 2. LULC map of South Gondar zone for the years 1990, 2005, and 2020.

Table 4. Types of biophysical conservation measures implemented in the SGZ from 2011 to 2022.

SN	Activities	Unit ('000)	Year											
			2011	2012	2013	2014	2015	2016	2017	2018	2019	2020	2021	2022
1	Bunds	Ha	51.4	65.7	74.6	78.8	67.2	54.5	31.6	32.6	20.3	16.7	21.9	20.5
2	Hillside terr.	Ha	6.0	15.7	15.2	9.8	5.1	5.4	3.6	4.6	2.3	1.8	3.6	2.6
3	Bench terrace	Ha	0.0	0.0	0.0	0.0	0.1	0.0	0.1	0.2	0.1	0.0	0.1	2.3
4	Moisture cons.	Ha	2.7	10.0	12.9	9.3	2.1	0.6	0.6	0.6	0.0	0.8	0.7	0.6
5	Gully rehab.	Ha	0.7	0.9	0.4	0.6	1.2	1.3	1.5	0.4	0.2	0.2	0.3	0.2
6	Cut of drains	m^3	3.9	2.6	0.7	0.6	0.2	0.1	0.2	0.1	0.1	0.0	0.2	0.1
7	Waterways	m^3	0.5	0.8	0.5	0.3	0.2	0.1	0.2	0.1	0.1	0.0	0.1	0.1
8	Area closure	Ha	16.0	49.8	43.8	43.4	34.1	26.0	39.4	38.9	24.1	22.9	21.0	13.8
9	Plantation	Ha	20.0	20.3	67.0	71.3	38.6	59.2	61.1	52.3	50.2	36.2	31.2	0.0

The magnitude of implementing these conservation measures varies from one district to the other. This is linked with the experts' inspiration to mobilize the community during the campaign. Figure 4 shows the total area of the districts covered by biophysical conservation measures. Some of the districts, such as Meketawa, Guna B/Midr, and Sedie Muja, are newly established districts showing a limited effort, whereas Fogera is an old district with low coverage that may be related to its plain landscape. Farmers in this area build on-plot moisture conservation structures for rice production. However, the Farta district completely covers its land mass with conservation measures. In addition, Tach Gayint, Andabet, and Estie left about 10% of the land to be covered with conservation measures (Figure 4).

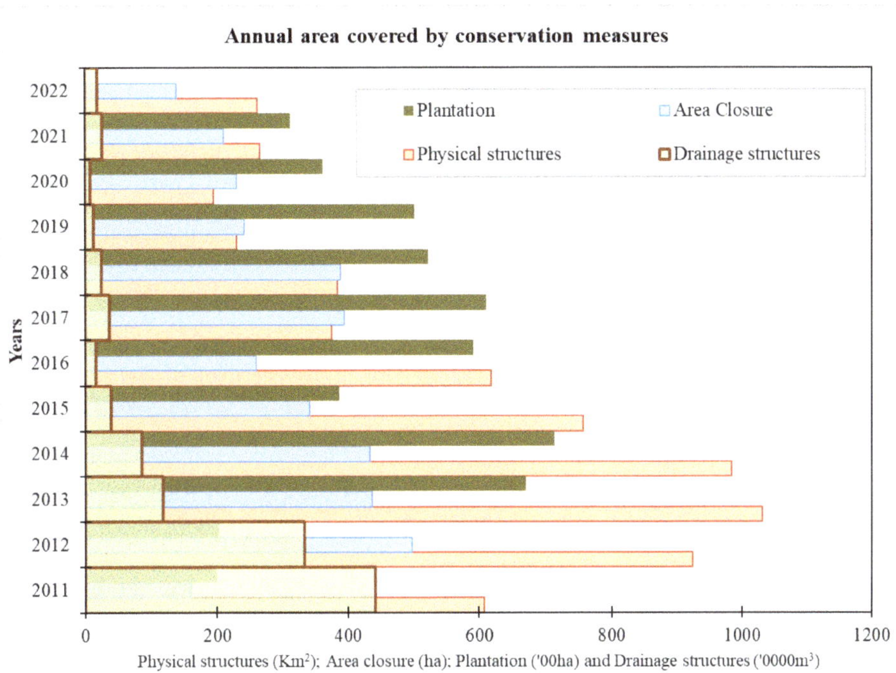

Figure 3. Annual biophysical conservation measures over the twelve-year period in the zone.

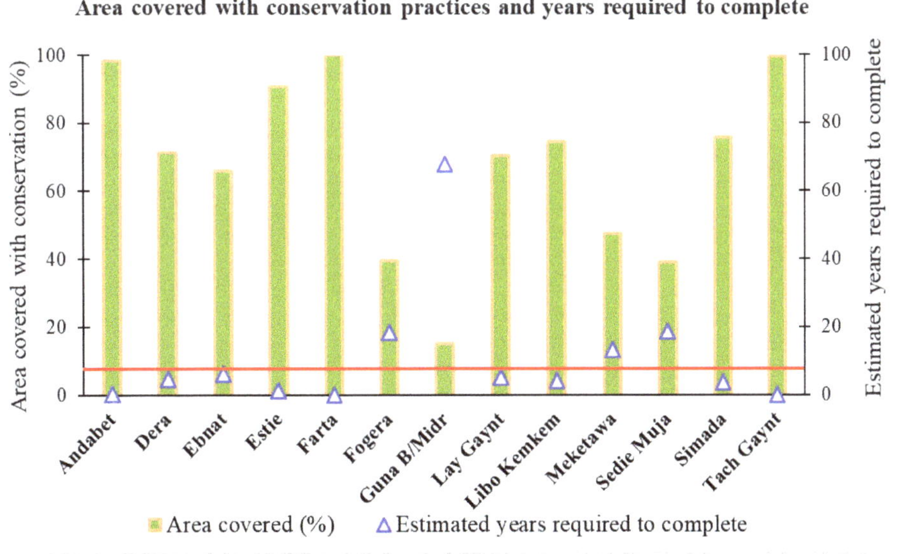

Figure 4. Area covered with biophysical conservation measures (%) at each district in the SGZ and years required to complete the remaining area. The pink color line indicates districts requiring fewer than 5 years.

4. Discussion

4.1. Linking LULC with Biophysical Conservation Efforts

Natural and anthropogenic LULC change has a significant impact and characterizes the level of land degradation. The type and distribution of surface cover affects surface runoff and sedimentation in the watershed [17]. Poor surface cover results in high soil erosion due to a quick response to rainfall. Biophysical soil and water conservation measures are linked with the type of land use and land cover in the area. For example, bunds are constructed on cultivated fields based on the available soil or rock resources. Hillside terraces are built on bushes and shrub land with medium soil depth; bench terraces are constructed on soils that are about 50 cm deep and hillslopes planned for cultivated land. Drainage structures are built on any land use type affected by upstream runoff. Gully control practices are implemented on land uses that are prone to gully erosion. The level of degradation in the land use reflects the suitable intervention proposed in the area. Biophysical conservation measures cover 73 percent of the zone's total land mass. Cultivated land accounts for 93 percent of the land use in the zone and is proportional to the magnitude of bunds constructed. Bund construction covers approximately 80% of cultivated fields (Table 4). A significant number of hillside terraces has been constructed along the contour, as a majority of the districts are characterized by steep slopes [18], except Fogera and Libo Kemkem. Lastly, meager bench terraces have been implemented in areas with thicker soil. Current land use may be suited for a certain conservation measure. Regrettably, the type of conservation measure has changed over time due to changes in land use. Over time, conservation planners have been challenged with this dilemma. This could be demonstrated by the predicted years required to cover the district's landscape with biophysical measures (Figure 4). For example, 61 percent of districts require fewer than five years to completely cover their landscape. However, as the land use system has changed, this may no longer be the case. Cultivated lands gradually expanded from gently sloping land into steeper slopes in the highlands with the subsequent clearing of forests and other vegetation. Despite this, the Watershed Development Guidelines advise against building stone bunds on slopes steeper than 35%; instead, terraces and bunds are built on hillslopes [19]. This finding is in agreement with the findings of Tsegaye [20], who state that land degradation, particularly soil erosion, was exacerbated by the influence of land use and land cover change because of agricultural land expansion on accessible steep slopes of forest, shrub, and grazing land.

Plantation is one of the biophysical measures where seedlings are planted on degraded areas, physical structures, and homesteads. The purpose of plantation as a conservation measure is sought in two ways. The planted materials reinforce constructed physical structures to prevent soil erosion and tree roots bind soil particles in situ. The study result shows biological measures cover approximately 92% of the bunds. However, it is challenging to find a bund covered with vegetation on the ground from each district for evidence. Reinforcing bunds with planation has the capacity to reduce runoff in cultivated fields by 49 percent [21]. According to Teshome [22], the profitability of SWC structures is greater for bunds with grasses than bunds without grasses. Integrated management of land resources requires both physical structures and multipurpose plantations for sustainable use of the resource. Furthermore, the plantation contributes to the area's forest cover. Forest coverage discrepancy was observed between the zonal report and satellite images. For example, the forest cover reported by the zone in 2020 is about 12 percent, whereas using the satellite images in the same year, it is only 5.1 percent. The reason may be the satellite images include artificial forests such as Eucalyptus at individual farm fields. Eucalyptus is a multipurpose fast-growing tree and has greater socio-economic benefits. Zonal experts define forest as two meters height, 20 percent canopy, and half a hectare of area coverage. More than 587 thousand hectares of land has been covered with more than 2.5 billion seedlings in the last eleven years, except 2022. This suggests that the area that the forest covers is expanding faster than before. However, as indicated in Figure 3, the rate of plantation is decreasing gradually, and the land cover expansion was not found as expected due to poor survival rate of seedlings, prolonged dry season, water scarcity, and unsustainable management of plantations [23–25]. Forest cover increased in some areas after watershed management programs launched in 2011 along with the establishment of eucalyptus plantations [26]. Farmers shift from cultivated land use to artificial forest land based on the income difference from crop and tree. This has a positive impact on the forest cover dynamics in the area. However, the overall trend of the forest cover is still declining over time, coupled with the increasing demand for firewood, construction, and other services [26–28].

Area closure is a land use practice essential to restore degraded areas, replace open grazing by a cut-and-carry system, and increase the source of income from the sale of grass and forest products [29]. The practice is difficult because it contradicts the widely used free grazing. Areas selected for closure are usually communal lands used for livestock grazing. The shortage of grazing land discourages the local people from practicing area closure. The free-grazing animals graze on crop residues, especially after crop harvest, and grazing lands, bushes, and shrub land. The free-grazing practice damages the land resources, and the livestock may destroy naturally regenerated and artificially planted seedlings. The advantage of practicing area closure is that it requires limited investment as compared with the other biophysical measures. However, the sustainability of the area closures can only be guaranteed if and only if the community has taken full responsibility for protection [30].

4.2. Trends of Conservation Measures

In the South Gondar zone, about 73% of the landscape is treated with biophysical conservation measures (Table 4). During the large-scale SWC campaign, between 2011 and 2016, large areas of cultivated fields were covered with bunds (Figure 4). Relatively, the investment on hillslopes and area closures in this period was also enormous. After 2016, the implementation of physical structures began to decrease, whereas the construction of area enclosures continued steadily in 2017 and 2018 (Figure 4). Table 5 and Figure 5 show the results of Mann–Kendall test and Sen Slope estimator for the four conservation measures, with statistical significance at the 5% level with increasing or decreasing trends. According to the Mann–Kendall trend test, a significant decreasing trend was observed from physical structures, drainage structures, and area closures (for the Sen Slope estimator, the same significant negative trend was detected). Plantation shows a negative non-significant decreasing trend ($p < 0.05$). The major reason for the decrement may be related to the limited participation of the community in physical structures, and experts focused on enclosures. In addition, as explained in Section 4.1, both natural and man-made factors contribute to its gradual reduction. The decreasing trend may also be more related with maintaining existing ones than constructing new structures. A maintenance report is not included in this study, though it is part of the annual action plan in some districts. According to Addisie and Molla [31] in the Gumara watershed, the local people are reluctant to participate in conservation campaigns, unlike in previous times. The reason is the success rate of efforts were below expectation, as farmers always hope to see immediate results [32]. Furthermore, zonal natural resource management experts explain that some of the farmers would prefer to stay on their farm fields rather than participate in watersheds outside their vicinity. This led to a decrease in the number of participating workers and area covered with conservation measures. The trend of plantation shows a decreasing trend; however, it is not a significant decrease (Figure 5). This could be related to the government's ambitious tree-planting target, which is a green legacy initiative that started in 2019 [33].

The sustainability of rehabilitation practices is fundamental and shall be taken in to account throughout the implementation period, particularly at the end of all efforts [34,35]. This could be achieved by incorporating maintenance as a planning element. As observed at selected fields, there is meager implementation of maintenance. Maintenance is an overlooked area of implementation [31]. Therefore, each district must continue to engage in maintenance in order for efforts to be used sustainably. A lack of maintenance activities for the installed conservation measures makes the land degradation problem worse, despite factors such as inadequate quality standards for installed physical structures [35,36] and poor livestock management systems that cause physical structures to be destroyed [37].

Table 5. Mann–Kendall trend and Sen Slope test summary results.

Series\Test	Kendall's Tau	*p*-Value	Sen Slope
Physical structures (ha)	−0.636	0.005	−7465.550
Drainage structures ('0000 m^3)	−0.727	0.001	−147,084.386
Area closure (ha)	−0.576	0.011	−2682.763
Plantation (ha)	−0.152	0.537	−1762.672
Alpha = 0.05			

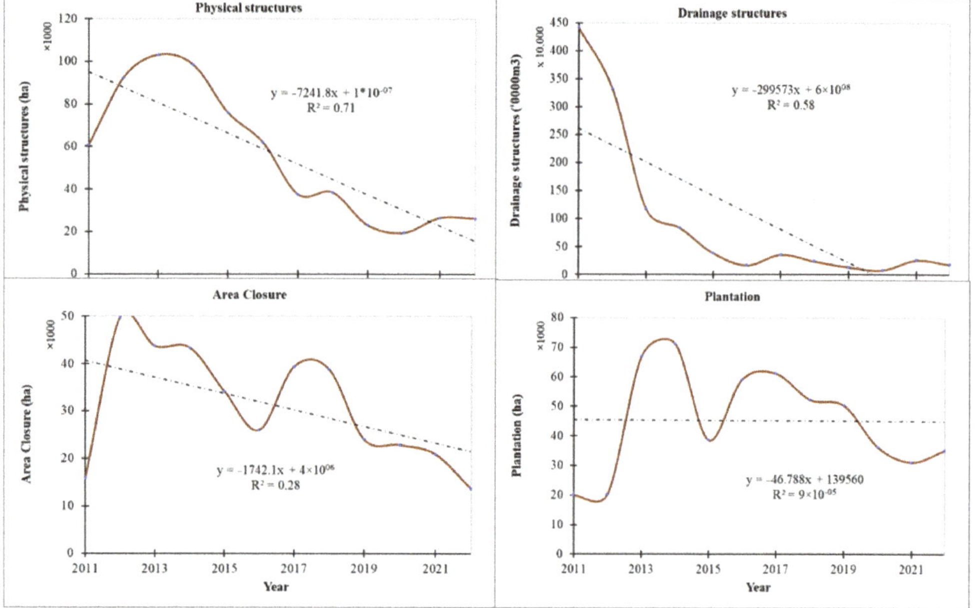

Figure 5. Trends of biophysical conservation measures over twelve years using the Mann–Kendall trend test graph and the regression. The dot lines are trend lines or lines of best fit indicating the general direction of points.

5. Conclusions

Evaluating the trends of biophysical soil and water conservation measures in relation to dynamic land use and land cover change is critical for the sustainable use of natural resources. Ethiopia has been implementing SWC practices on a watershed scale for the last five decades. However, the program maintains its current course, devoting vast amounts of both human and financial capital. As land use and land cover change affects the areas covered by specific controlling mechanisms, the implemented practices have yet to be completed. Data from satellite images and official reports show a discrepancy in the findings of the study. The study is intended to produce baseline information that helps conservation planners to monitor and evaluate real-time conditions in biophysical conservation measures, as well as changing land use and land cover in the future. The coverage of biophysical conservation measures in the majority of the districts in the zone indicates better achievement. However, the trend was significantly decreasing. The landscape experiences rapid land use and land cover change, hence it will require more structures in the future. Plantation was implemented at a larger scale, though it is unable to attribute it to the sustainability of physical soil, water conservation practices, and climate change anomalies. The findings of this study will contribute to developing interventions in the manner in which future watershed development programs are planned in response to sustainable natural resource management. Finally, the research recommends that, despite considerable investments in biophysical erosion control measures, integrated management efforts should be maintained.

Supplementary Materials: The following supporting information can be downloaded at: https://www.mdpi.com/article/10.3390/land11122187/s1, Table S1: Types of sensor used, acquisition date and percent cloud cover; Table S2: Land use classes and accuracy assessment results (1999–2020).

Author Contributions: M.B.A. and G.M. conceptualization, data analyses, and writing and editing the draft; M.T. and G.T.A. reviewing and editing; M.B.A., G.M., M.T. and G.T.A. writing and reviewing and editing the methodology; G.M. mapping. All authors have read and agreed to the published version of the manuscript.

Funding: Gebiaw T. Ayele covered the APC and received funding from Griffith Graduate Research School, the Australian Rivers Institute and School of Engineering, Griffith University, Queensland, Australia.

Data Availability Statement: All the data are presented in tables and figures in the manuscript.

Acknowledgments: Gebiaw T. Ayele acknowledges Griffith Graduate Research School, the Australian Rivers Institute and School of Engineering, Griffith University, Queensland, Australia.

Conflicts of Interest: The authors declare no conflict of interest.

References

1. Berry, L. *Land Degradation in Ethiopia: Its Extent and Impact*; Commissioned by the GM with WB Support; 2003; pp. 2–7. Available online: https://rmportal.net/library/content/frame/land-degradation-case-studies-03-ethiopia/at_download/file (accessed on 1 November 2022).
2. Yitbarek, T.; Belliethathan, S.; Stringer, L. The onsite cost of gully erosion and cost-benefit of gully rehabilitation: A case study in Ethiopia. *Land Degrad. Dev.* **2012**, *23*, 157–166. [CrossRef]
3. Erkossa, T.; Wudneh, A.; Desalegn, B.; Taye, G. Linking soil erosion to on-site financial cost: Lessons from watersheds in the Blue Nile basin. *Solid Earth* **2015**, *6*, 765–774. [CrossRef]
4. Sop, T.; Oldeland., J. Local perceptions of woody vegetation dynamics in the context of a 'greening Sahel': A case study from Burkina Faso. *Land Degrad. Dev.* **2013**, *24*, 511–527. [CrossRef]
5. Meshesha, D.T.; Tsunekawa, A.; Tsubo, M.; Ali, S.A.; Haregeweyn, N. Land-use change and its socio-environmental impact in Eastern Ethiopia's highland. *Reg. Environ. Chang.* **2014**, *14*, 757–768. [CrossRef]
6. Pech, S.; Sunada, K. Population growth and natural resources pressures in the Mekong River Basin. *AMBIO A J. Hum. Environ. Stud.* **2008**, *37*, 219–224. [CrossRef]
7. Desalegn, T.; Cruz, F.; Kindu, M.; Turrión, M.; Gonzalo, J. Land-use/land-cover (LULC) change and socioeconomic conditions of local community in the central highlands of Ethiopia. *Int. J. Sustain. Dev. World* **2014**, *21*, 406–413. [CrossRef]
8. Yesuf, H.M.; Assen, M.; Alamirew, T.; Melesse, A.M. Modeling of sediment yield in Maybar gauged watershed using SWAT, northeast Ethiopia. *Catena* **2015**, *127*, 191–205. [CrossRef]
9. Teshome, A.; Halefom, A. Mapping of soil erosion hotspot areas using GIS based-MCDA techniques in South Gondar Zone, Amhara Region, Ethiopia. *World News Nat. Sci.* **2019**, *24*, 218–238.
10. Selassie, Y.G. Problems, efforts and future directions of natural resources management in western amhara region of the Blue Nile basin, Ethiopia. In *Social and Ecological System Dynamics*; Springer: Berlin/Heidelberg, Germany, 2017; pp. 597–613.
11. Hurni, H. Degradation and conservation of the resources in the Ethiopian highlands. *Mt. Res. Dev.* **1988**, 123–130. [CrossRef]
12. Grunder, M. Soil conservation research in Ethiopia. *Mt. Res. Dev.* **1988**, *8*, 145–151. [CrossRef]
13. Herweg, K.; Ludi, E. The performance of selected soil and water conservation measures-case studies from Ethiopia and Eritrea. *Catena* **1999**, *36*, 99–114. [CrossRef]
14. Guzman, C.D.; Zimale, F.A.; Tebebu, T.Y.; Bayabil, H.K.; Tilahun, S.A.; Yitaferu, B.; Rientjes, T.H.; Steenhuis, T.S. Modeling discharge and sediment concentrations after landscape interventions in a humid monsoon climate: The Anjeni watershed in the highlands of Ethiopia. *Hydrol. Process.* **2017**, *31*, 1239–1257. [CrossRef]
15. Kindu, M.; Schneider, T.; Teketay, D.; Knoke, T. Land use/land cover change analysis using object-based classification approach in Munessa-Shashemene landscape of the Ethiopian highlands. *Remote Sens.* **2013**, *5*, 2411–2435. [CrossRef]
16. Sen, P.K. Estimates of the regression coefficient based on Kendall's tau. *J. Am. Stat. Assoc.* **1968**, *63*, 1379–1389. [CrossRef]
17. Alkharabsheh, M.M.; Alexandridis, T.; Bilas, G.; Misopolinos, N.; Silleos, N. Impact of land cover change on soil erosion hazard in northern Jordan using remote sensing and GIS. *Procedia. Environ. Sci.* **2013**, *19*, 912–921. [CrossRef]

18. Deng, C.; Zhang, G.; Liu, Y.; Nie, X.; Li, Z.; Liu, J.; Zhu, D. Advantages and disadvantages of terracing: A comprehensive review. *Int. Soil Water Conserv. Res.* **2021**, *9*, 344–359. [CrossRef]

19. Desta, L.; Carucci, V.; Wendem-Agenehu, A.; Abebe, Y. *Community Based Participatory Watershed Development*; A Guideline; Ministry of Agriculture and Rural Development: Addis Ababa, Ethiopia, 2005.

20. Tsegaye, B. Effect of land use and land cover changes on soil erosion in Ethiopia. *Int. J. Agric. Sci.* **2019**, *5*, 26–34.

21. Sultan, D.; Tsunekawa, A.; Haregeweyn, N.; Adgo, E.; Tsubo, M.; Meshesha, D.T.; Masunaga, T.; Aklog, D.; Ebabu, K. Analyzing the runoff response to soil and water conservation measures in a tropical humid Ethiopian highland. *Phys. Geogr.* **2017**, *38*, 423–447. [CrossRef]

22. Teshome, A.; Rolker, D.; de Graaff, J. Financial viability of soil and water conservation technologies in northwestern Ethiopian highlands. *Appl. Geogr.* **2013**, *37*, 139–149. [CrossRef]

23. Petros, W.; Tesfahunegn, G.B.; Berihu, M.; Meinderts, J. Effectiveness of water-saving techniques on growth performance of mango (*Mangifera indica* L.) seedlings in Mihitsab-Azmati watershed, Rama Area, northern Ethiopia. *Agric. Water Manag.* **2021**, *243*, 106476. [CrossRef]

24. Abrha, G.; Hintsa, S.; Gebremedhin, G. Screening of tree seedling survival rate under field condition in Tanqua Abergelle and Weri-Leke Weredas, Tigray, Ethiopia. *J. Hortic. For.* **2020**, *12*, 20–26. [CrossRef]

25. Demisachew, T.; Awdenegest, M.; Mihret, D. Impact evaluation on survival of tree seedling using selected in situ rainwater harvesting methods in Gerduba Watershed, Borana Zone, Ethiopia. *J. Hortic. For.* **2018**, *10*, 43–51. [CrossRef]

26. Alemayehu, A.; Melka, Y. small-scale eucalyptus cultivation and its socio-economic impacts in ethiopia: A review of practices and conditions. *Trees For. People* **2022**, *8*, 100269. [CrossRef]

27. Hailu, A.; Mammo, S.; Kidane, M. Dynamics of land use, land cover change trend and its drivers in Jimma Geneti District, Western Ethiopia. *Land Use Policy* **2020**, *99*, 105011. [CrossRef]

28. Alemu, B.; Garedew, E.; Eshetu, Z.; Kassa, H. Land use and land cover changes and associated driving forces in north western lowlands of Ethiopia. *Int. Res. J. Agric. Sci. Soil Sci.* **2015**, *5*, 28–44.

29. Aerts, R.; Nyssen, J.; Haile, M. On the difference between "exclosures" and "enclosures" in ecology and the environment. *J. Arid Environ.* **2009**, *73*, 762. [CrossRef]

30. Mulugeta, G.; Achenef, A. Socio-economic challenges of area exclosure practices: A case of Gonder Zuria Woreda, Amhara region, Ethiopia. *J. Nat. Sci. Res.* **2015**, *5*, 123–132.

31. Addisie, M.B.; Molla, G. Trends of community-based interventions on sustainable watershed development in the Ethiopian highlands, the Gumara watershed. *Sustain. Water Resour. Manag.* **2021**, *7*, 87 [CrossRef]

32. Mekuriaw, A.; Heinimann, A.; Zeleke, G.; Hurni, H. Factors influencing the adoption of physical soil and water conservation practices in the Ethiopian highlands. *Int. Soil Water Conserv. Res.* **2018**, *6*, 23–30. [CrossRef]

33. Jalleta, A.K. The Legal Protection of Forests: Ethiopian Green Legacy vs. International Environmental Regimes. *Beijing Law Rev.* **2021**, *12*, 725. [CrossRef]

34. Pretty, J.N.; Shah, P. Making soil and water conservation sustainable: From coercion and control to partnerships and participation. *Land Degrad. Dev.* **1997**, *8*, 39–58. [CrossRef]

35. Mitiku, H.; Herweg, K.G.; Stillhardt, B. Sustainable Land Management: A New Approach to Soil and Water Conservation in Ethiopia. Land Resources Management and Environmental Protection Department Mekelle University, Ethiopia, and Centre for Development and Environment (CDE), Swiss National Centre of Competence in Research (NCCR) North-South University of Bern, Switzerland. 2006. Available online: https://boris.unibe.ch/19217/1/e308_slm_teachingbook_complete.pdf (accessed on 1 November 2022).

36. Addisie, M.B.; Langendoen, E.J.; Aynalem, D.W.; Ayele, G.K.; Tilahun, S.A.; Schmitter, P.; Mekuria, W.; Moges, M.M.; Steenhuis, T.S. Assessment of practices for controlling shallow valley-bottom gullies in the sub-Humid Ethiopian Highlands. *Water* **2018**, *10*, 389. [CrossRef]

37. Belachew, A.; Mekuria, W.; Nachimuthu, K. Factors influencing adoption of soil and water conservation practices in the northwest Ethiopian highlands. *Int. Soil Water Conserv. Res.* **2020**, *8*, 80–89. [CrossRef]

Review

Taking/Compensations or Regulations? Balancing Landscape Conservation and the Development of Renewable Energy Facilities in Japan

Satomi Kohyama

Faculty of Social Sciences, Academic Assembly, University of Toyama, Toyama 930-8555, Japan; kohyama@eco.u-toyama.ac.jp; Tel.: +81-76-445-6415

Abstract: The application of regulations for the development of renewable energy facilities is one of the key environmental conservation strategies being implemented in Japan. However, regulations are only applied if the degree of environmental degradation falls below the "reference point." Thus, impacts of project development that are remarkably limited to scenic values of landscapes are largely overseen in Japan. On the other hand, establishing standards for the "reference point" is challenging, and existing scientific approaches and legal frameworks for conserving "daily landscapes" are largely absent. Therefore, it is necessary to establish a set of standards for "reference points" or indicators to classify landscape inventories, particularly those with scenic values. This study explored the potential of development-compensated implementation in Japan by scrutinizing relevant compensation measures in other countries. The results revealed that adding the aesthetic degradation of landscapes as an object in development compensation is challenging, as its value is difficult to monetize. Further, the evaluation of landscape degradation may be insufficient. Hence, there is a need for objective-driven indicators and methods that measure landscape degradation, particularly the effect of renewable energy facilities on the scenic values of "daily landscapes."

Keywords: landscape; scenic values; renewable energy; compensated measures; legal framework; Japan

Citation: Kohyama, S. Taking/Compensations or Regulations? Balancing Landscape Conservation and the Development of Renewable Energy Facilities in Japan. *Land* **2023**, *12*, 51. https://doi.org/10.3390/land12010051

Academic Editors: Kleomenis Kalogeropoulos, Andreas Tsatsaris, Nikolaos Stathopoulos, Demetrios E. Tsesmelis, Nilanchal Patel and Xiao Huang

Received: 28 November 2022
Revised: 19 December 2022
Accepted: 21 December 2022
Published: 24 December 2022

1. Introduction

Issues surrounding development persist in developed and developing countries despite continued discussions since the Brundtland report and other historical documents [1–3]. the most contested cases in contemporary developed countries include the installation of renewable energy facilities and natural resource management, such as forestry management and related regulations (i.e., tree thinning), as countermeasures against global warming [4–7]. These development activities are often accompanied by environmental changes, frequently in the form of trade-offs between environmental functions and social conflicts, altering the hydrology, biodiversity, culture, and/or landscape of the region in installation sites [8]. Moreover, there are trade-offs between social and institutional dimensions and among different stakeholders involved, which are often driven by research questions such as "Who pays for such conservation costs?"; "Who is exempted from paying conservation costs?"; and "Who will be responsible for environmental conservation?" [9–13]. These questions are particularly challenging in landscape conservation contexts, in which various stakeholders are involved. One way forward in addressing these concerns is through scientific and objective investigations.

Existing science-based methods consider the cost and benefit-sharing of landscape conservation, such as the impact assessment of scenic resources, aiming to objectively evaluate them and measure the degree of landscape degradation in question. Certain methods are currently being implemented in the UK and the US, such as the Landscape Character Assessment in the UK that aims to identify and describe variations in landscape characteristics [14]. In the US, when wind power is generated, developers or business operators are required to enter into a Host Community Agreement to return profits to the region [15]. Similar requests have been implemented in Japan; when business operators abandon development, there is no benefit or return from the region, and they will be forced to conserve landscapes free of charge for the region. These efforts to eliminate imbalances are globally sought after as climate change countermeasures and landscape conservation measures. The legal framework of Japan addresses such discrepancies. For instance, the "Economic Measures to Prevent Hindrances to Environmental Conservation" of the Basic Act on the Environment (*Kankyo-Kihon-Ho,* Act No 91 of 1993) states, in theory, that it is possible to introduce "economic instruments that differ from traditional regulatory instruments as an instrument of environmental policies" [16]. The law further encompasses the possibility of developing effective policies for environmental issues that are difficult to address with conventional regulatory approaches [17].

Economic instruments can be applied in multiple ways. For instance, business operators need to respect the so-called "landscape profits" or *Keikan Rieki,* where local communities contribute to constructing and maintaining landscapes [18]. The legal framework of Japan shows "in return for 'compensation' (*Hosyoh*) for landscape conservation, compensation from the public sector must be paid for business operators to abandon development, because of considering the regional consensus of the community despite the legal possibility of development. [19]" Since this concept is not practically connected, business entities are not recognized for the economic benefit or honor of withdrawing from the business for the sake of landscape conservation, so as a result, they hope that the land they own will be managed as efficiently as possible as a business. As a result, the public sector must use taxes at the national or prefectural level burden on residents to conserve the landscape.

In Japan, there are two types of "compensation." The first type is the idea of expropriating development rights from business operators (i.e., taking) rather than paying compensation (*Kaihatsu-Hosyoh*), and the second type is the government funds for business entities' development projects that are harmonized with landscape and biodiversity conservation [20]. For the former, monetary compensations are provided to business operators or developers for lost opportunities from abandoned business plans. For the latter, monetary compensations are paid by the government to operators to abandon their planned development as a conservation effort. To avoid confusion, in this manuscript, both types are referred to as "compensation"; however, this study deals with the former type.

In Japan, if the environmental standard of the current situation is maintained as is, that is, if it is not developed, it tends to be regarded as a regulation (in this case, compensations are not required). There are similarities with other countries, especially developed countries that are faced with declining birthrates and aging populations [21] or uneven distributions of the population [22]. It is instrumental to compare regulations in different contexts. Thus, this study inspected the regulations and legal frameworks of Japan and the existing approaches in the UK and the US. Section 2 presents the case of Japan, focusing on the boundary between private goods and public interests. Japan has the strongest landowner's (private individuals) rights among the developed countries with a compensation system, whereas the public sector has weak regulatory tools and specific laws including effective environmental conservation [23]. Section 3 overviews specific legal instruments that objectively address landscape degradations in the US and the UK, and Section 4 examines whether they can be applied in Japanese contexts or current legal frameworks. If the application for Japan can be demonstrated, these instruments could also be utilized in other countries with similar settings. Finally, Section 5 summarizes the key findings and emphasizes the need for inventories and indicators for scenic impact assessment.

2. Boundary between Private Goods and Public Interests

2.1. Research Overview

The relationship between personal property rights and environmental protection is an issue that has long been debated. Lubens clarified the relationship between social restrictions and property rights in German and US laws, indicating that property rights are the core part of the surrounding areas [24]. The study found that German law has a constitutionally guaranteed sphere of rights that exists as a "core field" (the *Kernbereich*), whereas U.S. law has the Taking Clause in the Fifth Amendment of the US Constitution [25] to protect the economic value and expectations related to realizing value in a property object.

The US Taking Law is underpinned by several common law judgments, and much debate surrounds this law. Essentially, expropriation of land ownership (land expropriation) is taking; however, it must be distinguished from other property rights regulations (property regulation by state police power with no compensation). Many property law researchers have investigated the Supreme Court case regarding taking [26] and explored the inherent quality of private property rights and public expropriation rights from their historical transition.

Findley and Farber [27] discussed the 1978 US supreme court judgment *Penn Cent. Transp. Co. v. New York City*. Moreover, there have been discussions on how to bring public interest to private land. The present study utilized Alexander's Social Obligation theory [28]. This theory is based on the analysis of 'taking', and the affirmative obligation of the property owner to the local community. He uses the "capability approach" proposed by Nussbaum and Sen to examine the community base and "functioning" of land. Penalver [29] and Wendel [30] defended this, while Smith [31] criticized it. Conversely, although not aggressively addressed in this study, Rule stated that if landowners are to be obliged to protect the natural environment, it is not a legal liability but an individual "norm" [32].

In Japan, if the current environmental standard is maintained (i.e., selected not to be developed), it tends to be regarded as a regulation (in this case, compensations are not required). While the current Japanese law has these characteristics, some similarities are shared with other countries (especially developed countries) that are faced with declining birthrates and aging populations [19] or uneven distributions of the population [20]. It is instrumental to compare regulations in different contexts and to estimate the agreement contents to obtain regional consensus.

Figure 1 shows the situation that this study aims to clarify and address. Currently, in Japan, the parties involved in the legal procedures of renewable energy projects include 'the Power Generation Business Operators/Developers', 'Public Sector: Ministry of Economy, Trade and Industry (METI) and Municipality', and 'Local Residents'. The Public Sector is where business operators apply for a business license and seek judgement. Local Residents represent the local environmental conservation spokespersons. The relationship and legal proceedings among these parties are unclear, and the issues on compensation and landscape assessment remain indefinite in the issuance of a business license by the Public Sector. For instance, when business operators apply for a license, the public sector examines the required documents based on the permit criteria and decides whether to grant a license or deny the application. If the permission criteria are appropriate and clear, the public sector can easily function as a regulation on the issuance of a business license. However, if the criteria are inappropriate and unclear, the decision of whether to approve or deny the application will be difficult. In such cases, regional consensus or agreement between local residents and operators is considered, as residents' intentions have a strong influence on the public sector's decision [33,34]. Particularly, if there is a strong opposition movement from local residents on the proposed renewable energy project, the public sector is likely to reject the license request. If such cases arise, regional consensus with the locals and municipality is achieved through the provision of benefits to the proposed project site. This includes compensation for environmental degradation and consideration for the economic value of landscape benefits, which are difficult to monetize.

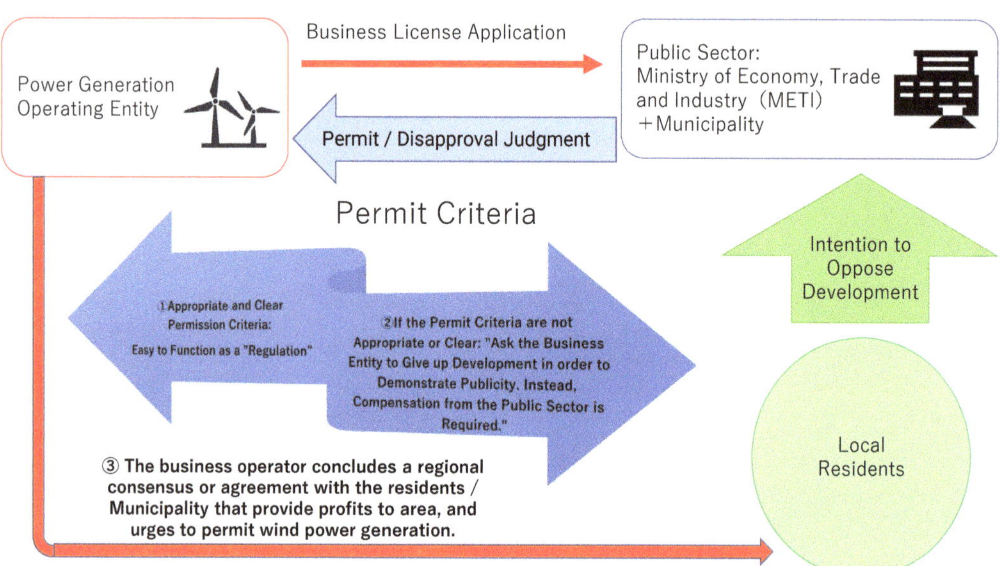

Figure 1. Parties involved in the issuance of business licenses for renewable energy projects in Japan.

As such, if the permit criteria are not clearly determined, it is not possible to provide compensation for businesses that abandon development and stop environmental deterioration and benefits provided to the region instead of developing and causing environmental deterioration. To achieve these points, it is necessary to have a clear indicator for converting the amount of damage caused by environmental deterioration.

Landscape deterioration is the primary reason for local residents' opposition to Japan's renewable energy power generation [35]. In promoting offshore wind power, Choshi offshore in Chiba Prefecture was selected as the only Renewable Energy Promotion Area (*Sokushin-Kuiki*) in the Kanto region [36]. Offshore Promotion Areas are designated by the government based on the act that establishes unified rules for offshore wind power generation (Act on Promoting the Utilization of Sea Areas for the Development of Marine Renewable Energy Power Generation Facilities (Shortened form: *Saiene-Kaiiki-Riyo-Ho*, Act No 89 of 2018)) after coordinating with the preceding users of the sea area, such as fishermen, and with the region. Choshi City is a fishing town; however, fishing yields have been declining. The promotion of this project had issues with compensating fishermen and handling the deterioration of the landscape, as this location has a scenic bay (*Byobugaura*) and a view of Mt. Fuji in the distance. Therefore, the landscape would change drastically if more than 30 wind turbines were constructed [37]. Consequently, fishermen received compensation, and Mitsubishi Corporation, which was selected as a business entity to promote offshore wind power generation, was asked to give "maximum consideration" to the local community [38]. The expression "maximum consideration" shows the lack of options for the modern landscape conservation criteria.

Therefore, if business entities can receive compensation for loss if they are prevented from development, it will be possible to prevent damage to the landscape to some extent. As such, I believe that clear standards (indicators) are necessary for everyday landscape conservation. Therefore, this study aimed to demonstrate the importance of indicator formulation and its continuous monitoring based on the cases of the UK and the US.

2.2. Internal and External Legal Obligations

Article 29 para 1 of the Constitution of Japan defines the property right as "the right to own or to hold property is inviolable" [39]. In para 2 of the same article, "property rights shall be defined by law, in conformity with the public welfare" and the exercise of property is accompanied by "internal legal obligations" [39]. Furthermore, para 3 stipulates that "private property may be taken for public use upon just compensation therefor," and "compensation (*Hosyho*)" are required legally [38]. In more concrete terms, para 2 indicates that property rights are restricted (subject to intrinsic restrictions) by law. In this regard, no compensation is required. Conversely, para 3 states that to achieve the public purpose, the state power may coercively expropriate or restrict the use of private property based on the French Declaration of Human Rights. In such cases, legitimate compensation is required (Figure 2) [40,41].

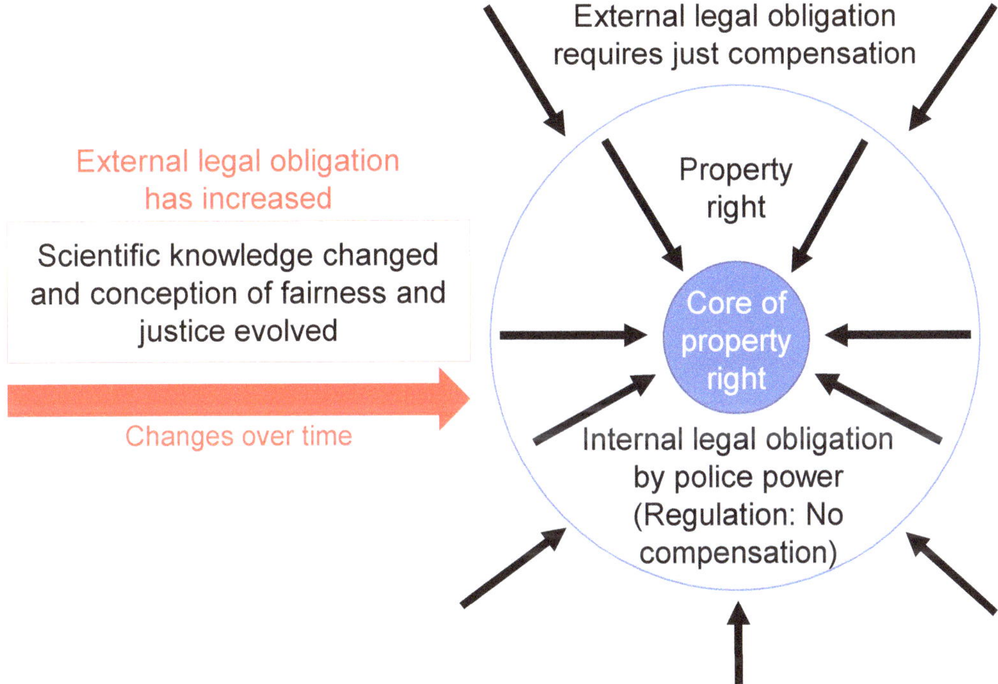

Figure 2. Internal and external legal obligations observed in Japan.

The restrictions outlined in para 2 are implicit and are subject to regulation by "police power." Police power means a fundamental power essential to government, and the inherent and plenary power of a sovereign to make all laws necessary without compensation (Figure 2). It is the "noxious-use doctrine" (*sic utere tuo*: to do this without harming others to use their things) and it is found in the realm of the provision of public goods, such as essential fields for human survival, including life, public health, national defense, law enforcement, and fire protection. This doctrine has gradually been applied to situations where public utility is expected by society, with no significant foreseen negative impacts. These are general norms of the regulation [31].

Alternatively, "taking" (public expropriation, regulated in para 3 of Article 29) aims at public use and does not take private property without appropriate compensation [39]. The "taking" clause is not a source of power for taking but rather a limitation [42]. For instance, "taking" must be for a public purpose, and compensation must be made because it entails special sacrifices for the benefit of the public. In addition, "taking" includes governmental or official actions that damage property or impair its use and physical appropriations [43].

2.3. Prioritizing Compensated Measures over Regulatory Actions

In the US, using "funding" measures for the conservation of endangered species was more effective than regulatory actions [44]. These "funding" measures are provided for land use rules, such as abandoning developments for the conservation of endangered species inhabiting private land [45]. Such trends can be applied in the context of landscape conservation as well. This study interpreted the term "funding" as synonymous with "compensation."

There are no existing criteria or legal protections for landscape degradation outside of national parks or designated areas in Japan. Moreover, one of the most frequent and opposing opinions of the local residents regarding renewable energy facility construction is the deterioration of "daily landscapes" (landscapes in daily settings without specific legal protections), making the issue a recurring problem in Japan from 2017 to 2020 [35]. Given the absence of legal frameworks and established criteria, the general trend of conflicts is on the rise, and issues surrounding landscape problems tend to rely on "residents' consent." This is because the construction activities of business operators are neither illegal nor unreasonable and frequently occur in unregulated sites, such as those that lack landscape protection. Moreover, target project sites are often perceived by people as "familiar" or "daily landscapes" in the region, and there are concerns that they will be affected or changed. For instance, a change in the landscape by cutting trees reminds residents of disasters such as floods that were once affected. Such "landscape change" is sensitive and affects the liberty of other economic activities, such as tourism in mountainous areas with strong historical and religious elements and idyllic hot spring areas with legitimate procedures. There is a need for objective-driven criteria to have a balanced system.

In the absence of clear regulations, it is imperative that residents provide sufficient compensation for business entities rather than forcing "residents' consent" by regulations or municipality ordinances to make business developers consider landscape elements. When there are signs that regulations will be imposed or when there is a small amount of compensation, business operators can develop the project legally [44]. Alternatively, if compensation is sufficient, development can be stopped and/or reexamined. In many development cases, local people want to protect familiar landscapes that affect their daily lives. The introduction of a "compensation" measure should be considered for the conservation of landscapes and historical and cultural heritage, which are not legally regulated.

2.4. Is it Taking or Regulation?

This study compiled descriptions from various sources that have changed along with the social and legal circumstances. Yoshida [46,47] noted that unlike "restrictions to avoid negative externalities (intrinsic restrictions)," there is an increasing number of "restrictions to ensure positive externalities (extrinsic restrictions)" based on certain real estate uses, such as conservation of green spaces and conservation of landscapes. This overseas restriction falls under Article 29 para 3 and requires compensation [38]. Sax [48] stated that "drawing the private/public line is not limited to physical harms." Furthermore, the balance between private desires and public obligations in the rules governing land ownership states may change over a longer period. Specifically, the development of scientific knowledge on hazards and concepts of fairness and justice affect such changes, which may include the possibility of both internal and external legal obligations. The full burden of maintaining that value falls on the relatively few who retained some natural value in their land, whereas those who have created the problem of potential species extinction bear no burden at all [48]. Alexander [28] argued that the owner should have active security obligation, in addition to "negative obligations" under "noxious-use doctrine." This idea is that landowners should be obliged not only to passively "not develop (prohibit)" but also actively "conserve (affirmative obligation)." This is a useful suggestion at present, when it is scientifically revealed that "care" is required in certain cases, in addition to "avoiding development" for environmental conservation.

It is difficult to determine whether the public nature of the government is "taking" (requiring payment of just compensation) or a regulation (with no requirements for compensation). There is no clear line or "standard" that can be distinguished [41,43,49,50]. This is partially due to changes over time, including scientific progress, and changes in the sense of equity and justice. In contrast to the compensation requirements overseas (as described later), the range of regulations is kept wide, as long as the content of the regulations is conserved in current Japanese laws [51]. For instance, under the Natural Parks Act Article 17 para 3 [52], a claim for compensation for loss was excluded, as the restriction on the use of the application for permission for the new construction of structures was an inherent restriction of property rights [53,54]. Another example is the case of local government-zoned green belts around urban areas. These include zoning that regulates the development of lands designated as green belts by landowners; however, litigation disputes that development indemnity would be sought in that regard. However, the Japanese Supreme Court determined that compensation was not required if conservation was forced [55]. Alternatively, if landowners' existing buildings are to be destroyed to install green belts, compensation is deemed necessary [51]. This means that not all landowners who currently have a natural environment to preserve are responsible for conserving the natural environment [48]. It is an academic doctrine and precedent that compensation for the loss under the Japanese Law enables the right to claim damages under the Constitution directly, even if there is no specific provision in the Individual Law [56].

In addition, it is difficult to consider the criteria for distinguishing between "taking" and regulation. There arises the need for scientific or agreed objective-driven criteria of landscape degradation for proper compensation. The amount of compensation does not indicate that development activities will be easier or unmonitored; rather, it promotes prudent and responsible development. To demonstrate this, under the US Endangered Species Act, payment of development compensation ("compensation for development control") rather than imposing regulation on endangered species has led to more controlled development activities [44]. Furthermore, the concept of development compensation is presented as a responsibility (indemnity) for local communities seeking responsibility from business operators.

There is a high demand that landowners assume the duty of conservation if they can play the role of conservation, as the number of actors in conservation is decreasing [28]. If the landowners are considered primarily responsible for conservation, it may be necessary to compensate the landowners for the additional efforts to conserve (which are higher standards than "not developing") [48]. Landowners may have the right to choose their preferences for conservation measures; however, for them to exercise their rights, a systematic operation, such as "ecological succession of cultivated land to wilderness area" over the long term, is required [57,58].

3. Criteria and Quantitative Approaches to Landscape Degradation

Existing compensation schemes in the UK and the US include measures to balance landscape conservation with developments including land-use changes. In addition, this study examined the legal settings related to ownership (private and public lands) in the two countries. In the current compensation schemes, both countries have a common context that "all property has a legally recognized owner, and if the owner cannot be found in a timely manner, then the government is presumed to be the owner" [59,60]. The landowners and the public sectors are responsible for maintaining the environment including landscapes.

The landscape character assessment (LCA), which is a process of identifying and describing variation in landscape characteristics, is used in the UK to evaluate the "visual despoliation of valued landscape" [14]. LCA documents identify and explain the unique combination of elements and features that make landscapes distinctive by mapping and describing character types and areas [61]. In addition, they show how the landscape is perceived, experienced, and valued by people [60]. Moreover, in the LCA approach, the landscape is evaluated whether it has "use value" or "non-use value" when wind turbines are installed or not [62,63]. It is logically deducted from "options to visit landscapes without wind turbines," "options to visit a landscape free of wind turbines," or simply "the presence of untouched landscapes" (that a "pristine" landscape exists as the "non-use value").

In many areas, wind farm development is located without protection areas, but close to these designations. Though, in these circumstances, the effects on the designated landscape remain a key consideration, LCAs do not place value on one landscape type over another. LCAs may point to the reasons why a landscape might be valued. LCA helps us understand what the landscape is like today, how it came to be like that and how it may change in the future. It also helps to ensure that change does not undermine whatever is valued or the characteristics of a particular landscape, and proposes ways to improve the landscape's character [63].

Looking at a specific case, the LCA methodology has proven beneficial for the Cape Wind Energy (CWE) and Ocotillo Wind Energy Facility (OWEF) projects in Scotland [64]. The LCA explained that "visual degradation" can be identified as a negative externality that may potentially affect human well-being and a realistic cost that needs to be considered in determining the economic viability of the project.

In the US, visual resource management (VRM) manages public lands in a manner that protects the quality of visual or scenic values. VRM is conducted under the Federal Land Policy and Management Act (FLPMA) by the Bureau of Land Management (BLM) The BLM has developed the VRM system for visual resource inventory, management, and impact assessment. BLM-administered lands are managed in accordance with approved resource management plans, which are developed with public participation and collaboration. The VRM classes set the objectives for lands in each class and describe the limits of allowable visual change in the landscape character with which proposed management activities must comply [65].

VRM is a methodical approach used for cataloging and managing scenic resources of public lands managed by the BLM. Specifically, all public lands managed by the BLM are classified into four categories based on the maintenance of natural landscapes (Class I), scenic quality (Class II), sensitivity level (Class III), and distance (Class IV). This classification has an impact on business approvals. In VRM classification, considerations are sought before undertaking development activities in the region and provide a methodical means to evaluate activities that they conformed with the approved objectives. There is a unique scale of landscape evaluation.

4. Potential of Development Compensation in the Legal Framework of Japan

Current Japanese regulations of environmental conservation are generally required only if they are below a "reference point" (a benchmark of being "neutral"). In Article 21 of the Japanese Basic Act on the Environment (Regulations to Prevent Hindrances to Environmental Conservation), regulatory methods are stipulated as "the State must take the following regulatory measures to prevent a hindrance to environmental conservation." The conceptual framework is outlined in Figure 3 [52,66]. For Case 1, development activities are above the reference point and have desirable effects on the environment. For Case 2, environmental conditions are deteriorating but are still above the reference point. In this case, regulation is not yet necessary, although the validity of the content is questioned, such as "Is the business operator taking and evading compensation payments?" Moreover, "soft methods" are desirable as incentive measures, awareness-raising, or educational activities. In contrast, in Case 3, the environmental conditions are below the reference point; thus, environmental conservation control measures are needed to avoid degradation. These public measures are implemented in the form of regulations.

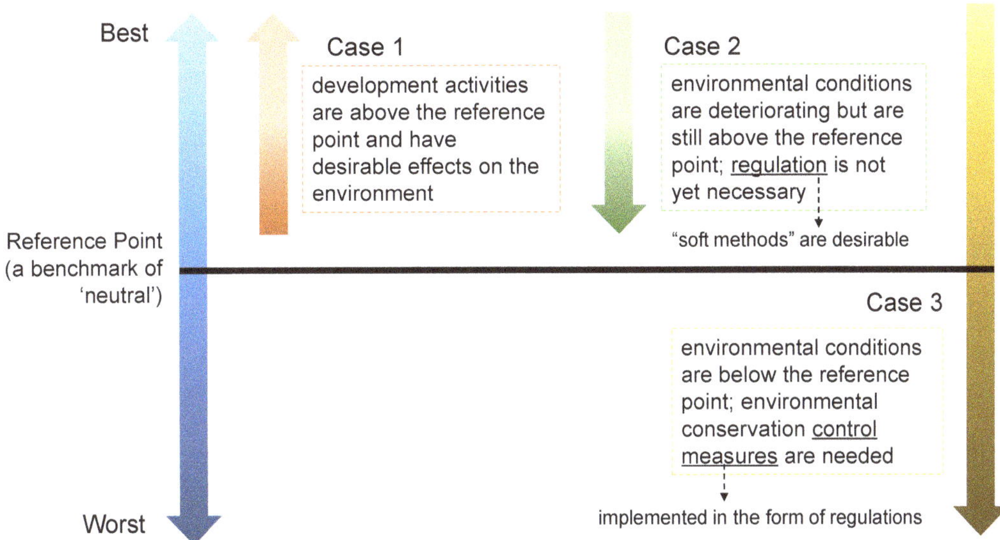

Figure 3. Conceptual framework of implementing Japanese regulations for environmental degradation control measures (conceptualized from MoE 2002).

Article 2 of the Japanese Basic Act on the Environment defined "environmental load" that leads to a worsening of environmental conditions as "negative effects of human activities on the environment causing hindrances to the environmental conservation." "Hindrances to environmental conservation" means environmental degradation to the extent that measures directly related to the rights and obligations of citizens, such as regulations, are implemented. Examples include damage to human health and living environments, such as pollution and failure to secure natural blessings that are indispensable for the public. Despite this regulation, Japan is facing the concern of deteriorating ecosystems and landscapes caused by inaction, such as the lack of care for artificial forests and the neglect of *satoyama* [21,67]. It is difficult for landowners to be legally liable for inaction, as they have the freedom to leave (or use) the land [21] and, in certain cases, due to cultural differences affecting people's perceptions and behavior [68]. In several cases, inactions can be "hindrances to the environmental conservation," as conservation activities in degraded ecosystems are an extra burden for the landowners [54].

In addition to the above regulation, there are also specific landscape-related measures. For instance, in Article 2 of the Landscape Act, the "good natural landscape" (*Ryoko-na-Kekan*) needs to be protected and is described as a "Basic Philosophy" [69]. Other measures regarding natural landscapes include Article 1 of the Natural Parks Act for excellent natural scenic areas [52] and Article 2 (5) of the Act on Protection of Cultural Properties for the cultural landscape (*Bunkateki-Kekan*) [70]. Although these measures for "good landscape" need to be followed and regulated, there are no specific indicators of landscape degradation or reference point indicators. Furthermore, the Japanese legal environmental assessment includes a procedure called "post-hoc investigations" (*Jigo-Cyōosa*), which are surveys conducted after the commencement of projects [71]. Business operators must consider the implementation of the follow-up survey in certain cases, and if the results show that the environmental impact is significant, they must take environmental conservation measures. This follow-up survey avoids environmental impact caused by unpredictable factors and prevents problems with local residents; however, investigations related to landscapes are lacking compared with surveys related to animals or noise which are steadily conducted [72]. The measurement and monitoring of such landscapes, including the development and operation of negative indicators of landscape degradation, are not regularly conducted, if not absent, despite efforts made to conserve "good landscapes."

If certain landscapes considered for development can be ranked or categorized, the degree of damage to the landscape should also be classified. The idea of subtracting from the condition without the damage is effective for the classification. There are three reasons to support approaches of quantifications for Japan (as in the UK's LCA and the US's VRM). First, the method can be applied to the degree of damage to natural landscapes in daily settings in addition to the current "good natural landscape," such as scenic places, which are the focus of current environmental impact assessments. The indicators further lead to the calculation of development compensation more proportionally, as they contribute to balancing the pros and cons with objective criteria reflecting arguments for developments and conservations by continuously accumulating and utilizing data in monitoring.

Second, such scenic resource indicators can be applied to areas or landscapes that are not yet protected by law. Before the enactment of the Landscape Act, there were landscape ordinances in some advanced municipalities [69]. In other words, the idea of these landscape ordinances becomes a policy at the national level, and what is further cultivated is called the Landscape Act. Some of the ideas of conserving the natural landscapes of various places that exist in municipalities have been inherited by the Landscape Act [73]. However, due to its national disposition (nationwide), the methods prepared by the Landscape Act are not intended to conserve local daily popular landscapes or popular natural landscapes. The trade-offs and conflicts are likely to intensify with the revisions of the renewable energy schemes, and deregulation of law-based environmental impact assessment standards. In Japan, each municipality is making policy decisions on renewable energy promotion areas [74]. The area can be classified into one of three categories: development promotion area, adjustment area, and excluded development area (Figure 4). The significance of the classification is progressing in Japan as the number of conflicts occurring in various places is increasing [75–77]. Examples of these conflicts are cases with opposition movements, even if the developments are neither illegal nor unreasonable (Figure 4) [57]. Such conflicts are likely to linger when the development promotion and adjustment areas are introduced based on legal procedures. There is a need to consider scenic resource indicators in advance and examine land-use policies and other development plans for the introduction of renewable energy. Furthermore, "residents' consent" tends to fall into subjective judgment; when consents or agreements tend to be difficult, they tend to eliminate the problem itself.

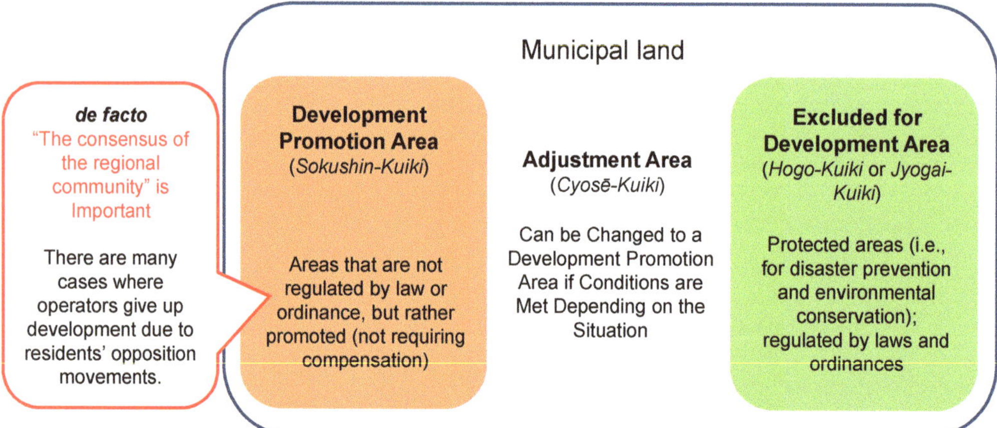

Figure 4. Current status of promoting areas for renewable energy sites in Japan (modified from Kohsaka and Kohyama 2022).

The third reason to support approaches of quantifications in Japan is the sense of aesthetics, which may change over time. An iconic example is the Eiffel Tower in the city of Paris. Initially, the tower was perceived as "out of place"; however, it is currently the most famous and beloved symbol in Paris [78,79]. In Japan, the Tokyo Metropolitan Government Office received negative perceptions in the earlier periods; however, over time, it changed to positive and consequently became one of Tokyo's landmarks. The evaluation of aesthetic values is possible by adding the damage of the development to the condition in which such transition was not damaged. For instance, if wind turbines are perceived as beautiful with the transition to the next generations, they can be evaluated as having positive effects in the longer term.

As another aesthetic, we have also noticed the existence of ethical aesthetics. We must pay attention to changes in public nature in an era when climate change countermeasures are required more and more. Renewable energy business entities tend to be treated as private businesses (not public nature) under Japanese law [80]. Therefore, when a business entity gives up its business to preserve the environment and landscape, it can be regarded as a withdrawal of a public nature. In this case, the beneficiaries are residents, etc. However, it is necessary to consider "whether the renewable energy business really does not have a public nature". This is because the role of the renewable energy business as a countermeasure against climate change and the provision of services in society in the form of electricity supply is truly important. However, the question here is 'the location'. Even if it is a project for such an important business, whether the project should be implemented in "this place" is being questioned again.

5. Conclusions

This study examined various approaches to scenic impact assessment. In the UK and US, the degree of damage to the natural environment is foreseen and evaluated, and such methodologies are instrumental in decision-making processes. Moreover, by applying necessary assessment, it is possible to measure or estimate who (including individuals and communities) and where will be affected, including benefits and adverse effects. This enables a better understanding of both positive and negative effects at the landscape level, allowing the decision regarding which level of effects the development is likely to result in. Based on these findings, classification allows the owners who are only developing to compensate residents as an offset for development. Items of scenic impact assessment are de jure mandatory issues but are de facto limited parts of the environmental impact assessment (EIA) when considering the effects of landscape degradation from development projects.

The ways to address landscape impacts in EIA, the impacts of community concerns on the consensus of the regional community, and the degradation of natural landscapes in large-scale wind power projects are increasing. Local residents are frequently concerned with wind turbines and facilities constructed at the ridges of mountains, which are separated from human living spaces. These concerns arise from the projected "changing living environment (hometown)" to the change of the distant natural landscape rather than the risk of biodiversity damage (including migratory birds) or health-related noise risks. In such cases, concerns regarding landscape deterioration are interlinked with concerns about disasters, including land modifications [81]. Thus, there is a need to develop compensation for the cost and benefit-sharing of landscape conservation. To balance landscape degradation with other functions and preferences of residents, there is a need to develop indicators for the calculation, concrete scenic resource inventory, and establishment of a methodical approach to management.

Existing methods were examined to balance landscape conservation with the increased construction of renewable energy facilities. This study outlined several aspects as a way forward. There is a need to encompass and evaluate landscapes in daily settings beyond those "good natural landscapes" in scenic places. With the increasing of conflicts, it is essential to evaluate and rank or categorize landscape resources for all landscapes. As such, methodical efforts are required to conserve areas beyond "good natural landscapes," including those in daily settings of rural areas, which are largely overseen and neglected.

Furthermore, there is a need for objective-driven approaches that can measure landscape degradation. It is relatively easy to evaluate the beauty of the landscape that needs to be protected; however, there are no established methods from various perspectives or social contexts. Adding landscape's aesthetic degradation as an object in development compensation is challenging, as its value is difficult to monetize, and it may not be possible to sufficiently conduct research such as aging evaluation regarding landscape degradation. As a methodical approach to advance landscape conservation, examinations balancing development projects are required.

The significance of this paper is that, for a country like Japan, which does not have clear, objective-driven indicators and methods that measure landscape degradation, it is important to establish scientific indicators and the sharing of responsibility for their development (to provide an opportunity to think about the relationship between developers and victims, potential victims and beneficiaries) including the importance of compensation.

Funding: This research received no external funding.

Data Availability Statement: Not applicable.

Acknowledgments: The author would like to thank Ryo Kohsaka (Graduate School of Agricultural and Life Sciences, The University of Tokyo) for his support in this study.

Conflicts of Interest: The author reports there are no competing interest to declare.

References

1. Brundtland, G. *World Commission on Environment and Development (WCED). Our Common Future;* Oxford University Press: Oxford, UK, 1987.
2. Worthington, E. *The Ecological Century: A Personal Appraisal;* Clarendon Press: Oxford, UK, 1983.
3. Weizsäcker, E. *ERDPOLITIK: Ökologische Realpolitik an der Schwelle zum Jahrhundert der Umwelt;* Wissenschaftliche Buchgesellschaft (wbg): Darmstadt, German, 1989.
4. Salcido, R. Rationing environmental law in a time of climate change. *Loyola Univ. Chic. Law J.* **2015**, *46*, 617–668.
5. Morris, A.; Owley, J. Mitigating the impacts of the renewable energy gold rush. *Minn. J. Law Sci. Technol.* **2014**, *15*, 293–388.
6. Michie, P. Billing credits for conservation, renewable, and other electric power resources: An alternative to marginal-cost-based electric power rates in the Pacific Northwest. *Environ. Law* **1983**, *13*, 963–1029.
7. Kohyama, S. A study on "law-environmentalization" of forest law. *Kyushu Int. Univ. Law Rev.* **2014**, *20*, 43–63.
8. Outka, U. Renewable energy siting for the critical decade. Univ. Kans. *Law Rev.* **2021**, *69*, 857–887.
9. Parkhurst, G.; Shogren, J. Evaluating incentive mechanisms for conserving habitat. *Nat. Resources J.* **2003**, *43*, 1093–1149.
10. Adler, J. Money or nothing: The adverse environmental consequences of uncompensated land use controls. *Boston Coll. Law Rev.* **2008**, *49*, 301–366. [CrossRef]
11. Stern, S. Encouraging conservation on private lands: A behavioral analysis of financial incentives. *Ariz. Law Rev.* **2006**, *48*, 541–583.
12. Shelton, H. Conservation in Texas: Bridging the gap between public good and private lands using landowner incentive programs. *Vt. J. Environ. Law* **2018**, *19*, 273–305.
13. Rinehart, L. Zoned for injustice: Moving beyond zoning and market-based land preservation to address rural poverty. *Georget. J. Poverty Law Policy* **2015**, *23*, 61–105. [CrossRef]
14. Tudor, C. *An Approach to Landscape Character Assessment*, 1st ed.; Natural England: York, UK, 2014. Available online: https://assets.publishing.service.gov.uk/government/uploads/system/uploads/attachment_data/file/691184/landscape-character-assessment.pdf (accessed on 17 November 2022).
15. Kohyama, S. A study on wind power generation business and noise: Focusing on U.S. cases. *Tomidai-Keizai-Ronsyu* **2019**, *64*, 557–600. [CrossRef]
16. Basic Act on the Environment (Act No 91 of 2018). Available online: https://www.japaneselawtranslation.go.jp/en/laws/view/3850 (accessed on 17 November 2022).
17. Kitamura, Y. *Environmental Law*, 5th ed.; Kobundo: Tokyo, Japan, 2020; pp. 157–162.
18. Otsuka, T. *Environmental Law*, 4th ed.; Yuhikaku: Tokyo, Japan, 2020; pp. 104–117.
19. Uga, K. *State Compensation;* Yuhikaku: Tokyo, Japan, 1997; pp. 387–494.
20. Kohyama, S. *Introduction to Administrative Disputes*, 2nd ed.; Bunshindo: Kure, Japan, 2021; p. 108.
21. Morgan, C. Demographic crisis in Japan: Why Japan might open its doors to foreign home health-care aides. *Pac. Rim Law Policy J.* **2001**, *10*, 749–779.

22. Dileo, A. Directions and dimensions of population policy in the United States: Alternatives for legal reform. *Tulane Law Rev.* **1971**, *46*, 184–231.
23. Kohyama, S. Amendment of basic civil legislation and ownership of forests ownership. *Sanrin* **2022**, *1653*, 2–10.
24. Lubens, R. The social obligation of property ownership: A comparison of German and U. S. law. *Ariz. J. Int. Comp. Law* **2006**, *24*, 389–449.
25. Constitution Annotated (USCS Const. Amend. 5.). Available online: https://constitution.congress.gov/constitution/amendment-5/ (accessed on 23 December 2022).
26. JUSTIA US Supreme Court, Penn Cent. Transp. Co. v. New York City, 438 U.S. 104, 98 S. Ct. 2646, 57 L. Ed. 2d 631, 1978 U.S. LEXIS 39, 8 ELR 20528, 11 ERC (BNA) 1801. Available online: https://supreme.justia.com/cases/federal/us/438/104/web (accessed on 23 December 2022).
27. Findley, R.; Faber, D. *Environmental Law in a Nutshell*, 2nd ed.; Inada, H., Translator; Bokutan-sya: Tokyo, Japan, 1992; pp. 216–217.
28. Alexander, G. The social-obligation norm in American property law. *Cornell Law Rev.* **2009**, *94*, 745–819.
29. Penalver, E. Land virtues. *Cornell Law Rev.* **2009**, *94*, 821–888.
30. Wendel, W. Legal ethics as "political moralism" or the morality of politics. *Cornell Law Rev.* **2008**, *93*, 1413–1436.
31. Smith, H. Mind the gap: The indirect relation between ends and means in American property law. *Cornell Law Rev.* **2009**, *94*, 959–989.
32. Ruhl, J.; Kraft, S.; Lant, C. *The Law and Policy of Ecosystem Services*; Island Press: Washington, DC, USA, 2007.
33. Kohsaka, R.; Kohyama, S. Contested renewable energy sites due to landscape and socio-ecological barriers: Comparison of wind and solar power installation cases in Japan. *Energy Environ.* **2022**, 0958305X221115070. [CrossRef]
34. .Kohyama, S.; Kohsaka, R. Wind farms in contested landscapes: Procedural and scale gaps of wind power facility constructions in Japan. *Energy Environ.* **2022**, 0958305X221141396. [CrossRef]
35. Yamasahita, N. Local Troubles of Solar Power Generation and Harmonization/Regulation Ordinances, toward Proper Promotion in the Future. Available online: https://www.meti.go.jp/shingikai/energy_environment/saisei_kano_energy/pdf/002_02_00.pdf (accessed on 17 November 2022).
36. Chiba-Prefecture. Offshore Choshi-City: Offshore Wind Power Generation Business. Available online: https://www.pref.chiba.lg.jp/sanshin/ocean-re/01_choshi.html (accessed on 17 November 2022).
37. NHK Chiba. Offshore Wind Power Generation Project underway off the Coast of Choshi-City, Chiba-Prefecture. Available online: https://www.nhk.or.jp/shutoken/chiba/article/000/35/ (accessed on 17 November 2022).
38. NIKKEI. Offshore Wind Power off the Coast of Choshi, Chiba, Consideration for the View of Mt. Fuji. Available online: https://www.nikkei.com/article/DGXZQOCC26D0Y0W2A420C2000000/ (accessed on 17 November 2022).
39. Constitution of Japan (Constitution 3 November 1946). Available online: https://elaws.e-gov.go.jp/document?lawid=321CONSTITUTION (accessed on 17 November 2022).
40. Nishino, A. *Loss compensation laws: Commental*; Keiso-Syobo: Tokyo, Japan, 2018; pp. 3–208.
41. Tsujimura, M. *The Constitution of Japan*, 7th ed.; Nihon-hyoron-Sya: Tokyo, Japan, 2021; pp. 247–254.
42. Abe, Y. *Research of the State Compensation Laws I*; Shinzan-Sya: Tokyo, Japan, 2019; pp. 365–408.
43. Epstein, R. *Takings: Private Property and the Power of Eminent Domain*; Harvard University Press: Cambridge, MA, USA, 1985.
44. Ferraro, P.; Mcintosh, C.; Ospina, M. The effectiveness of the US endangered species act: An econometric analysis using matching methods. *J. Environ. Econ. Manag.* **2007**, *54*, 245–336. [CrossRef]
45. Hoshino, M.; Tanaka, H. *Empirical Analysis by R: From Regression Analysis to Causality*; Ohmusha: Tokyo, Japan, 2016; pp. 169–188.
46. Yoshida, K. *Modern Land Property Rights Theory: Legal Issues Surrounding Land of Unknown Ownership and a Declining Population*; Shinzan-Sya: Tokyo, Japan, 2019; pp. 3–142.
47. Yoshida, K. Today's Challenges for Real Estate Ownership. *NBL* **2019**, *1152*, 4–12.
48. Sax, J. Environmental disruption and landowner obligations—Time for some new thinking. *Res. Environ. Disrupt.* **2010**, *40*, 2–7.
49. Ministry of Construction. *Loss Compensation Standards Associated with the Acquisition of Public Land*; Ministry of Construction: Tokyo, Japan, 1962.
50. Ministry of Land, Infrastructure, Transport and Tourism (MLIT). Loss Compensation Standards Associated with the Acquisition of Public Land (2020 Ver.). Available online: https://www.mlit.go.jp/common/001338606.pdf (accessed on 17 November 2022).
51. Shiono, H. *Administrative Law*, 6th ed.; Yuhikakku: Tokyo, Japan, 2019; pp. 383–389.
52. Natural Parks Act (Act No. 161 of 1957). Available online: https://www.japaneselawtranslation.go.jp/en/laws/view/3060 (accessed on 17 November 2022).
53. Tokyo District Court. *A Case Showing Certain Criteria as to Whether the Regulations Under Articles 17 and 35 of the Natural Parks Act Are Special Sacrifices beyond the Scope of Internal Legal Obligations*; Tokyo District Court: Tokyo, Japan, 1990.
54. Kohyama, S. Legal study to solve the problem of poorly maintained forests—from the viewpoint of clarifying the obligations of owners. *J. Environ. Law Policy* **2012**, *15*, 278–291.
55. Japanese Supreme Court. *Defendant Case of Violation of the River Neighborhood Restriction Order*; Japanese Supreme Court: Tokyo, Japan, 1968.
56. Japanese Supreme Court. *City Road Area Decision Disposition (1938, Cancellation Request Case; Shiono n. 51; Kohyama n.20)*; Japanese Supreme Court: Tokyo, Japan, 2005.

57. Kohsaka, R.; Kohyama, S. State of the art review on land-use policy: Changes in forests, agricultural lands and renewable energy of Japan. *Land* **2022**, *11*, 624. [CrossRef]

58. European Centre for Nature Conservation. ECNC's Participate in Activities to Regenerate Abandoned Cultivated Land in Europe into a Natural Wilderness Region. Available online: https://tenbou.nies.go.jp/news/fnews/detail.php?i=4639 (accessed on 17 November 2022).

59. West Coast Stock Transfer, Inc. Escheatment. Available online: https://westcoaststocktransfer.com/escheatment/ (accessed on 15 November 2022).

60. Kohyama, S. A comparative study on escheatment laws of Japan and U. S. *TJAIBL* **2020**, *5*, 87–101.

61. Landscape and Seascape Character Assessments. Available online: https://www.gov.uk/guidance/landscape-and-seascape-character-assessments (accessed on 15 November 2022).

62. Transport Scotland. Landscape and Visual Effects, Chapter 8. Available online: https://www.transport.gov.scot/media/46461/a720-es-chapter-8-landscape-and-visual-effects.pdf (accessed on 15 November 2022).

63. Scottish Natural Heritage. Siting and Designing Wind Farms in the Landscape, 3a ed. Available online: https://www.nature.scot/sites/default/files/2017-11/Siting%20and%20designing%20windfarms%20in%20the%20landscape%20-%20version%203a.pdf (accessed on 17 December 2022).

64. Dussias, A. Room for a (sacred) view? American Indian tribes confront visual desecration caused by wind energy projects. *AILR* **2014**, *38*, 333–420.

65. U.S. Department of the Interior, Bureau of Land Management. The Federal Land Policy and Management Act. Available online: https://www.blm.gov/sites/blm.gov/files/documents/files/FLPMA2016.pdf (accessed on 17 November 2022).

66. Ministry of the Environment. *Explanation of the Basic Act on the Environment*; Revised ed.; Gyousei: Tokyo, Japan, 2022.

67. Kohsaka, R.; Ito, K.; Miyake, Y.; Uchiyama, Y. Cultural ecosystem services from the afforestation of rice terraces and farmland: Emerging services as an alternative to monoculturalization. *FEM* **2021**, *497*, 119481. [CrossRef]

68. Kohsaka, R.; Handoh, I. Perceptions of "close-to-nature forestry" by German And Japanese groups: Inquiry using visual materials of "cut" and "dead" wood. *JFR* **2005**, *11*, 11–19. [CrossRef]

69. Landscape Act. Available online: https://www.japaneselawtranslation.go.jp/en/laws/view/2533 (accessed on 17 November 2022).

70. Act on Protection of Cultural Properties. Available online: https://elaws.e-gov.go.jp/document?lawid=325AC010000021 (accessed on 17 November 2022).

71. Environmental Impact Assessment Act. Available online: https://www.japaneselawtranslation.go.jp/en/laws/view/3375 (accessed on 17 November 2022).

72. Study Group on the Ideal Way of EIA for the Proper Introduction of Renewable Energy. The Report of the Study Group on the Ideal Way of EIA for the Proper Introduction of Renewable Energy. Available online: https://www.meti.go.jp/shingikai/safety_security/renewable_energy/pdf/20210331_1.pdf (accessed on 17 November 2022).

73. Ito, S. *Policy innovation from Municipalities: From Landscape Ordinance to Landscape Act*; Bokutansya: Tokyo, Japan, 2006.

74. Ministry of the Environment. Ordinance/Special Zone System Regarding Renewable Energy. Available online: https://www.env.go.jp/policy/local_keikaku/ (accessed on 17 November 2022).

75. Obane, H.; Nagai, Y.; Asano, K. Assessing land use and potential conflict in solar and onshore wind energy in Japan. *Renew. Energy* **2020**, *160*, 842–851. [CrossRef]

76. Kitano, M.; Shiraki, S. Estimation of bird fatalities at wind farms with complex topography and vegetation in Hokkaido, Japan. *Wildl. Soc. Bull.* **2013**, *37*, 41. [CrossRef]

77. Environmental Impact Assessment Network. Enforcement Status of the Environmental Impact Assessment Act. 2021. Available online: http://assess.env.go.jp/2_jirei/2-4_toukei/index.html (accessed on 17 November 2022).

78. Brisman, A. The Aesthetics of Wind Energy Systems. *NYU Environ. Law J.* **2005**, *13*, 1–133.

79. Wickersham, J. Sacred landscapes and profane structures: How offshore wind power challenges the environmental impact review process. *Boston Coll. Environ. Aff. Law Rev.* **2004**, *31*, 325–347.

80. Kohyama, S. The Study of an Administrative Case: In Tahara City. Aichi Prefecture; Injunction demand for operation of wind power generation facility. *Tomidai-Keizai-Ronsyu* **2015**, *64*, 183–208. [CrossRef]

81. Kohyama, S.; Kohsaka, R. Japan's mountainous region changed by the promotion of de-carbonization policy—regional dynamics over decarbonization and renewable energy site debate. *Jichi-Soken* **2022**, *523*, 1–37.

 *land*

Review

Rural Energy Communities as Pillar towards Low Carbon Future in Egypt: Beyond COP27

Ahmed Abouaiana

Department of Planning, Design and Technology of Architecture, Sapienza University of Rome, Via Flaminia 72, 00196 Rome, Italy; ahmed.abouaiana@uniroma1.it

Abstract: Egypt pays extraordinary attention to climate action, which is gaining momentum, coinciding with reaching the peak of the status quo by hosting the 2022 United Nations Climate Change Conference, Conference of Parties (COP27). Renewable energy sources are one of the principal axes of the state's plan to combat climate change and open new horizons toward decarbonization. Rural commons act as a food basket and are essential to function in urban areas and enhance ecosystem services, even though currently they are facing extraordinary environmental challenges. Therefore, this study aims to restore the function of the rural commons from consumerism to productivity as an energy basket and create a tendency and momentum toward a self-sufficiency dogma by promoting the rural energy community concept from a top-down approach in Egypt. Two steps can articulate this: First, defining the legal key concept and showing its roots in European policies to provide a direction to this research. Second, by analyzing the current Egyptian legalization, laws, efforts, and best practices, those could address, allow, and encourage the concept's core. The results proved that this is the first research to discuss the concept from the climate–energy–land use perspective, integrated with a previous bottom-up intervention. Meanwhile, it explains the current state of knowledge and a better understanding of the institutional context, showing the high level of coordination of cross sectors and proving that rural energy communities are presented in the bottom-up practices. This can support decisionmakers and paves the way for researchers, academic bodies, and energy experts to explore other insights.

Keywords: climate action; land–energy–climate nexus; multi-function land use; policy evaluation; renewable energy sources; sustainable rural development

Citation: Abouaiana, A. Rural Energy Communities as Pillar towards Low Carbon Future in Egypt: Beyond COP27. *Land* **2022**, *11*, 2237. https://doi.org/10.3390/land11122237

Academic Editors: Kleomenis Kalogeropoulos, Andreas Tsatsaris, Nikolaos Stathopoulos, Demetrios E. Tsesmelis, Nilanchal Patel and Xiao Huang

Received: 11 November 2022
Accepted: 7 December 2022
Published: 8 December 2022

Publisher's Note: MDPI stays neutral with regard to jurisdictional claims in published maps and institutional affiliations.

1. Introduction

1.1. Research Background

Mediterranean rural and agricultural areas play a substantial role in attaining sustainable development and food security, particularly in Egypt. The nature of rural commons is characterized by productivity and self-sufficiency. They act as a food basket and are essential to the functioning of urban areas, enhancing ecosystem services and providing aesthetic value and recreational services.

Rural areas have witnessed significant transformation due to consecutive factors since the past mid-century. Producing modern built environment patterns and different anthropological activities leads to shifting them from productivity to consumerism and irreversible adverse environmental impacts, in addition to confronting exclusive environmental challenges as one of the most fragile spots due to climate change (CC) surpassing the global norms and the annual decrease of the farmlands.

The global community has intensified its efforts since the Paris Agreement in 2015 (international treaty on CC) to fulfill the commitment at the national scales to mitigating climate actions in urban and rural contexts, in which energy is a pivotal element in this equation. Among these efforts is the energy communities concept. It has been gaining global tendency and interest, especially in Europe, since 2018 as one of the successful strategies for enhancing sustainable development goals, as emphasized in References [1,2].

European regulatory frameworks are already recognized and clearly explained to develop many energy community projects as possible in the different European countries (Section 2.2). Recently, the European Commission launched the Rural Energy Community Advisory Hub (REACH) initiative in June 2022 to set up energy communities in European rural areas, officially launching direct technical assistance in September 2022 for developing concrete energy community initiatives across the EU. In light of this, the welfare of rural communities (the heartbeat of the EU) is a priority [3].

In the same vein, two recent bottom-up studies by the author support this direction and boost energy communities and agrivoltaic key concepts in rural contexts, as has been concluded by them in two rural agricultural-based settlements in Egypt and Italy. On top of that, a recent field study concluded that low domestic electricity consumption in rural buildings in Egypt showed the potential to achieve zero energy building targets [4,5] as well as achieving positive energy ones.

Based on the integration of the introduced facts and these outcomes, this study is an attempt to provide a comprehensive foundation bringing attention to restoring the function of the rural commons from consumerism to productivity and acting as energy baskets, in addition to creating a tendency and momentum to self-sufficiently dogma by promoting the rural energy community concept in Egypt. This can be addressed by investigating the following research questions (RQ):

- RQ1: What is the current status of the concept's core in the legal framework from a top-down standpoint?
- RQ2: Is the concept implemented in Egypt, and can it be localized?

It is noteworthy that, in an African context, Ambole et al. [6] provided a similar approach, where they aimed to identify the potential of energy communities in sub-Saharan countries. However, their intervention was different from this study for two reasons. First, the majority of these countries face a lack of energy supplies and energy poverty. Second, they focused on the role of bottom-up practices; namely, they promoted a collaborative design method to deliver a platform to involve multiple stakeholders that enables citizens' incorporation during the life cycle of establishing energy communities, showing how local societies can push innovation and development in the direction of sustainable energy systems.

1.2. Study Hypothesizes

The author argues that the rural energy community and agrivoltaic concepts can enable green transition in Egypt and enhance land and food security for many reasons, summarized as follows:

- The concepts are gaining momentum, which can benefit the local community and decisionmakers alike and preserve natural resources in the light of the land–energy–climate nexus. The development scene in Egypt and climate and energy policies are also witnessing significant improvement (Sections 3.1 and 3.2) post the socio-political fluctuations in the past few years;
- The extraordinary support to regenerate rural built environments within the national presidential project Decent Life *"Haya Karima."* (Section 3.3.1);
- In the national agenda to achieve sustainable development Vision 2030, Egypt supports innovation and scientific research (Goal 4) as essential pillars of development, showing how this research can support decision-making;
- Egypt qualifies to become a regional hub for producing and exporting renewable and low-carbon energy to Europe [7], the Middle East, and North Africa [8]. For instance, the EU–Egypt partnership priorities formally endorsed on 19 June 2022, to work jointly, focusing on renewable energy (RE) and energy efficiency actions based on implementing ambitious climate policies and targets. Meanwhile, its pivotal role is in representing Africa during COP27;

- The rural energy communities concept promotes a self-sufficiency dogma that can mitigate external uncertainties that human society is being exposed to in recent times, which have reached unheard-of levels, such as the coronavirus pandemic and regional wars, which brings to mind the specter of the 1970's energy crisis. Beyond that, it is an issue of existence.

Figure 1 illustrates the rationale behind promoting the rural energy community key concept in the legal framework, from global to local perspectives, as an effective strategy toward green transition and a low-carbon future in Egypt.

Micro Perspective

Author's research line in retrofitting rural built environments

Extraordinary support to researchers for innovation (Goal 4, Egypt Vision 2030)

Extraordinary support to regenerate the rural built environments (*Haya Karima*)

The rapid urban development scene in Egypt since 2014

Egypt's qualification to become a regional energy hub

The energy community concept is gaining a global trend

Macro Perspective

Figure 1. The argument's rational background, from a global to national scale.

2. Materials and Methods

2.1. Reserch Structure and Metodology

To articulate how the aim was accomplished and the research questions addressed, this study blends diverse methods:

2.1.1. Part 1—Theoretical Method

The aim is to provide a general direction to the Egyptian context via:

- Defining the rural energy community's key concept and highlighting the concept's roots, progress, and current legal framework in the European context (Section 2.2).

2.1.2. Part 2—Analytical Method

The aim is to provide a comprehensive foundation on the topic and explain the current situation to support decision-making and pave the way for the scientific community in Egypt towards a better understanding of the concept in order to answer RQ1 via:

- Review of the current Egyptian efforts related to climate actions and renewable energy that could address the core of energy communities; namely: climate actions (Section 3.1) and energy policies (Section 3.2), in addition to highlighting the rural and agricultural policies and related national projects (Section 3.3.1).

2.1.3. Part 3—Field Method

The aim is to investigate bottom-up studies, to define whether the energy community concept (as its present European form) or its core exists or not, and to provide holistic insights showing to what extent the concept is localized on-ground (to answer RQ2), in addition to supporting the promoted top-down approach, via:

- Discuss local relevant best practices on different scales based on the previous literature (Section 3.3.2) and on-site investigations (Section 3.3.3).

Figure 2 summarizes the methodology and research structure.

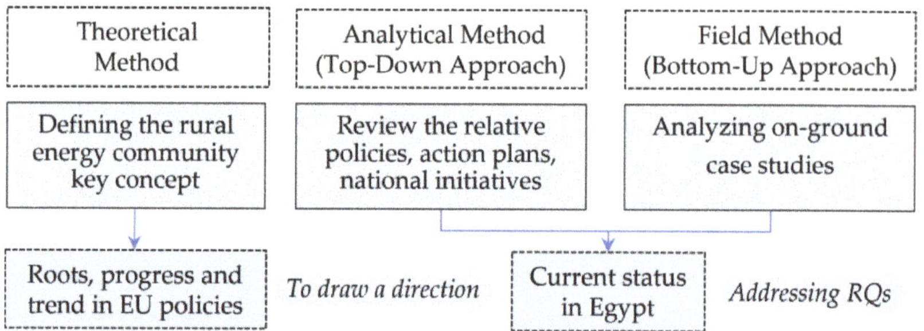

Figure 2. A visual summary of the article's methodology and design.

2.2. Rural Energy Community Key Concept

2.2.1. European Legal Framework: At Glance

Many literature reviews have discussed the concept's roots in the European regulatory framework. For example, references [9,10] have reviewed the direct and indirect regulations that addressed the energy community concept in European policies until full legal recognition. They presented how the concept rooted since the 1990s, since the first energy wave (Directive 96/92/EC), and the white paper *"Energy for the Future: Renewable Sources of Energy"* in 1997, the need for "Community Strategy" mechanisms of coordination of the broad policies engaged in promoting renewable generation process. Both can be considered the first phase of liberalization of the electricity sector.

Then, the second energy package in the 2000s [11] consisted of policies, such as Directives (2001/77/EC) and (Directive 2003/54/EC) to organize the European en-ergy sector in the RE field and to secure a competitive retail market. This was followed by the third energy package in 2009, including Directives (2009/72/EC) and (2009/73/EC) to increase the independence of national regulatory agencies and open fair retail markets. In addition, (Directive 2009/28/EC) allows the concept of local energy structures of a civic character such as energy cooperatives. Finally, the fourth (clean) energy package [12] added new electricity market rules to enable Europe to meet its climate and energy targets and attract investments (leading to energy communities).Self-consumption energy communities have two categories: the renewable energy community (REC) endeavors to expand the role of renewables (energy communities and self-consumers) in line with the revised renewable energy directive (EU) 2018/2001 [13]; and the citizen energy community (CEC), which aims to make energy communities accessible for citizens as active contributors to blend in the electricity system effectively, in line with the directive on shared rules for the internal electricity market (EU) 2019/944 [14].

2.2.2. Definition and Forms

The European Commission defined energy communities as "Citizen-driven energy actions that contribute to the clean energy transition, advancing energy efficiency within local communities." [15]. Cappellaro et al. [16] added the self-consumption of energy as an association of users through voluntary membership of a legal entity who collaborate to produce, consume, and manage energy through one or more local energy plants for self-consumption and collaboration. Each community has particular characteristics, but they all share the same goal: to self-produce and provide affordable renewable energy to its members. A study by de Simón-Martín et al. [17] defined them as generic terms describing a set of consumers and producers that belong to the same legal entity (e.g., cooperative, consortium, and association).

The concept itself is adaptable, granting a variety of initiatives that have numerous practices and forms [18,19], such as the clean energy community [20] and local energy initiatives [21]. All aim to decentralize energy production.

In the same vein, numerous studies have discussed the implementations of the concept and its forms at the European level. Their practices have been growing gradually since the 2000s as significant actors in the energy transition by improving social acceptance and enabling citizen participation. In general, Europe demonstrates an exceptional benefit in investments of the REC [22]. Nevertheless, the concept still has vast growth potential, particularly in rural areas in northwest Europe [23].

For instance, Heaslip et al. [24] promoted sustainable energy communities, and they stated that a lot of European studies are still needed to introduce the concept in rural areas and islands. They suggested an inclusive consideration of how to transform obstacles into enabling tools to enhance the successful development of energy communities in rural areas. Chamorro et al. [25] defined the rural energy community as delivering low-cost and reliable electricity to supply households and prevalent farming and agriculture business models to fill the gap between rural and urban communities while preserving individual characteristics. Furthermore, the Future of Rural Energy in Europe initiative (FREE) was established in 2010 to promote sustainable energy use within rural communities and aid in the objective of climate neutrality by 2050 [26].

Although Frieden et al. [27] stated that although most European countries had established regulatory frameworks to implement EU legislation, implementing efficient policies to foster energy communities seems challenging [28], they do not have a unified and mature definition. They are still "highly heterogeneous" at many levels, such as business patterns, operational geographical range, members' attributes, and numbers, in addition to the adopted forms of policy and regulation [29].

Tarpani et al. [30] mentioned that European regulation is complex as it depends strictly on the countries and their management of national policies. They classified the European countries based on establishing energy community projects into "early adopters" and "laggard" countries, such as Italy (however, it has launched its national regulatory framework [16]). Whereas in a country such as Poland (the most coal-dependent economy in the EU), Jasiński et al. [31] highlighted the absence of rural energy communities, despite having existed in 2019 in a similar form: energy cooperative as a policy to enable a collaborative approach (no on-ground cooperative established yet).

Otherwise, Abouaiana and Battisti discussed the juxtaposition of the energy communities concept's core with relevant regulatory frameworks, Figure 3. Briefly, for example, the energy community concept juxtaposed with the reformed common agriculture policy (CAP) by four out of ten keys. The EU rural pact makes EU rural areas more resilient, green, connected, and robust. Some initiatives provide support and keep the rural communities' locals involved. It is noteworthy that the covenant of mayor initiative on its official website indicated that it is for cities. However, it includes the action plans for rural settlements, so it has been included. The author is implementing a similar study focusing on the Pontinia rural settlement in Italy that will support the action plan of Pontinia [32].

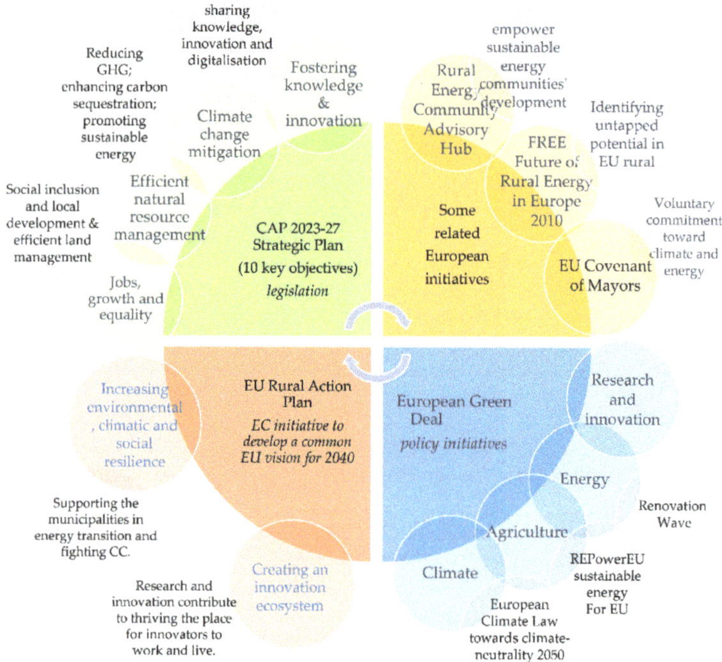

Figure 3. Mapping the core of the energy communities concept in current European policies and initiatives.

3. Results

3.1. Egypt's Global and Local Climate Commitment

Egypt has intensified its efforts since the Paris Agreement, represented by developing its national sustainable development agenda Egypt Vision 2030 in 2016 to fulfill its commitment, for which energy is a pivotal element in this equation. Therefore, climate and energy policies push for carbon neutrality in general, in addition to supporting citizens' quality of life, well-being, local economies, and ecosystems.

Hosting the United Nations Climate Change Conference, Conference of Parties (COP27) of November 2022 in Sharm El-Sheikh represents the status quo's peak milestone in climate action, which will undoubtedly affect Egypt's development scene towards decarbonization in the near future. Hence, climate and energy policies' objectives are backed via explicit targets, policies, restructuring laws, and the development of coordinated cross-sectoral national action plans.

Cross-Sectorial Coordination of Climate Action

In 2015, due to Prime Minster (PM) Decree 1912, Formation of the National Council on Climate Change (NCCC), supported by PM Decree 1129 to draw up the state's general policies in dealing with climate change via linking, for the first time in Egypt, policies, plans, and strategies for climate change with sustainable development at the state's sectorial levels were implemented. The ministries and concerned authorities assigned one of their relevant units to work on CC issues to provide implementation and follow-up plans to compact climate change (Article 11).

In 2016, the national agenda of sustainable development (Egypt Vision 2030) was launched, based on Egypt's commitment to providing and ensuring a good life for inhabitants, in line with the sustainable development goals and the Agenda of Africa 2063. It represents the governing structure for all projects and development programs that will be put into operation until 2030.

In 2018, the report *"Egypt's Voluntary National Review"* (VNR) was launched to present the country's progress toward achieving the sustainable development goals. The second VNR was established in 2021 [33]. In 2020, PM Decree 193 identified the tasks and competencies of the Ministry of Planning and Economic Development (MPED), which includes following up and assessing the state's performance to implement the sustainable development agenda.

Meanwhile, in 2021, MPED launched the *"Environmental Sustainability Standards Guide: The Strategic Framework for Green Recovery"* in association with the Ministry of Environment and all relevant governmental entities [34]. The guideline aims to raise awareness of interventions that positively impact the environment, guiding government and private sectors towards investment and showing the strategic objectives and priorities in all sectors. For example, in the energy sector, solar and wind energy production came first among the priority projects and activities in green financing.

In 2022, the NCCC requested to prepare a benchmark, the National Climate Change Strategy (NCCS) 2050, to incorporate CC components across sectors and combat its impacts [35]. In addition, reducing greenhouse gas (GHG) emissions, considering that the contribution of the energy and agriculture sectors is 75% of the total, shows the potential of the rural energy community to decrease emissions.

The benchmark consists of five pillars. Each has a set of objectives and enabling directions (tools and policies). In a nutshell: (i) achieve low-emission development (e.g., promoting micro-scale decentralizing RE systems and increasing the RE share towards energy transition); (ii) resilience to CC (e.g., conserving and increasing the agricultural land, developing infrastructure in rural communities); (iii) supporting climate action governance (e.g., aligning CC units across sectors); iv) green finance units (e.g., promoting green jobs and climate funding opportunity; (v) knowledge management (e.g., boosting the scientific research's role in CC in all fields).

3.2. Energy and Renewable Energy in Egypt

The Egyptian Electricity Holding Company manages the electricity sector on behalf of the Ministry of Electricity and Renewable Energy (Formerly named Ministry of Electricity, changed in 2014). It is regulated by the Egyptian Electric Utility and Consumer Protection Regulatory Agency (EgyptERA), responsible for implementing policy decisions and administering licenses.

The electricity sector has significantly improved since 2014, after a few years of domestic electricity shortage. Generally, electricity production increased by 128% in the past decade. Regarding the Electricity Holding Company's annual reports, the installed capacity was 25,705 MegaWatt (MW) in 2012 and 58,818 MW in 2021. However, the associated GHG emissions require significant steps to increase the RE share to mitigate CC impacts, as emphasized in reference [36] (solar and wind represented 5% of the total in 2021).

Does Regulatory Framework Liberalize the Sector?

Egypt pays remarkable attention to achieving energy efficiency at many levels. The first attempt in 1986, Law 102, was to establish the New and Renewable Energy Development and Use Authority that followed the Ministry of Electricity. In 2012, the national energy efficiency action plan (NEEAP)'s phase one was established to improve energy efficiency (e.g., by saving public lighting, improving electrical appliances, and improving energy plants). It provided a bottom-up approach with limited applications, such as the missing energy-saving component in the agriculture and transportation sectors [37].

For these reasons, in 2016, phase two was released with a holistic approach considering the laws and numerous buildings, such as industrial and educational ones. The main aims are to be an energy efficiency (EE) hub leading the region toward low carbon dioxide

(CO$_2$) policies, such as enhancement of energy use efficiency into energy, environment, and economic policies to mitigate the carbon footprint [38]. In this domain, the "Integrated and Sustainable Energy Strategy for Egypt 2035" was adopted to achieve energy security and make suitable renewable energy source development conditions engaging all sectors to reach the ambitious share of RE by 42% in 2035 [39]. Figure 4 depicts the NEEAP implementation milestones, reflecting the clear intention to open a competitive energy market that may contribute to fully liberalizing the sector within the next decade and enabling the application of energy communities or more advanced innovative solutions.

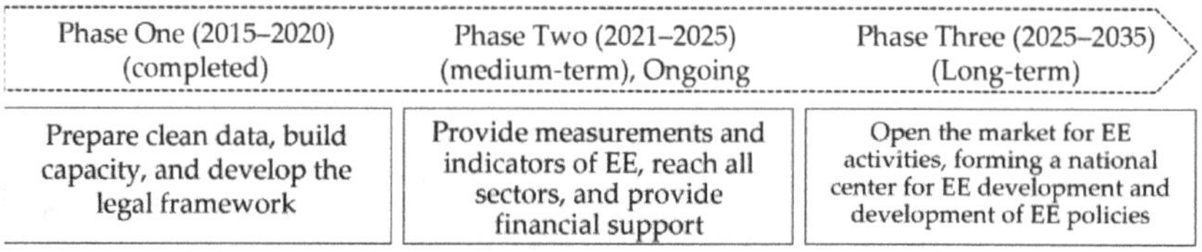

Phase One (2015–2020) (completed)	Phase Two (2021–2025) (medium-term), Ongoing	Phase Three (2025–2035) (Long-term)
Prepare clean data, build capacity, and develop the legal framework	Provide measurements and indicators of EE, reach all sectors, and provide financial support	Open the market for EE activities, forming a national center for EE development and development of EE policies

Figure 4. NEEAP's implementation and time frame.

In 2014, Law 203 determined the purchase prices of electrical energy supplied to electricity distribution companies from electricity production plants used for RE sources (sun and wind) to be contracted with the feed-in tariff (FiT) scheme. The electricity prices from solar and wind energy were determined by PM decree 1947. Additionally, Republican Decree 135 allows the RE authority to invest in RE.

In 2015, Law 87 aimed to restructure the electricity sector. It is a crucial step toward liberalizing the electricity production and distribution market. It encourages private investments in the energy sector, where the transition period is from ten years up to twelve. In 2016, the Republican Decree 116 allocated governmental lands for RE projects by the authority. In 2019 PM Decree 183 identified the electricity prices from biomass energy to diversify RE sources.

In 2017, EgyptERA issued Decree 3/2017's amendment of the regulatory rules for encouraging the exchange and use of electric energy produced from solar energy by the net metering system; the client should be the solar plant owner. This was followed by decree 3/2018: Amending Contracts Forms of Net Metering, which enabled making an energy purchasing agreement between a qualified third-party company (operating and managing the plant) and a client who benefits from the purchased energy and making a contract with the electricity distributor. Figure 5 visualizes the mechanism.

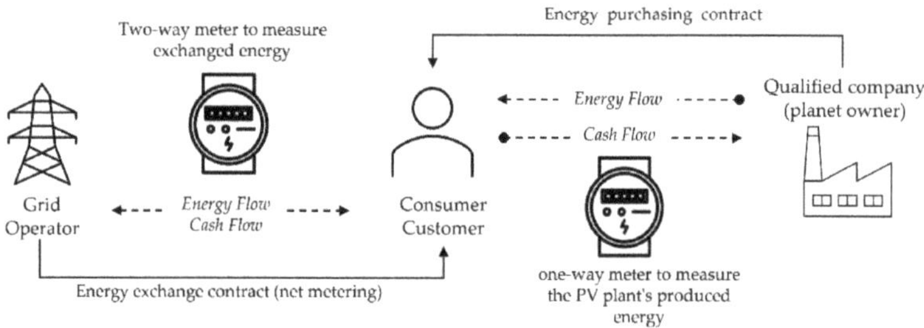

Figure 5. Contractual mechanism of energy purchasing agreement within the net metering scheme.

This was followed by Decree 2/2020, which set regulations on governing the net metering system, and its update Decree 6/2022 [40], which increased the permissible capacity limit for the total solar energy projects to be operated on a net metering system to 1000 MW (it was 400 MW). The entire solar capacity owned by any licensed party or one of the corporate clients should not exceed a total of 35 MW (it was 25 MW) with a shorter limit of 30 MW (it was 20 MW) for one project [41]. The legalization was amended to contribute to the preparation for hosting COP27. This shows one of its indirect impacts on the regulatory framework and supports this study's argument. Reference [42] shows more about investment mechanisms.

3.3. Rural Development and Rural Energy Communities

3.3.1. Top-Down Implementations

The government reflects all policies, particularly energy and climate, into national urban development, chiefly in rural commons, that host 58% of the total population (104 million inhabitants in October 2022) and host the majority of building stock by 70% (11.4 million buildings) of a total 16.2 million buildings [43].

This occurs within two main prominent streams. First, to regenerate the existing rural built environments. Second, by penetrating the desert (93% of the total area) to double the inhabited areas and agricultural lands (from 7% to 14.5%), such as with the New Delta Project, Figure 6a, which aims to achieve food self-sufficiency. It is noteworthy that the state controlled the phenomena of infringement on agrarian land through intensive penalties (Law 164/2019), which peaked in the aftermath of the January Revolution of 2011.

(**a**)

(**b**)

Figure 6. (**a**) An example of increasing the reclaimed land areas: New Delta national Project between 2012 and 2020 (Google Earth Engine, 2022) Available online: https://earthengine.google.com/timelapse#v=30.14018,30.36834,9.591,latLng&t=3.63&ps=100&bt=19840101&et=20201231&startDwell=0&endDwell=0 (accessed on 15 October 2022). (**b**) Progress on a project's canal lining, implemented to improve road quality and enhance water efficiency and climate adaptation (taken by Ahmed Abouaiana).

On the other side, the government aims increase the amount of reclaimed agricultural land by showing the potential of considering them as energy baskets and employing energy communities since the early planning. Table 1 summarizes the state trend in supporting agriculture and food security.

Table 1. Amount of annual infringement on agricultural land and reclaimed ones, by roughly one thousand acres—"feddan" [44].

Year	2012	2013	2014	2015	2016	2017	2018	2019	2020
Agricultural land Infringements	15.6	14.8	12.9	9	8.1	8.7	4.2	2.5	2.9
Reclaimed lands	39	22.9	22.6	14.5	38.5	38.9	59.2	115.7	81

The presidential initiative Decent Life *"Haya Karima"* is the central pillar to achieving sustainable rural development in Egypt. The main objectives are to improve the quality of life and regenerate the existing built environments through intensive and instant interventions. For instance, renovating public buildings and dwellings, offering job opportunities, agriculture development, and developing infrastructure, Figure 6b. It is split into three phases, according to the poverty rate: greater than 70% (2019–2021, completed), 50–70% (from 2021 until three years after, currently running), and less than 50% (the completion date 2030) [45,46]. Reference [47] visualizes examples of these interventions.

3.3.2. Bottom-Up Practices: Literature Review

Many practices have been implemented in rural settlements in Egypt to achieve sustainability in general and meet energy demands from RE at different scales, such as mono-buildings, multi-buildings, and farms. Scholars have discussed these issues for more than three decades [48]. For instance, Ahmad [49] provided a technoeconomic assessment of rural dwellings; their life cycle costs are competitive with other conventional energy sources.

Abdelsalam et al. [50] implemented decentralized photovoltaic (PV) solar cells on the rooftops of five buildings in two villages, with a total capacity of 11 kW. Shouman [51] provided a stand-alone solar system (off-grid) for a dwelling, with a daily consumption of 4.555 kWh, in remote areas as an optimum solution for villages with no access to centralized electricity power. She stated that solar energy systems are sustainable technological solutions for rural electrification and are cost-effective compared to expensive grid extensions in remote areas.

Nemr et al. [52] have provided technoeconomic optimization for alternatives of stand-alone RE systems applied in rural areas to cover the energy load of domestic, agricultural, greenhouse system, and fish farm loads (a total of 365 kWh/day). They recommended a hybrid system according to the cost and fulfillment of the required loss of power supply probability term.

Mohamed [53] mentioned that solar energy is efficient in remote areas, and the return on investment is much better-recommended than conventional power supplies. He stressed the importance of providing economic incentives and equipment and encouraging agricultural investors to rely on solar energy solutions by providing the system's components and financial support.

Ibrahim et al. [54] recently provided technoeconomic optimization of proposed decentralizing solar energy in an industrial building, school, and farm in remote rural areas (a total of 10.8 kWh/day). They focused on developing storage systems (batteries) that significantly increased the efficiency of the PV system in all case studies and decreased the cost in both buildings by an average of 65%.

At a large scale, Heliopolis University [55] discussed the impact of their projects; beyond harvesting the RE and meeting their demands from RE, they showed how their sustainable agricultural practices, mainly using bioenergy, reduced 11,930 tCO2e of Egypt's share. In this light, they proposed developing "a national framework for emissions reduction and carbon credits from the agricultural sector." They discussed that the promising future of the "Agricultural Carbon Credit" could follow two paths: certified emission reductions via registering for the development mechanism or voluntary emission reductions. The author emphasizes the last pattern that exists in Egypt but still needs more visibility.

Fortunately, on 6 November 2022, the Egyptian Exchange established the Carbon Certificates Trading Company, in partnership with the Agricultural Bank, as the first platform to enable the carbon certificate trading mechanism [56]. It came true after a few months of discussion [57], reflecting a dramatic change at the national level.

3.3.3. Bottom-Up Practices: Field Study

Recently, the author has implemented a field trip to a nearly zero-energy private farm in Ismailia Governorate, Suez Canal Region in eastern Egypt, located at longitude 30.498597 and latitude 32.076638, with a total area of 15 feddan. The owners aimed to provide a prototype of a self-sufficient agriculture community. The project offers typical farming activities: livestock, fisheries, and farming. Consequently, it has the necessary equipment (e.g., pumps, refrigerators, heaters, and washers) and a few buildings (accommodation, stores, industry, and a lecture hall).

Different RE sources were established to meet the annual electricity demand (100,680 kWh), namely solar energy (on-grid and off-grid), wind turbines (horizontal and vertical), and hybrid systems, Figure 7a. This system is more efficient than the mono one (supporting the results of Reference [58]). This mix produces 96,372 kWh annually, which covers 96% of the total, considering that pumps and the associated equipment consume the majority (58%), followed by ventilation (19%), other equipment (17%), and lighting (6%)—showing the importance of energy-efficient irrigation and pumps in similar realms.

(a)

(b)

(c)

(d)

Figure 7. Different implementations promote the REC concept in rural Egypt. (**a**) A hybrid off-grid system. (**b**) Multi-function land use: parking shading and producing electricity to balance consumption in the adjacent activity (solar pump). (**c**) The manager describes a micro-scale practice to generate green hydrogen. (**d**) The first self-consumption social building in the village and surrounding ones. (Images taken by Ahmed Abouaiana.)

In contrast, biogas units were implemented to meet the annual gas demand of 14,364 m^3 (21% for kitchens and 79% for animal wards and barns) and manage waste (25.8 tons). The units produce gas with a surplus of 15% and produce 1.3 tons of compost yearly.

It is noteworthy that the author discussed multi-function land use (Figure 7b) with the manager, specifically, the agrivoltaic concept. The manager was willing to apply the idea to a small agricultural area that requires shading (it already exists and uses a conventional shading device). Moreover, the farm developed a micro-scale implementation to produce green hydrogen, and the system is under development, Figure 7c.

On a different scale, the author implemented a self-consumption building (service building) of a non-governmental organization (NGO) that consumes 100% of generated RE (3 kW plant capacity) in January 2022 [5]. The author declares the project is not feasible economically, in terms of the long payback period, because of the high cost of the batteries and annual maintenance fees. However, this project was implemented as a pilot project, with a clear intention to create momentum for the core of the renewable energy communities concept.

Considering that the author discussed (the NGO founder) a possible volunteer collaboration to develop the vast area in front of a building (Figure 7d) through temporary outdoor activities for kids with a client, the fence of the building will have lighting, which will be operated from the installed PV, to light both the building and the surrounding area, showing another prospective positive impact of this intervention. It is noteworthy that the author was invited to an interview on national TV to discuss this intervention a few days after implementing it [59], indicating the special attention of decisionmakers to scientific research and bottom-up practices as well, which also supports the argument of this manuscript.

Figure 8 shows an imaginative example of an actual case study in Lasiafr Albalad village in northern Egypt, showing prospective investment in PV cells by families within the existing legal framework. Each family within the same building internally arranges their consumption and monthly electricity prices. This ambitious proposal may succeed, especially in extended family buildings (e.g., father and sons), with the argument that they can maintain a cooperative level. The author aims to implement this rudimental idea within his research line to provide a positive social impact on his original community.

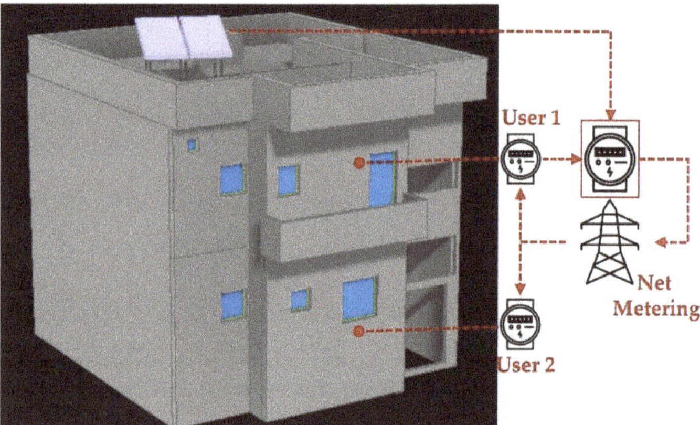

Figure 8. An imaginative vision of a collective PV production (net metering scheme). The model was generated by the author using 3ds Max.

4. Discussion

In Europe, the progress of liberalizing the electricity sector and renewable energy communities was initiated over two decades until it settled into the legal framework (Directives (EU) 2018/2001 and (EU) 2019/94). Energy Communities are the heart of innovation in the energy sector. Despite the associated vagueness at the national EU level, according to Rossetto, the concept's core covers broad domains in European policies (Figure 3). The obvious takeaway from the European experience is the focus on citizen engagement as an active contributor through the value creation of RE.

This study contributes to the local context by generating momentum to support decision-making in Egypt—which is witnessing intensive and rapid development (new projects and retrofitting the existing ones) in addition to hosting COP27, which will indeed affect the scene and increase opportunities for the paper's argument, as well as having already positively impacted RE laws (EgyptERA decree 6/2022)—via promoting the rural energy community as a pillar to support land, agriculture, and rural development and accelerating the transition towards low GHG emissions from the agriculture and energy sectors. Both represent 75–80% of the total (352 MtCO2e in 2019), which increased by 19% compared to 2009 [60]. Thus, the concept is optimal for revitalizing the productive nature of rural commons, as recommended by Abouaiana and Battisti, besides the introduced rational argument in Figure 1. Moreover, rural commons can be a safe haven for the country under global uncertainties and their associated threats.

4.1. Key Concept Juxtaposition in Legal Framework (RQ1 Answer)

The overview of climate–energy–land policies proved that they are intermingled at a sectorial level, and that the government has started a high level of coordination in recent years among the sectors from the top of the planning pyramid, beginning by arranging institutional frameworks since launching Egypt Vision 2030, where MPED plays the pivotal role in achieving that, in addition to establishing the dedicated entity NCCC. On the other side, it proposes the inclusion of, for the first time, carbon policies with energy action plans, creating climate benchmark NCCS policies, and developing key performance indicators to measure the progress towards the sustainable development goals. Despite this, seeing the effects of these policies on the bottom levels still requires time.

Electricity policies have improved since the introduction of Law 87/2015, which, with NAEEP, aids in liberalizing the sector. However, Kamal [61] stated that the transition to a genuinely competitive market in Egypt still requires more time. Thusly, she highlighted that blockchain applications in sector management could speed up the transition, such as electric energy exchanges through smart electronic contracts to allow customers to choose between many energy suppliers who provide electricity at competitive prices.

In the same context, there is a meaningful improvement in electricity production. However, RE transition in Egypt is sluggish, being below the target of the RE share in electricity generation (20% in 2020), possibly because of the massive expansion of other conventional power plants [62]. Salah et al. [63], in their holistic investigation of RE transition in Egypt, emphasized that the current regulations slow down the development of the RE sector, such as the "non-harmonized regulations between the government sectors." Notwithstanding, the author may differ from them, as this study showed the coordination level. The author believes there is "no roadmap without a narrative" [64], and this study tries to contribute to this belief.

Consequently, localizing RE industries can inevitably reduce the high cost or lack of competencies radically, and it is the role of the state to provide sufficient enabled policies. Meanwhile, the economic incentive is insufficient, needs clear strategies, and should be more visible and competitive to the local community and individuals.

The state pays considerable attention to rural agriculture development, which is almost centralized and intensive, represented in reclaiming new lands (Table 1), and implementing new projects. For instance, the New Delta Project represents nearly 0.9% of Egypt's area (2.2 million feddan). Likewise, in regenerating existing environments, the philosophy is clear enough in defining the intensive interventions and rapid correction actions that burden the state, in light of the well-known global economic challenges, and Egypt primarily. Each may affect the optimal implementations of these trends (settled in regulations) at the micro-levels. Therefore, when developing the deteriorated contexts, the priorities would be to rapidly achieve the essential aspects of human needs, even if in a conventional way.

For instance, decent houses instead of deteriorated ones, clean water, infrastructure, enhanced internet, and public buildings to decentralize governmental services will surely enable employing RE gradually in further steps. As supported by the Food and Agriculture Organization, the gradual increase in the use of RE and considering the natural resources nexus can make agriculture and rural development more energy-smart and energy-efficient [65,66]; this is on one side. On the other, developing energy infrastructure and networks accelerate liberalizing the market, as emphasized in Reference [67].

Despite this, it cannot be dispensed that reversing adaptation and mitigation efforts [68] and renewable energy and low-emissions policies, proportionately with this rapid development's pace, is inevitable, and it should be carefully planned to be considered on the grounds of maximizing the development impacts [69]. Concomitantly, providing national sustainability guidelines for agriculture-based urban communities is absolutely essential, as about 90% of Egypt's area is desert that can still be innovatively and sustainably developed.

Indirectly, the core of REC is included in the policies, like in the European case (Figure 3). In contrast, the core of CEC and effective citizen engagement in development, in general, are absent in the scene. For example, *Haya Karima* follows a participatory approach among the stakeholders: institutional entities, civil society, and the private sector are contributors to the implementation and monitoring of their activities. At the same time, the local communities and citizens participate in determining the requirement. However, it could be a user-centered development [70]. The author believes it is essential but is limited. In the meantime, it seems that it is too early to target the citizen until the model gets settled into regulatory framework.

The author would say that citizens should be more engaged as active contributors, which is one of the vital difference points with the EU policies [71]. Because they can provide collaborative and innovative solutions, both are inevitable in sustainable rural agricultural development in Egypt [72], especially in energy issues, as promoted in Ibrahim et al.'s study and as emphasized in Ambole et al.'s experiment.

The agricultural cooperatives should be enabled to invest in RE with clear support from the government. They can contribute as a first step toward rural energy communities, such as in France [73], Germany [74], Portugal [75], and Spain [76], or in less developed European countries such as Poland's case. Logistic and economic support from the state in rural energy transition is essential in developed countries such as the United Kingdom [77], a fortiori in Egypt.

4.2. Concept Localization On-Ground (RQ2 Answer)

From a bottom-up perspective, the REC's core exists on-ground, represented by micro-scale projects. The vast decentralized interventions are mainly for rural, remote areas and agricultural supplies, mostly off-grid solutions [78]. This remains an efficient solution despite the high cost of batteries and storage systems. The author implemented a zero-energy building in a village in Egypt (Figure 7d), utilizing a transdisciplinary participatory approach [79]; the locals determined the target building, supported establishing the project, and appreciated it. Notwithstanding, a similar process was implemented in rural Italy [80]. Both reflect that the local community can contribute if they have the chance. Accordingly, combining top-down with bottom-up approaches is inescapable.

On the other side, the role of the private sector and PV suppliers is to pay more attention to these business opportunities, as well as to carbon certificates that are circulating these days. Meanwhile, negotiations with them (from the researchers or the local authorities) shall support these projects (at an individual scale). In other words, prospecting for private sector entities with a corporate social responsibility policy would provide financial and technical support. Vice versa, this can provide them with benefits, such as job opportunities, revenue, and a good reputation in the local market. Interested stakeholders in energy, climate, and sustainable rural and agriculture development domains should seize the anticipated advantage post-COP27 to support decision-making and enhance our built environment.

5. Conclusions and Implications

In a nutshell, the rural energy community is a new actor that can decarbonize the entire energy sector [81], accelerate reaching Vision 2030 goals, support quality of life, and mitigate poverty in Egypt, as found by Belaïd [82]. This is the first study discussing the concept of rural energy communities in Egypt in their legal EU form. It presented a macro-level discussion of this new actor by providing insights to explain the current state of knowledge and a better understanding of the institutional context. In addition, it highlighted the potential of developing the concept, which is a fact in Egypt, as proven by the presented micro-scale practices, such as the self-sufficient community (Ismailia farm) (Section 3.3.3). Furthermore, multi-functional land use can support this and maximize the government's ongoing targets to double the inhabited areas and expand agricultural land, which has already started in recent years.

Eventually, this study—distinctly—confirms it is not an attempt at Europeanization of the local context, but it attempts to highlight the cutting-edge, transferring technologies from developed countries [83] and maximizing the researchers' role as knowledge brokers and acting as mediators between the local society and authorities, which can support the decision-making and raising of locals' awareness, which is crucial to preparing for social acceptance of energy transition [84,85], especially in agricultural and rural communities [86,87].

The knowledge highlighted in this study can be a good foundation point for future work. Although the discussion of the top-down approach and limited case studies may not generalize the results on a wide scale in this initial study, they were comprehensive in nature. The results may also provide a direction in which to generalize the approach to urban contexts. Until the energy sector reaches complete liberalization, this study recommends that other investigations be conducted in the aftermath of COP27 to assess its influences and provide additional insights that can support the agricultural security and energy transition leading to a low-carbon future in Egypt.

Funding: This research received no external funding.

Informed Consent Statement: Informed consent was obtained by the author from all subjects involved in the study.

Data Availability Statement: Not applicable.

Acknowledgments: The author acknowledges the support from Engineer Waseem El Hefnawy, Triple M Farm ("REEF") https://www.facebook.com/REEFFarm (accessed on 10 November 2022), in providing technical support related to consumption and RE production patterns in the farm data presented.

Conflicts of Interest: The author declares no conflict of interest.

References

1. Candelise, C.; Ruggieri, G. Status and evolution of the community energy sector in Italy. *Energies* **2020**, *13*, 1888. [CrossRef]
2. Ceglia, F.; Marrasso, E.; Pallotta, G.; Roselli, C.; Sasso, M. The State of the Art of Smart Energy Communities: A Systematic Review of Strengths and Limits. *Energies* **2022**, *15*, 3462. [CrossRef]
3. European Commission. A Long-Term Vision for the EU's Rural Areas Building the Future of Rural Areas Together. Available online: https://ec.europa.eu/info/strategy/priorities-2019-2024/new-push-european-democracy/long-term-vision-rural-areas_en (accessed on 10 November 2022).
4. Abouaiana, A.; Battisti, A. Multifunction Land Use to Promote Energy Communities in Mediterranean Region: Cases of Egypt and Italy. *Land* **2022**, *11*, 673. [CrossRef]
5. Abouaiana, A. Agile Methodology as a Transdisciplinary Retrofitting Approach for Built Environment in Traditional Settlements in Mediterranean Region. Ph.D. Thesis, Sapienza University of Rome, Rome, Italy, 2022.
6. Ambole, A.; Koranteng, K.; Njoroge, P.; Luhangala, D.L. A review of energy communities in sub-saharan Africa as a transition pathway to energy democracy. *Sustainability* **2021**, *13*, 2128. [CrossRef]
7. European Commission. EU-Egypt Joint Statement on Climate, Energy and Green Transition—STATEMENT/22/3703. 2022. Available online: https://ec.europa.eu/commission/presscorner/detail/en/STATEMENT_22_3703 (accessed on 10 November 2022).
8. Esily, R.R.; Chi, Y.; Ibrahiem, D.M.; Amer, M.A. The potential role of Egypt as a natural gas supplier: A review. *Energy Rep.* **2022**, *8*, 6826–6836. [CrossRef]
9. Sokolowski, M.M. European Law on the Energy Communities: A Long Way to a Direct Legal Framework. *Eur. Energy Environ. Law Rev.* **2018**, *27*, 60–70. [CrossRef]
10. Biresselioglu, M.E.; Limoncuoglu, S.A.; Demir, M.H.; Reichl, J.; Burgstaller, K.; Sciullo, A.; Ferrero, E. Legal provisions and market conditions for energy communities in Austria, Germany, Greece, Italy, Spain, and Turkey: A comparative assessment. *Sustainability* **2021**, *13*, 11212. [CrossRef]
11. Ciucci, M. *Fact Sheets on the European Union—INTERNAL ENERGY MARKET*; European Parliament: Strasbourg, France, 2021.
12. European Commission. Clean Energy for All Europeans Package. Available online: https://energy.ec.europa.eu/topics/energy-strategy/clean-energy-all-europeans-package_en (accessed on 10 November 2022).
13. European Commission. Directive (EU) 2018/2001 of the European Parliament and of the Council of 11 December 2018 on the promotion of the use of energy from renewable sources (Text with EEA relevance). *Off. J. Eur. Union* **2018**, *61*, 82–209.

14. European Commission. DIRECTIVE (EU) 2019/944 OF THE EUROPEAN PARLIAMENT AND OF THE COUNCIL of 5 June 2019 on common rules for the internal market for electricity and amending Directive 2012/27/EU (recast) (Text with EEA relevance). *Off. J. Eur. Union* **2019**, *62*, 125–199.
15. European Commission. European Commission—Energy—Topics—Markets and consumers—Energy communitie. Available online: https://energy.ec.europa.eu/topics/markets-and-consumers/energy-communities_en (accessed on 23 February 2022).
16. Cappellaro, F.; Palumbo, C.; Trincheri, S. *La Comunità Energetica—Vademecum 2021 [The Energy Community-Vademecum 2021]*; ENEA: Roma, Italy, 2021.
17. de Simón-Martín, M.; Bracco, S.; Piazza, G.; Pagnini, L.C.; González-Martínez, A.; Delfino, F. The Role of Energy Communities in the Energy Framework. In *Levelized Cost of Energy in Sustainable Energy Communities*; Springer: Berlin/Heidelberg, Germany, 2022; pp. 7–29. ISBN 978-3-030-95932-6.
18. van Bommel, N.; Höffken, J.I. Energy justice within, between and beyond European community energy initiatives: A review. *Energy Res. Soc. Sci.* **2021**, *79*, 102157. [CrossRef]
19. European Commission. Organisational Form and Legal Structure. Available online: https://rural-energy-community-hub.ec.europa.eu/energy-communities/organisational-form-and-legal-structure_en (accessed on 10 November 2022).
20. Gui, E.M.; MacGill, I. Typology of future clean energy communities: An exploratory structure, opportunities, and challenges. *Energy Res. Soc. Sci.* **2018**, *35*, 94–107. [CrossRef]
21. Ghorbani, A.; Nascimento, L.; Filatova, T. Growing community energy initiatives from the bottom up: Simulating the role of behavioural attitudes and leadership in the Netherlands. *Energy Res. Soc. Sci.* **2020**, *70*, 101782. [CrossRef]
22. Echave, C.; Ceh, D.; Boulanger, A.; Shaw-Taberlet, J. An ecosystemic approach for energy transition in the mediterranean region. In Proceedings of the 2019 1st International Conference on Energy Transition in the Mediterranean Area (SyNERGY MED), IEEE, Cagliari, Italy, 28–30 May 2019; pp. 1–5.
23. ENRD. *Smart Villages and Renewable Energy Communities*; The European Network for Rural Development: Bruxelles, Belgium, 2020.
24. Heaslip, E.; Costello, G.J.; Lohan, J. Assessing good-practice frameworks for the development of sustainable energy communities in Europe: Lessons from Denmark and Ireland. *J. Sustain. Dev. Energy Water Environ. Syst.* **2016**, *4*, 307–319. [CrossRef]
25. Chamorro, H.R.; Yanine, F.F.; Peric, V.; Diaz-Casas, M.; Bressan, M.; Guerrero, J.M.; Sood, V.K.; Gonzalez-Longatt, F. Smart Renewable Energy Communities-Existing and Future Prospects. In Proceedings of the 2021 IEEE 22nd Workshop on Control and Modelling of Power Electronics (COMPEL), IEEE, Cartagena, Colombia, 2–5 November 2021; pp. 1–6.
26. Rural-Energy About FREE. Available online: https://www.rural-energy.eu/about-free/ (accessed on 9 October 2022).
27. Frieden, D.; Tuerk, A.; Roberts, J.; D'Herbemont, S.; Gubina, A.F.; Komel, B. Overview of emerging regulatory frameworks on collective self-consumption and energy communities in Europe. In Proceedings of the 2019 16th International Conference on the European Energy Market (EEM), IEEE, Ljubljana, Slovenia, 18–20 September 2019; pp. 1–6.
28. Martens, K. Investigating subnational success conditions to foster renewable energy community co-operatives. *Energy Policy* **2022**, *162*, 112796. [CrossRef]
29. Rossetto, N.; Verde, S.F.; Bauwens, T. A taxonomy of energy communities in liberalized energy systems. In *Energy Communities Customer-Centered, Market Driven, Welfare-Enhancing?* Löbbe, S., Sioshansi, F., Robinson, D., Eds.; Elsevier: Cham, Switzerland, 2022; pp. 3–23. ISBN 978-0-323-91135-1.
30. Tarpani, E.; Piselli, C.; Fabiani, C.; Pigliautile, I.; Kingma, E.J.; Pioppi, B.; Pisello, A.L. Energy Communities Implementation in the European Union: Case Studies from Pioneer and Laggard Countries. *Sustainability* **2022**, *14*, 12528. [CrossRef]
31. Jasiński, J.; Kozakiewicz, M.; Sołtysik, M. Determinants of Energy Cooperatives' Development in Rural Areas—Evidence from Poland. *Energies* **2021**, *14*, 319. [CrossRef]
32. Associazione, G.A.P.; Brighenti, J.; Cerroni, A.; Corradini, F.; Della Rocca, M.; De Prosperis, F.; Di Girolamo, M.; Di Pastena, S.; Franco, A.; Libralato, G.; et al. *PONTINIA2020 Il Piano d'Azione per l'Energia Sostenibile [PONTINIA 2020 The Sustainable Energy Action Plan]*; Patto dei Sindaci: Pontinia, Italy, 2015.
33. MPED. *Egypt's 2021 Voluntary National Review*; The Ministry of Planning and Economic Development, Egypt: Cairo, Egypt, 2021.
34. MPED. *Environemntal Sustainability Standards Guide: "The Strategic Framework for Green Recovery"*, 1st ed.; Ministry of Planning and Economic Development, Egypt: Cairo, Egypt, 2021. (In Arabic)
35. EEAA. *Egypt National Climate Change Strategy (NCCS) 2050*; Ministry of Environment, Egypt: Cairo, Egypt, 2022.
36. Mondal, M.A.H.; Ringler, C.; Al-Riffai, P.; Eldidi, H.; Breisinger, C.; Wiebelt, M. Long-term optimization of Egypt's power sector: Policy implications. *Energy* **2019**, *166*, 1063–1073. [CrossRef]
37. Elrefaei, H.; Khalifa, M.A. A Critical Review on the National Energy Efficiency Action Plan of Egypt. *JNRD-J. Nat. Resour. Dev.* **2014**, *4*, 18–24.
38. MOEE. National Plan to Improve Electricity Energy Efficiency (2018–2020). 2018. Available online: http://www.moee.gov.eg/test_new/DOC/p.pdf (accessed on 10 November 2022). (In Arabic)
39. IRENA. *Renewable Energy Outlook: Egypt*; International Renewable Energy Agency: Abu Dhabi, United Arab Emirates, 2018; ISBN 978-92-9260-069-3.
40. EgyptERA. Regarding the Amendments to Net Metering Controls as an Incentive to Support and Encourage Solar Energy Projects in a Way that Contributes to the Preparation for Hosting the Conference of Parties (COP27) in Sharm El-Sheikh. 2022. Available online: http://egyptera.org/ar/download/journal/2022/6_2022.pdf (accessed on 10 November 2022). (In Arabic)

41. EgyptERA. Periodical Books of Net Metering System. Available online: http://egyptera.org/ar/RegulatedRules.aspx# (accessed on 10 November 2022). (In Arabic)
42. NREA. *New and Renewable Energy Authority Annual Report 2021*; New and Renewable Energy Authority, Ministry of Eletricity and Renewable Energy: Cairo, Egypt, 2021. (In Arabic)
43. CAPMAS. *Egypt Population, Housing, and Establishments Census 2017*; Central Agency for Public Mobilization and Statistics: Cairo, Egypt, 2017. (In Arabic)
44. CAPMAS. *Annual Bulletin of Lands Reclamation 2019/2020*; Central Agency for Public Mobilization and Statistics: Cairo, Egypt, 2020. (In Arabic)
45. UN. Decent Life (Hayah Karima: Sustainable Rural Communities (Government) #SDGAction43597. Available online: https://sdgs.un.org/partnerships/decent-life-hayah-karima-sustainable-rural-communities (accessed on 10 November 2022).
46. Haya Karima Haya Karima. Available online: https://www.hayakarima.com/index_en.html (accessed on 10 November 2022).
47. DMC. The Documentary Film "A decent Life … for All Egyptians" … The Journey of the Largest National Project in the Governorates of Egypt. 2021. Available online: https://www.youtube.com/watch?v=E5GQW8ivGS8 (accessed on 10 November 2022). (In Arabic)
48. Arafa, S. Renewable energy solutions for development of rural villages and desert communities. In Proceedings of the 2011 International Conference on Clean Electrical Power (ICCEP), IEEE, Ischia, Italy, 14–16 June 2011; pp. 451–454.
49. Ahmad, G.E. Photovoltaic-powered rural zone family house in Egypt. *Renew. Energy* **2002**, *26*, 379–390. [CrossRef]
50. Abdelsalam, T.I.; Darwish, Z.; Hatem, T.M. Solar Energy for Rural Egypt. In Proceedings of the International Congress on Energy Efficiency and Energy Related Materials (ENEFM2013) Proceedings, Antalya, Turkey, 9–12 October 2013; Oral, A.Y., Bahsi, Z.B., Oze, M., Eds.; Springer: Cham, Switzerland, 2014; pp. 363–370, ISBN 978-3-319-05521-3.
51. Shouman, E.R. International and national renewable energy for electricity with optimal cost effective for electricity in Egypt. *Renew. Sustain. Energy Rev.* **2017**, *77*, 916–923. [CrossRef]
52. El-Nemr, M.K.; Elgebaly, A.; Ghazala, A.I. Optimal Sizing of Standalone PV-Wind Hybrid Energy System in Rural Area North Egypt. *J. Eng. Res.* **2021**, *5*, 46–56. [CrossRef]
53. Mohamed, E.M.S. Cost-benefit analysis and environmental impact of the use of solar energy in the Egyptian agricultural sector (A case study of the Al-Maghra Oasis in Matrouh Governorate, within the framework of the One and a Half Million Feddan Project). *Sci. J. Bus. Environ. Stud.* **2022**, *13*, 89–128. (In Arabic) [CrossRef]
54. Ibrahim, K.H.; Hassan, A.Y.; AbdElrazek, A.S.; Saleh, S.M. Economic analysis of stand-alone PV-battery system based on new power assessment configuration in Siwa Oasis–Egypt. *Alex. Eng. J.* **2023**, *62*, 181–191. [CrossRef]
55. Heliopolis University Concept Note: Creating Agricultural Carbon Credits for a Sustainable Future in Egypt. 2022. Available online: https://www.sekem.com/wp-content/uploads/2022/05/Concept-Note-Agriculture-for-Climate-Agricultural-Carbon-Credits.pdf (accessed on 20 October 2022).
56. Egypt Today Staff. Egypt to Establish Carbon Certificates Trading Company. Available online: https://www.egypttoday.com/Article/3/120419/Egypt-to-establish-Carbon-Certificates-Trading-Company (accessed on 10 November 2022).
57. CNN Egypt Is Studying the Establishment of a Platform for Trading Carbon Bonds and Experts Explain the Reasons. Available online: https://arabic.cnn.com/business/article/2022/02/09/egypt-platform-carbon (accessed on 18 October 2022). (In Arabic)
58. Akrami, M.; Gilbert, S.J.; Dibaj, M.; Javadi, A.A.; Farmani, R.; Salah, A.H.; Fath, H.E.S.; Negm, A. Decarbonisation Using Hybrid Energy Solution: Case Study of Zagazig, Egypt. *Energies* **2020**, *13*, 4680. [CrossRef]
59. Egyptian Channel One. The First Service Building Fully Powered by Clean Energy Inside the Village of Kafr El-Sheikh. Arch.Ahmed Abouaiana Reveals the Details. Available online: https://www.youtube.com/watch?v=CmfAKqkLVfE (accessed on 28 June 2022). (In Arabic)
60. ClimateWatch Egypt. Available online: https://www.climatewatchdata.org/countries/EGY?end_year=2019&start_year=1990#ghg-emissions (accessed on 20 October 2022).
61. Kamal, N. The most important messages of the International Conference of the National Planning Institute "Energy and Sustainable Development" Cairo 21–22 November 2020. *Egypt. J. Dev. Plan.* **2020**, *28*, 149–154. (In Arabic) [CrossRef]
62. Shaaban, M.; Scheffran, J.; Elsobki, M.S.; Azadi, H. A Comprehensive Evaluation of Electricity Planning Models in Egypt: Optimization versus Agent-Based Approaches. *Sustainability* **2022**, *14*, 1563. [CrossRef]
63. Salah, S.I.; Eltaweel, M.; Abeykoon, C. Towards a sustainable energy future for Egypt: A systematic review of renewable energy sources, technologies, challenges, and recommendations. *Clean. Eng. Technol.* **2022**, *8*, 100497. [CrossRef]
64. Escribano, G. Toward a Mediterranean energy community: No roadmap without a narrative. In *Regulation and Investments in Energy Markets*; Elsevier: Amsterdam, The Netherlands, 2016; pp. 117–130.
65. FAO. *Energy, Agriculture and Climate Change towards Energy-Smart Agriculture, Ref: I6382En/1/11.16*; Food and Agriculture Organization of the United Nations: Rome, Italy, 2016.
66. Cheo, A.E.; Adelhardt, N.; Krieger, T.; Berneiser, J.; Sanchez Santillano, F.A.; Bingwa, B.; Suleiman, N.; Thiele, P.; Royes, A.; Gudopp, D. Agrivoltaics across the water-energy-food-nexus in Africa: Opportunities and challenges for rural communities in Mali. (20 April 2022). Discussion Paper Series, Wilfried Guth Endowed Chair of Constitutional Political Economy and Competition Policy, University of Freiburg, No. 2022-03 2022. Available online: https://ssrn.com/abstract=4088544 (accessed on 20 October 2022).

67. Andoura, S.; Hancher, L.; Van der Woude, M. *Towards a European Energy Community: A Policy Proposal*; Notre Europe, Jacques Delors Institute: Paris, France, 2010.
68. Kassem, H.S.; Bello, A.R.S.; Alotaibi, B.M.; Aldosri, F.O.; Straquadine, G.S. Climate change adaptation in the delta Nile Region of Egypt: Implications for agricultural extension. *Sustainability* **2019**, *11*, 685. [CrossRef]
69. McGookin, C.; Mac Uidhir, T.; Gallachóir, B.Ó.; Byrne, E. Doing things differently: Bridging community concerns and energy system modelling with a transdisciplinary approach in rural Ireland. *Energy Res. Soc. Sci.* **2022**, *89*, 102658. [CrossRef]
70. Braiki, H.; Hassenforder, E.; Lestrelin, G.; Morardet, S.; Faysse, N.; Younsi, S.; Ferrand, N.; Léauthaud, C.; Aissa, N.B.; Mouelhi, S. Large-scale participation in policy design: Citizen proposals for rural development in Tunisia. *EURO J. Decis. Process.* **2022**, *10*, 100020. [CrossRef]
71. REScoop. *A Supportive EU Legal Framework for Energy Communities in Energy Efficiency*; REScoop.eu: Brussels, Belgium, 2021.
72. Al-Makhzangy, A.S.M. Al-Ibtikar ka Aliyah Litahkik Altanmia Almostadama fi Misr [Innovation as a Mechanismto to Achieve Sustainable Development in Egypt]. *Arab J. Adm.* **2022**, *42*, 359–378. [CrossRef]
73. SAS. Ségala Agriculture Énergie Solaire [Ségala Agriculture Solar Energy]. Available online: https://energie-partagee.org/projets/segala-agriculture-energie-solaire/ (accessed on 5 January 2022).
74. European Union. Jühnde Bio-Energy Village in Germany. Available online: https://www.eesc.europa.eu/en/news-media/presentations/juhnde-bio-energy-village-germany (accessed on 20 October 2022).
75. Coopernico. We Own Our Energy. Available online: https://www.coopernico.org/ (accessed on 5 January 2022).
76. IRENA Coalition for Action. *Community Energy Toolkit: Best Practices for Broadening the Ownership of Renewables*; International Renewable Energy Agency: Abu Dhabi, United Arab Emirates, 2021; ISBN 978-92-9260-366-3.
77. Clausen, L.T.; Rudolph, D. Sustainable rural development and rural energy communities in a post-Brexit UK: Paralysis or broader visions in uncertain times? In *Rural Governance in the UK*; Routledge: London, UK, 2022; pp. 140–161. ISBN 1003200206.
78. IRENA. *Future of Solar Photovoltaic: Deployment, Investment, Technology, Grid Integration and Socio-Economic Aspects (A Global Energy Transformation: Paper)*; International Renewable Energy Agency: Abu Dhabi, United Arab Emirates, 2019; ISBN 978-92-9260-156-0.
79. Abouaiana, A. A Transdisciplinary Framework for Developing Energy Efficiency in Rural Communities- at Lasaifar Al-balad. Available online: https://www.facebook.com/ahmad.abdulmalek/posts/pfbid02QpMBGzv4J6JRb9uh9PUKBk9AqRKqNwkXeuRRqn8e6KnodbYXv8j2GgN6XtzpeVkhl (accessed on 20 October 2022). (In Arabic).
80. Comune di Pontinia. Available online: https://www.facebook.com/permalink.php?story_fbid=143295724790254&id=109416818178145 (accessed on 11 March 2022).
81. Golla, A.; Röhrig, N.; Staudt, P.; Weinhardt, C. Evaluating the impact of regulation on the path of electrification in Citizen Energy Communities with prosumer investment. *Appl. Energy* **2022**, *319*, 119241. [CrossRef]
82. Belaïd, F. Mapping and understanding the drivers of fuel poverty in emerging economies: The case of Egypt and Jordan. *Energy Policy* **2022**, *162*, 112775. [CrossRef]
83. Hongyun, H.; Radwan, A. Economic and social structure and electricity consumption in Egypt. *Energy* **2021**, *231*, 120962. [CrossRef]
84. Duarte, R.; García-Riazuelo, Á.; Sáez, L.A.; Sarasa, C. Analysing citizens' perceptions of renewable energies in rural areas: A case study on wind farms in Spain. *Energy Rep.* **2022**, *8*, 12822–12831. [CrossRef]
85. OECD. Managing environmental and energy transitions in rural areas. In *Managing Environmental and Energy Transitions for Regions and Cities*; OECD Publishing: Paris, France, 2022; pp. 95–121. ISBN 9789264182660.
86. Pascaris, A.S.; Schelly, C.; Burnham, L.; Pearce, J.M. Integrating solar energy with agriculture: Industry perspectives on the market, community, and socio-political dimensions of agrivoltaics. *Energy Res. Soc. Sci.* **2021**, *75*, 102023. [CrossRef]
87. Pascaris, A.S.; Schelly, C.; Rouleau, M.; Pearce, J.M. Do agrivoltaics improve public support for solar? A survey on perceptions, preferences, and priorities. *Green Technol. Resil. Sustain.* **2022**, *2*, 1–17. [CrossRef]

MDPI
St. Alban-Anlage 66
4052 Basel
Switzerland
www.mdpi.com

Land Editorial Office
E-mail: land@mdpi.com
www.mdpi.com/journal/land